Ecohydrology and Environmental Watershed Management

Ecohydrology and Environmental Watershed Management

Edited by **Herbert Lotus**

SYRAWOOD
PUBLISHING HOUSE

New York

Published by Syrawood Publishing House,
750 Third Avenue, 9th Floor,
New York, NY 10017, USA
www.syrawoodpublishinghouse.com

Ecohydrology and Environmental Watershed Management
Edited by Herbert Lotus

International Standard Book Number: 978-1-68286-151-6 (Hardback)

Printed in the United States of America.

Contents

Permissions

List of Contributors

Preface

In my initial years as a student, I used to run to the library at every possible instance to grab a book and learn something new. Books were my primary source of knowledge and I would not have come such a long way without all that I learnt from them. Thus, when I was approached to edit this book; I became understandably nostalgic. It was an absolute honor to be considered worthy of guiding the current generation as well as those to come. I put all my knowledge and hard work into making this book most beneficial for its readers.

The sustainable management of water resources has become an increasingly prominent field of study and research. Ecohydrology and water management are crucial topics of interest in this field. This book explores the cohesive relationship between water resources and their respective ecosystems. The chapters included herein trace the progress of these fields and highlight some of the key concepts such as ecological engineering, catchment hydrology, biogeochemical processes, mathematical analysis, management of different water resources, etc. Scientists and students will find this book full of crucial and unexplored concepts.

I wish to thank my publisher for supporting me at every step. I would also like to thank all the authors who have contributed their researches in this book. I hope this book will be a valuable contribution to the progress of the field.

Editor

Multi-scale analysis of bias correction of soil moisture

C.-H. Su and D. Ryu

Department of Infrastructure Engineering, University of Melbourne, 3010 Victoria, Australia

Correspondence to: C.-H. Su (csu@unimelb.edu.au)

Abstract. Remote sensing, in situ networks and models are now providing unprecedented information for environmental monitoring. To conjunctively use multi-source data nominally representing an identical variable, one must resolve biases existing between these disparate sources, and the characteristics of the biases can be non-trivial due to spatio-temporal variability of the target variable, inter-sensor differences with variable measurement supports. One such example is of soil moisture (SM) monitoring. Triple collocation (TC) based bias correction is a powerful statistical method that is increasingly being used to address this issue, but is only applicable to the linear regime, whereas the non-linear method of statistical moment matching is susceptible to unintended biases originating from measurement error. Since different physical processes that influence SM dynamics may be distinguishable by their characteristic spatio-temporal scales, we propose a multi-timescale linear bias model in the framework of a wavelet-based multi-resolution analysis (MRA). The joint MRA-TC analysis was applied to demonstrate scale-dependent biases between in situ, remotely sensed and modelled SM, the influence of various prospective bias correction schemes on these biases, and lastly to enable multi-scale bias correction and data-adaptive, non-linear de-noising via wavelet thresholding.

1 Introduction

Global environmental monitoring requires geophysical measurements from a variety of sources and sensors to close the information gap. However, different direct and remote sensing, and model simulation can yield different estimates due to different measurement supports and errors. Soil moisture (SM) is one such variable that has garnered increasing interest due to its influences on atmospheric, hydrologic, geomorphic and ecological processes (Rodriguez-Iturbe, 2000; GLACE Team et al., 2004; Legates et al., 2011). It also represents an archetype of the aforementioned problem, where in situ networks, remote sensing and models jointly provide extensive SM information.

In situ networks usually provide point-scale measurements; satellite retrieval of shallow SM at a mesoscale footprint of 10–50 km must resort to a homogeneity or dominant-feature assumption, whereas modelled SM depends on the simplified model parameterization, and the quality, resolution and availability of forcing data. Subsequently, the spatial (lateral and vertical) variability of SM can lead to systematically different measurements regarded as *biases*. Descriptive or predictive spatial SM statistics can be used to relate point-scale to mesoscale estimates (Western et al., 2002), but in situ data are often limited in describing the spatial heterogeneity of SM. However, without bias correction, it is not possible to conduct meaningful comparisons between in situ, satellite-retrieved and modelled SM for validation (Reichle et al., 2004) and optimal data assimilation (Yilmaz and Crow, 2013). Standard bias correction methods are now increasingly being applied to SM assimilation in land models (Reichle et al., 2007; Kumar et al., 2012; Draper et al., 2012), numerical weather prediction (Drusch et al., 2005; Scipal et al., 2008a) and hydrologic models (Brocca et al., 2012). Reichle and Koster (2004) proposed matching statistical moments of the data, while linear methods based on simple regression and matching dynamic ranges have also been considered (e.g. Su et al., 2013a). But these methods can induce artificial biases in the signal component of the corrected data as the error statistics were ignored; this also suggests a connection that the issue of bias correction is inseparable from that of error characterisation (Su et al., 2014a).

Triple collocation (TC) (Stoffelen, 1998), which is a form of instrument-variable regression (Wright, 1928; Su et al.,

2014a), is increasingly being used to address these issues in oceanography (Caires and Sterl, 2003; Janssen et al., 2007) and hydrometeorology (Scipal et al., 2008b; Roebeling et al., 2013). In particular, it was used to estimate spatial point-to-footprint sampling errors (Miralles et al., 2011; Gruber et al., 2013), and correct biases in SM (Yilmaz and Crow, 2013). Based on an affine signal model and additive orthogonal error model, it assumes that representativity differences are manifested as additive and multiplicative biases. But these assumptions may have limited validity, as the temporal behaviour of SM may vary across different spatial scales, driven by a continuum of localised and mesoscale influences (e.g. Entin et al., 2000; Mittelbach and Seneviratne, 2012). Specifically, the coupling of SM with precipitation and evaporative losses (controlled by temperature, humidity, wind speed) varies across spatial scales. This can be more pronounced at places where surface hydrological features (e.g. topography, infiltration rate and storage capacity) are highly heterogeneous. Thus, the biases are likely to be non-systematic across short and long timescales on different spatial scales and errors are non-white, undermining the utility of the affine model. One possible remedy is to apply bias correction, either TC or statistical-moment matching, only to anomaly time series (Miralles et al., 2011; Liu et al., 2012; Su et al., 2014a), but it remains unclear how these methods affect the signal and noise components in the corrected data. Alternatively a moving time window can be used to examine the time-varying statistics of time series (Loew and Schlenz , 2011; Zwieback et al., 2013; Su et al., 2014a).

Given the possible (time)scale dependency in biases and errors, we propose an extension to TC analyses to include wavelet-based multi-resolution analysis (MRA) (Mallat, 1989) as a framework to (1) provide a fuller description of the temporal scale-by-scale relationships between coincident data sets; (2) study the influence of various prospective bias correction schemes; and (3) achieve multi-scale bias correction. To avoid excessive changes in the noise characteristics upon correction, TC can be further combined with the wavelet thresholding (Donoho and Johnstone, 1994) to (4) achieve non-linear, data-adaptive de-noising, with contrast to existing linear schemes (Su et al., 2013b). The techniques were applied to SM data from an in situ probe, satellite radiometry and land-surface model, but the proposed methods are generally enough to be applied to other geophysical variables.

The paper is organised as follows. Section 2 presents the study area over Australia and the SM data sets used in our pilot studies. Section 3 explains the theoretics behind MRA and applies it to SM, following by examination of scale-by-scale statistics in Sect. 4. Section 5 presents a new joint MRA-TC analysis framework, which is then applied to examine the influence of different bias correction schemes in Sect. 6. Importantly, both Sects. 4 and 6, using wavelet correlation, wavelet variance and scale-level TC analyses, provide evidence to support the need to extend traditional bulk

and anomaly based analyses. Section 7 demonstrates the use of wavelet thresholding to de-noise satellite SM. Section 8 offers our concluding remarks.

2 Study areas and data sets

We consider in situ, satellite-retrieved and modelled SM over Australia. For an in-depth study, we consider point-scale and pixel-scale SM estimates at K1 monitoring site (147.56° longitude, −35.49° latitude) situated at Kyeamba Creek catchment, southeastern Australia (Smith et al., 2012; Su et al., 2013a). The in situ SM (INS as shorthand) was sampled at 30 min intervals, 0–8 cm depth using a time-domain interferometer-based Campbell Scientific 615 probe during November 2001–April 2011. The region experiences a temperate (Cfb) climate characterised by seasonally uniform rainfall but variable evapotranspiration forcing, so that SM varies between dry in summer (December–February) to wet in winter (June–August). The creek is located on gentle slopes with rain-fed cropping and pasture, and the soil varies from sandy to loam. Figure 1 illustrates the land cover, elevation, monthly rainfall accumulation (from 2002 to 2011), and clay content over the region.

The satellite SM was retrieved by AMSR-E (Advanced Microwave Scanning Radiometer for Earth Observing System; AMS) of the AQUA satellite. The retrieval is based on an inversion of the forward radiative transfer model of a vegetation-masked soil surface, relating observed brightness temperature to soil dielectric constant estimates. A dielectric mixing model is then used to related the dielectric constant to volumetric SM. The combined C/X-band $1/4° \times 1/4°$ gridded, half-daily (~ 1.30 a.m./p.m. LT – local time) version 5 product (July 2002–October 2011) is based on the Land Parameter Retrieval Model (Owe et al., 2008). C-band (X-band) has a shallow sampling depth of ~ 1–2 cm (~ 5 mm), although it is mostly C-band data over Australia due to relatively small radio frequency interference. Given the 1–2 day revisit times of the satellite, there is a significant number of missing values in the AMS data. However, we found that (not shown) over 99 % (95 %) of the gaps over Australia are ≤ 1.5 day (≤ 1 day) long. For use in wavelet analysis (Sect. 4), a one-dimensional (1-D in time) interpolation algorithm (Garcia, 2010) based on discrete cosine transform (Wang et al., 2012) was applied to infill gaps of lengths ≤ 5 days in AMSR-E. Other interpolation methods were trialled; e.g. linear interpolated AMSR-E shows great similarities to the DCT interpolated data, while cubic spline interpolation leads to spurious peaks.

The modelled SM is taken from MERRA (Modern Era Retrospective-analysis for Research and Applications) – Land produced by the Catchment land surface model GEOS version 5.7.2. The MERRA atmospheric re-analysis is driven by a vast collection of in situ observations of atmospheric and surface winds, temperature, and humidity, and remote sens-

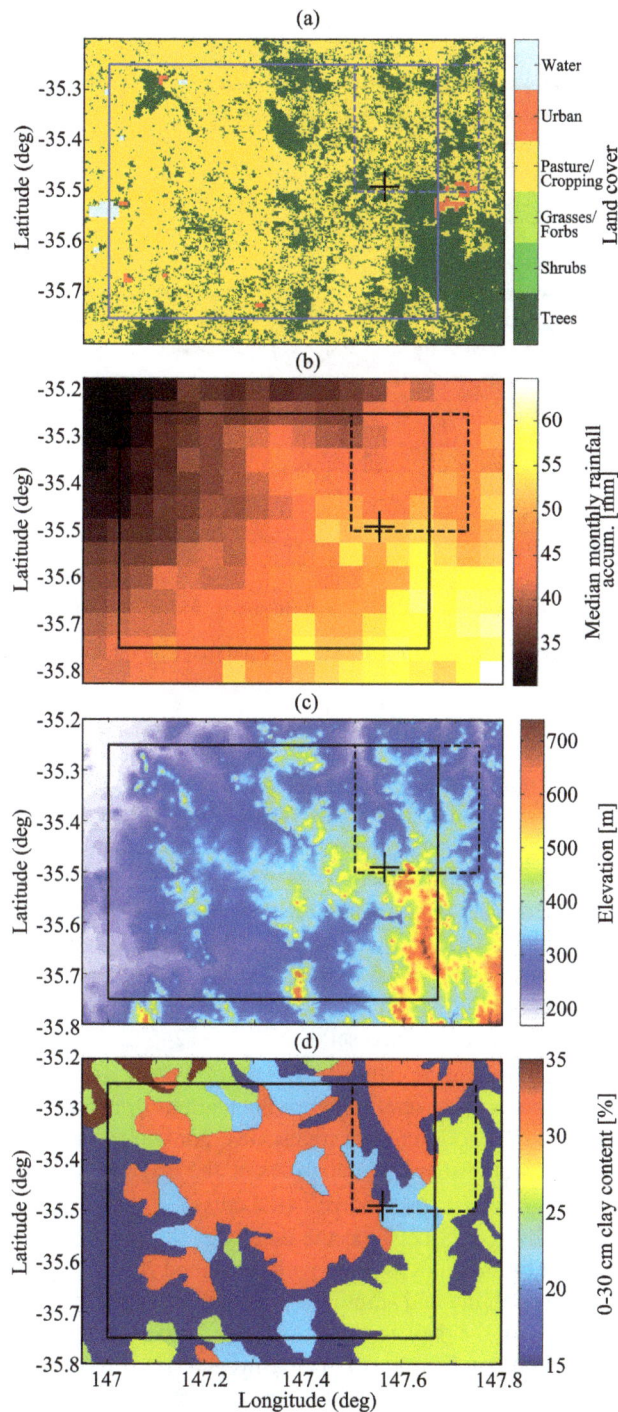

Figure 1. Spatial variability of land surface and rainfall over Kyeamba Creek. The cross denotes the location of the K1 monitoring station, and the dashed (solid) box is the pixel area of AMS (MER).

ing of precipitation and radiation (Rienecker et al., 2011). The MERRA land-only fields were post-processed by reintegrating a revised Catchment model with more realistic precipitation forcing to produce the MERRA-Land (MER as shorthand) data set (Reichle et al., 2011). The resultant SM field corresponds to the hourly averages of the uppermost layer (0–2 cm) and is gridded on a $2/3° \times 1/2°$ grid.

The three data are co-located spatially via nearest neighbour and temporally at around the satellite overpass times of 1.30 a.m./p.m. LT. Their time series are plotted in blue in the first panels of Fig. 2. While co-located, the three methods observed SM dynamics over different locations and areas of the catchment (Fig. 1), due to differences in their pixel resolutions and alignments.

Continental-scale AMS and MER data over Australia are also considered. The continent has great variability in climatic and land surface characteristics. Most of the northern regions experience a tropical savannah (Aw) Köppen–Geiger climate as classified by Peel et al. (2007), central Australia is largely arid desert (BWh), and eastern mountainous areas have a temperate climate with no dry seasons (Cf). The southwestern regions similarly have a temperate climate, but with dry summers (Cs). These temperate regions have higher vegetation compared to the tropical north with moderate vegetation cover.

3 Multi-scale decomposition of soil moisture

The observed Kyeamba SM (denoted by blue curves p in Fig. 2) exhibits a long-term cycle of wet and dry years due to the El Niño–Southern Oscillation and seasonal and diurnal cycles originating from the fluctuations in vegetation and solar radiation, and experiences transient decay from various loss mechanisms, and abrupt increase from individual rainfall events. Their influences on observed SM can vary with the measurement methods. To unravel these differences, we turn to wavelets as the analysing kernels to study variability on individual broad-to-fine timescales. The *scale* under investigation is *temporal* for the rest of the paper, unless stated otherwise.

The 1-D orthogonal discrete wavelet transform (DWT) enables MRA of a time series $p(t)$ of dyadic length $N = 2^J$ and a regular sampling interval Δt by providing the mechanism to go from one resolution to another via a recursive function

$$p_{j-1}^{(a)}(t) = p_j^{(a)}(t) + p_j(t), \tag{1}$$

with an expectation value $E(p_j^{(a)}) = E(p) = p_J^{(a)}(t) = \mu_p$ and $E(p_j) = 0$, where the superscript (a) labels approximated representations. The integer $j \in [1, J]$ labels the scale of analysis, with $j = 1$ (J) denoting the finest (coarsest) scale, and serves to define a spectral range in a spectral analysis. The recursion therefore relates an approximation or coarse representation $p_j^{(a)}$ of the signal at one resolution to

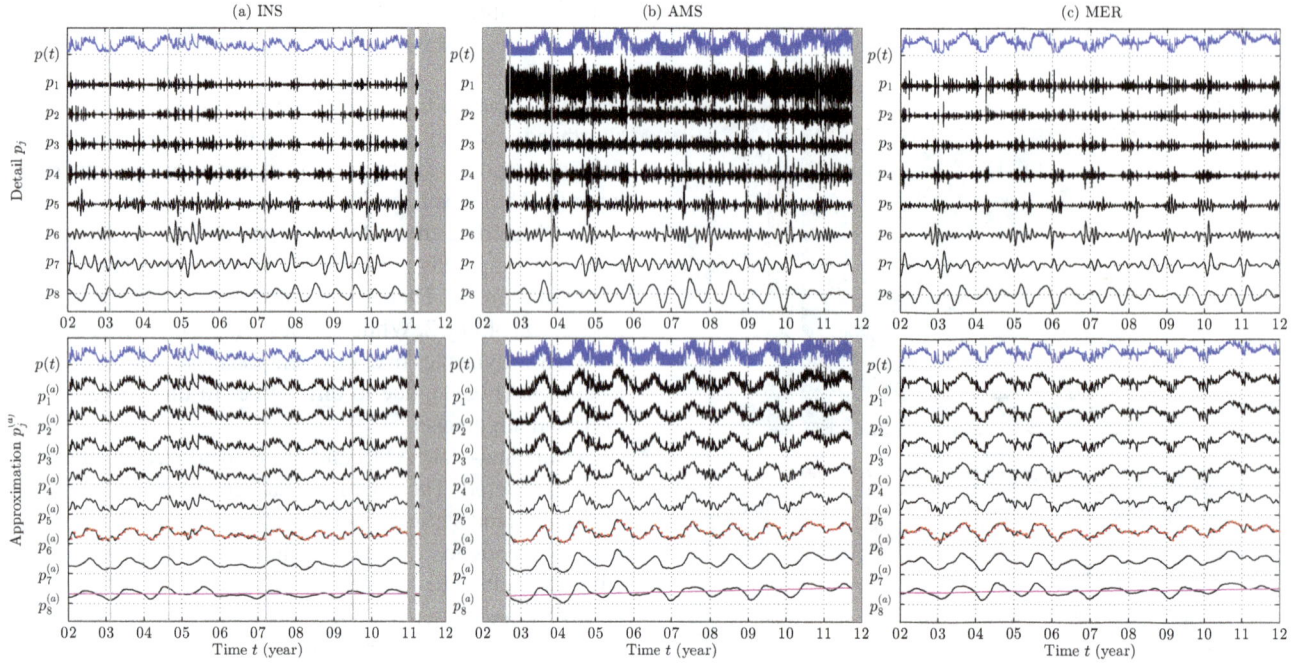

Figure 2. MRA of INS, AMS and MER SM at Kyeamba. p denotes the original time series, p_j the detail time series, and $p_j^{(a)}$ the approximation time series. Grey shadings are > 5 day data gaps, red dots superimposed in $p_6^{(a)}$ are monthly means of p, and magenta lines are trend lines fitted to $p_8^{(a)}$.

that at a higher resolution $p_{j-1}^{(a)}$ by adding some fine-scale detail denoted by p_j. The end of the recursion chain leads to reconstruction of the original time series with the equality $p_0^{(a)}(t) = p(t)$, and a multi-resolution decomposition of p as

$$p(t) = p_{j_0}^{(a)}(t) + \sum_{j=1}^{j_0} p_j(t) \quad (2)$$

$$= \sum_{k=1}^{n_{j_0}} p_{j_0 k}^{(a)} \phi_{jk}(t) + \sum_{j=1}^{j_0} \sum_{k=1}^{n_j} p_{jk} \psi_{jk}(t) \quad (3)$$

under j_0 levels of decomposition. Loosely speaking, for a half-daily time series, the *detail time series* p_j for $j = 1, 2, 3, \ldots$ corresponds to (fine-scale) dynamics observed on 1 day (1 d), 2 d, 4 d, etc., timescales, while the *approximation time series* $p_j^{(a)}$ for $j = 1, 2, 3, \ldots$ contains (broad-scale) dynamics on scales longer than 1 d, 2 d, 4 d, etc.

In Eq. (3), each of these components is further decomposed into a linear summation of $n_j = N/2^j$ number of basis functions ϕ_{jk} and ψ_{jk} with scale of variability $2^j \Delta t$ and temporal location $k \, 2^j \Delta t$. The weighting or wavelet coefficients, determined via DWT of p, measure the similarity between p and the bases via the inner products $p_{jk}^{(a)} \equiv \langle p, \phi_{jk} \rangle$ and $p_{jk} \equiv \langle p, \psi_{jk} \rangle$. Hence the coefficients indicate changes on a particular scale and location, and enable the above scale-by-scale decomposition. Note that the bases are defined in $L^2(R)$ space and satisfy orthonormality conditions

prescribed by $\langle \phi_{jk}, \phi_{j'k'} \rangle = \delta_{jj'} \delta_{kk'}$, $\langle \psi_{jk}, \psi_{j'k'} \rangle = \delta_{jj'} \delta_{kk'}$, $\langle \phi_{jk}, \psi_{j'k'} \rangle = 0$, where δ is the Kronecker delta function. For detailed expositions of the mathematical theory of wavelets and MRA, consult Daubechies (1992) and Mallat (1989).

The detail and approximated time series of Kyeamba's SM are illustrated in subsequent panels of Fig. 2, analysed using the Daubechies $D(4)$ wavelet for $j_0 = 8$. On the finest scales $j = 1$–2 (1–2 d), the details show variability due to rainfall wetting, and over the next set of scales $j = 2$–5 (2–16 d) they describe transient moisture loss. The $p_6^{(a)}$ (≥ 32 d) component accounts for several scales of fluctuations over seasonal, inter-annual, and long-term timescales. For comparison, the standard monthly average analyses of the original time series p are superimposed on $p_6^{(a)}$ (red dots).

The differences between the details of the three SM are apparent on the finest scales, with AMS and MER showing greater variability and amplitude compared to INS. However, the similarity of their temporal patterns, in both details and approximations, grows with increasing scales $j > 3$ (see also Fig. 3). Fitting a trend line to their coarsest scale approximation series suggests $p_8^{(a)}$ that the trends (magenta lines) in the three data show different gradients, with the trend in INS showing the smallest positive gradient. The differences in dynamic ranges of their detail and approximation time series, together with their mismatch in shape and trend, are indicative of multiplicative biases and noise.

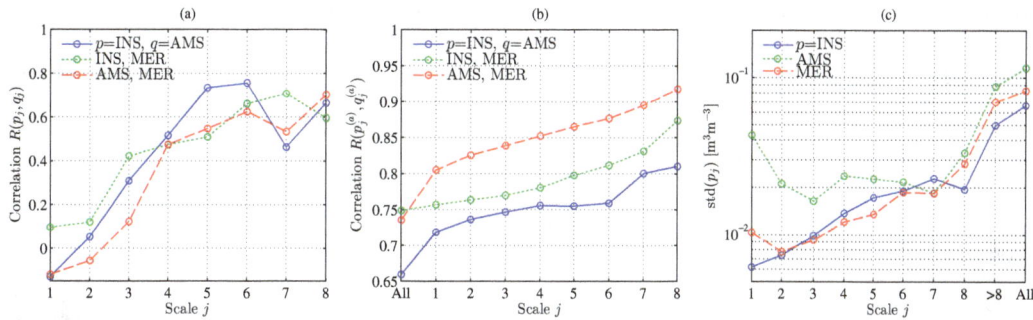

Figure 3. Comparisons of correlation R and SD between INS, AMS and MER at scale levels. **(a)** compares the correlation between their detail time series p_j, and **(b)** compares between their approximation time series $p_j^{(a)}$. Scale $j > 8$ corresponds to $p_8^{(a)}$, and "All" refers to statistics of the original time series.

4 Multi-scale statistics

MRA enables direct comparisons between any two representations $p = \{X, Y\}$ of a given variable f (e.g. SM) on various temporal scales independently, owing to the orthonormal properties of wavelet bases. It also offers an additional degree of freedom in temporal positions (using the index k) to allow better representation of local variability. By subsetting the wavelet coefficients over certain range of k values, non-stationary statistics can also be examined. However, in this work, we consider only variability across j and assume stationarity on each scale. Pearson's linear correlation R and variance analyses (see Appendix A) are performed on the Kyeamba's INS, AMS and MER SM (as p in Eq. 2) detail (p_j) and approximation ($p_j^{(a)}$) time series in Fig. 3. The strength of MRA is that since the detail time series p_j on a given scale j does not contain variations on timescales greater than j, the weak-sense stationarity conditions can be better met.

Before proceeding, we recall that weak R indicate the presence of noise and/or the presence of non-linear correlation between any pairs of the data, while differences in standard deviation can also indicate the presence of noise, but also an extraneous signal and/or multiplicative bias. Typically one invokes a linearity assumption and assumes an affine relation between the signal components of the different data and an additive noise model (more later in Sect. 5), so that the differences between the data are attributed only to an overall additive bias $E(X) - E(Y)$, multiplicative biases, and noise. While we adopt this simplistic viewpoint here, its limitations to properly account for variable lateral and vertical measurement supports should be noted. For instance, short-timescale SM dynamics show increasing attenuation in amplitude, but are also delayed in time in deeper soil columns (e.g. Steelman et al., 2012). Additionally, SM is physically bounded between field capacity and residual content and these thresholds can vary with soil texture, location and depths. These effects can give rise to temporal autocorrelation in errors and undermine the linearity assumption be-

tween coincident measures. Finally, the non-stationary characteristic of noise in satellite SM (Loew and Schlenz , 2011; Zwieback et al., 2013; Su et al., 2014a) due to e.g. dynamical land surface characteristics such as soil moisture (Su et al., 2014b), is not treated here.

With these considerations, we first examine the correlations between the three data. For the detail time series (Fig. 3a), their correlations are lowest on the finest scales ($R < 0.2$) but generally improves with scale ($R > 0.5$), as noted previously. There is however no data pair that shows consistently higher R than other pairs: $R(\text{INS}_j, \text{AMS}_j) > R(\text{INS}_j, \text{MER}_j)$ on coarser scales $j = 4$–6, 8, whereas $R(\text{INS}_j, \text{MER}_j)$ is highest on other scales. Comparing their approximation time series (Fig. 3b), R between AMS and MER are higher than the other two pairs, ranging from ($j = 2$) 0.8 to 0.92 ($j = 8$), largely due to the strong correlation between their respective p_8 and $p_8^{(a)}$. In other words: on one hand, AMS and MER both show skill in representing some aspects of the in situ SM temporal variability; on the other hand, stronger AMS-MER correlations on the coarsest (temporal) scales and their mesoscale spatial resolutions would indicate lesser representativeness of in situ measurement on these spatio-temporal scales.

Furthermore, we observe that $R(p_j^{(a)}, q_j^{(a)})$ reduces with decreasing j as more components are added to the reconstruction of $p_j^{(a)}$ and $q_j^{(a)}$. The inclusion of noisy AMS_1 in the makeup of AMS leads to a drop in $R(\text{INS}, \text{AMS})$ and $R(\text{AMS}, \text{MER})$. Aside from including more noise in the approximation time series, adding components with different multiplicative biases (more later in Sect. 6) can also diminish the correlations. The scale dependence of multiplicative biases and added noise can contribute to the contrasting results of applying TC to raw versus anomaly SM time series in Draper et al. (2013). In particular, given the presence of noise in p_j for $j \geq 7$, error analysis of the anomaly SM (i.e. in p_j for $j \leq 6$) will under-estimate the total error in the raw data p.

Next, Fig. 3c plots their wavelet spectra that decompose total variance $\text{var}(p)$ into individual scales

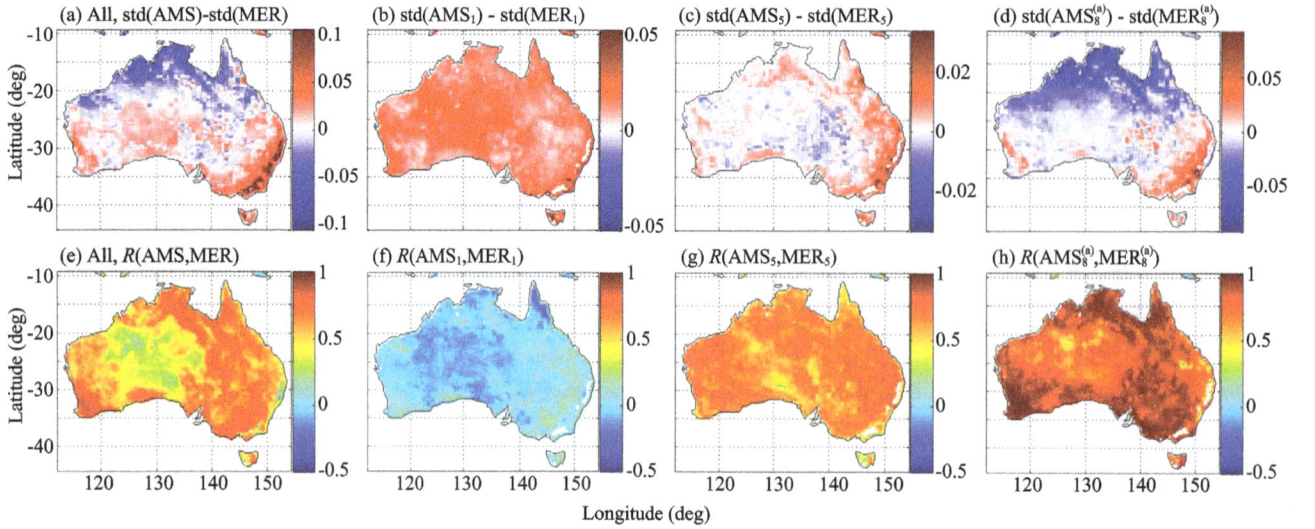

Figure 4. Difference in SD (in units of $m^3\,m^{-3}$) and correlation R between AMS and MER for **(a, e)** all and (rest) on selected timescales.

$var(p_j) \equiv std(p_j)^2$. The three data show clear differences in their standard deviation (SD) profile, both in the fine and coarse scales. As already noted, both noise and/or multiplicative biases are possible contributing factors such that noise can inflate the variance, while biases can cause suppression or inflation. Following the visual inspection of Fig. 2 and the noted weak correlations $R(INS_j, AMS_j)$ and $R(INS_j, MER_j)$ at small j, it can be argued that there is significant noise in AMS (for $j = 1$–3) and MER ($j = 1$). This in turn leads to their larger SD cf. INS. On coarser scales where R values are significantly higher, the differences in SD may be attributed more to multiplicative biases. For instance for their p_8 and $p_8^{(a)}$ components, AMS and MER shows larger SD and thus positively biased relative to INS.

Figure 4 extends the variance and correlation analyses between AMS and MER to the Australian continent using their coincident data from the period July 2002–October 2011. The spatial maps of SD differences (ΔSD) and correlations show significant variability in the statistics with timescales and spatial locations. On the finest scale $j = 1$, the similarity between the difference map (Fig. 4a) and the TC-derived error map of AMSR-E (see Fig. 6a in Su et al., 2014a) in terms of spatial variability and the low AMS-MER correlations (Fig. 4f) support our observation that the detail time series AMS_1 is noise dominated. Weak negative correlation between AMS_1 and MER_1 can also be observed over arid regions. By contrast, owing to the strong correlation $R \sim 0.6$–0.9 (Fig. 4g and h) on the coarse scales, the causes of ΔSD (Fig. 4c and d) are related to biases. In particular, at $j > 8$, the ΔSD map in Fig. 4d also suggests a possible association between biases and climatology or land cover characteristics, with negative biases dominating northern tropical (Aw) and semi-arid (BS) regions, and positive biases in temperate, vegetated regions (Cs and Cf) over southeastern and southwest-

ern Australia. The visual comparisons between scale-level ΔSD and bulk ΔSD enable stratification of the continent to central arid regions of higher noise identified in $j = 1$ and 2 and temperate (tropical) regions, with a positive (negative) bias seen on coarser scales.

5 Joint MRA-TC analysis

In order to quantify observed differences between the data, we propose a *scale-dependent* linear model: a multi-scale (MS) model that distinguishes the signal components of the two data X and Y via an overall additive bias and a set of positive scaling coefficients $\alpha_{p,j}, \alpha'_p$, and assumes an additive and zero-mean independent but non-white noise model $\epsilon_p(t)$. Focusing on the zero-mean signal and noise components, the "structural relationship" model reads

$$p'(t) = \alpha'_p f'(t) + \epsilon'_p(t), \tag{4}$$

$$p_j(t) = \alpha_{p,j} f_j(t) + \epsilon_{p,j}(t), \tag{5}$$

for $p' = p_{j_0}^{(a)} - E(p)$ and $f = f' - E(f)$, where the signal and noise components have been decomposed into their multi-resolution forms. The standard assumptions of orthogonal and mutually uncorrelated errors are used, so that the covariance $cov(f_j, e_{p,j}) = 0$, $cov(f', e'_p) = 0$, $cov(e_{p,j}, e_{q,j}) = 0$, $cov(e_{p,j}, e'_q) = 0$ and $cov(e'_p, e'_q) = 0$ for $p \neq q$, $p, q, \in \{X, Y\}$. The differences in the values of the scaling coefficients between data, i.e. $\alpha_{X,j} \neq \alpha_{Y,j}$, signify multiplicative biases on individual scales. To see this, we express their mean-squared deviation $MSD \equiv E[(Y - X)^2]$ in terms of variables in Eqs. (4) and (5) to arrive at

$$\text{MSD} = (\mu_X - \mu_Y)^2 + \sum_{j}^{J} \left[\left(\alpha_{Y,j} - \alpha_{X,j} \right)^2 \text{var}\left(f_j \right) \right.$$
$$\left. + \text{var}\left(\epsilon_{X,j} \right) + \text{var}\left(\epsilon_{Y,j} \right) \right]. \tag{6}$$

The first term is the additive bias, and the summation consists of scale-specific multiplicative biases proportional to $(\alpha_{X,j} - \alpha_{Y,j})^2$ and noise contributions from each datum. The interpretation of the discrepancies between X and Y can vary depending on the time period of the data and the analysis, and the adopted signal/noise model. By using the entire 9 year record of INS, AMS and MER data in MRA, the MS model does not observe a time-varying additive bias (e.g. from using the moving-window approach of Su et al., 2014a) or autocorrelated errors (from using the lagged covariance in Zwieback et al., 2013). Rather, MRA and the MS model enable a description of the systematic differences based wholly in terms of multiplicative biases at individual timescales, and the random differences in terms of additive noise. Specifically, this contrasts with the short time-window approach (e.g. ≤ 32 d), where multiplicative biases existing on coarse scales ($p_6^{(a)}$) will manifest as both time-varying additive and multiplicative biases.

Importantly, the model allows for different scaling coefficients between scales, i.e. $\alpha_{p,j} \neq \alpha_{p,j'}$ for $j \neq j'$, as a form of non-linearity with f. The equality $\alpha_{p,j} = \alpha'_p = \alpha_p$ is therefore a special case of (bulk) linearity. As our focus of the above model is the multiplicative biases and noise, for convenience of notation, we remove the mean of the X and Y prior to MRA and bias correction. Furthermore, without the loss of generality, we choose X as the reference henceforth and let $\alpha_{X,j}, \alpha'_X = 1$.

By using a third independently derived representation (Z) of f, TC enables estimation of the required scaling coefficients and noise std($\epsilon_{p,j}$) (Appendix B). As we will see later, these estimates are needed for bias correction and de-noising. Within the operating assumptions of TC, TC estimates are unbiased and consistent; that is, the estimated $\hat{\alpha}_{Y,j} = \alpha_{Y,j}$ as the asymptotic limit. However, TC's superiority is dependent on the availability of a strong instrument and a large sample for statistical analyses (Zwieback et al., 2012; Su et al., 2014a). Standard linear estimators, namely ordinary least-square (OLS) regression and variance matching (VAR), can be considered as substitutes, although they are biased estimators of α when X and Y are both noisy (Yilmaz and Crow, 2013; Su et al., 2014a), e.g. OLS yields $\hat{\alpha}_{Y,j} < \alpha_{Y,j}$. In summary, we propose that combining these estimators with MRA via the MS linear model enables investigation into the distribution of the multiplicative biases and additive noise over j, and their response to various bias correction schemes.

6 Multi-scale analysis of bias correction

Consider now the bias correction of Y to produce a corrected datum Y^* that "matches" X. Different interpretations of a "match" and assumptions about signal and noise statistics lead to different bias correction schemes. To describe matching, there are different choices of optimality criterion. The first is based on matching the statistics of the signal-only component of Y^* to that of X. This approach requires consistent estimation of slope parameters α's and the resultant statistics of X and Y^* may differ due to different noise statistics. The second is based on the matching of the statistical moments between Y^* and X (e.g. VAR matching), although the statistics of their constitutive signal components may differ for the same reason. The third is based on the minimum-variance principle of minimizing the least-square difference between Y^* and X (i.e. the OLS estimation), but as already noted, the estimator becomes inconsistent when there are measurement errors in X and Y.

Following our theoretical model in Sect. 5, we define our optimality criterion based on the first criterion of matching the first two moments of the signal components in X and Y so that Y^* is suitable for bias-free data assimilation. In particular, Yilmaz and Crow (2013) have shown that residual multiplicative biases due to a sub-optimal bias correction scheme will cause filter innovations to contain residual signal and sub-optimal filter performance. Thus, within the paradigm of the MS model, our goal of bias correction is to minimise the difference $|\alpha_{Y^*,j} - 1|$ for $\alpha_{X,j} = 1$, so that the multiplicative bias terms in Eq. (6) are eliminated.

- *Bulk linear rescaling* assumes bulk linearity between X and Y so that the correction equation is

$$Y^* = \frac{Y}{\hat{\alpha}_Y}, \tag{7}$$

  where $\hat{\alpha}_Y$ is given by TC for our objective. When the bulk linearity is satisfied, this approach ensures that the statistical properties (SD and higher moments) of the signal components in X and Y^* are identical. Linear rescaling using $\hat{\alpha}_Y$ values estimated by OLS and VAR matching have previously been considered by e.g. Su et al. (2013a); but due to error-in-variable biases, they can induce artificial biases in the signal component of Y^* even if the bulk linearity condition is valid.

- *Bulk cumulative distribution function (CDF) matching* assumes non-linearity between X and Y and transforms Y^* so that (Reichle and Koster, 2004)

$$\text{cdf}\left(Y^* \right) = \text{cdf}(X), \tag{8}$$

  where cdf($\circ$) computes the CDF. This ensures that the mean, SD, and higher statistical moments of X and Y^* are identical, but the statistical properties of their signal and noise components that make up X and Y^* are not necessarily identical. In particular, when the relative signal and noise statistics in the two data are different, CDF matching leads to artificial biases between the signal components in X and Y^*. As with VAR matching of first two moments, the CDF counterpart is expected to contain extraneous contribution of the noise variances in the mapping of the second moment, as well as at higher moments (Su et al., 2014a). The issue can be exacerbated by variable signal and noise statistics on different scales.

– *Anomaly/seasonal (A/S) linear rescaling* allows biases between X and Y to be different on two scales of variation. In practice, the useful information content in observations is primarily based on their representation of anomalies, where observations are assumed in a particular land surface model's unique climatology (Koster et al., 2009). The correction is therefore limited to the anomalies, although other components (e.g. seasonal fluctuation and long-term trend) may be preserved to validate model prediction. Here the linear correction using TC estimator is applied to match the characteristics of each component – anomaly ($i = A$) and seasonal (S) – separately, so that the corrected Y has the form

$$Y^* = Y_S^* + Y_A^*, \tag{9}$$

with $Y_i^* = Y_i/\hat{\alpha}_{Y_i}$ for $i \in \{S, A\}$. In one approach, p_S is computed using moving window averaging of multi-year data within a window size of 31 days centered on a given day of year (Miralles et al., 2011; Su et al., 2014a), so that inter-annual cycles and long-term trends are retained in p_A. In an alternative approach (Albergel et al., 2012), a sliding 31 day window is used such that

$$p_A \approx \sum_{j=1}^{6} p_j$$ for half-daily time series. In this work, the former, more conventional approach was taken.

– *A/S CDF matching* applies CDF matching to anomaly and seasonal components separately as per Eq. (9) but with $\mathrm{cdf}(Y_i^*) = \mathrm{cdf}(X_i)$. The application of CDF matching to the anomaly component of soil moisture data was considered by Liu et al. (2012).

– *Multi-scale (MS) rescaling* is the direct consequence of the MS model where information in Y is rescaled at individual scales,

$$Y^* = \frac{Y'}{\hat{\alpha}_Y'} + \sum_{j=1}^{j_0} \frac{Y_j}{\hat{\alpha}_{Y,j}}. \tag{10}$$

In relation to Eq. (6), this approach obviously eliminates that the multiplicative terms in the summation. The bulk and A/S linear correction schemes can be considered as special cases of MS rescaling where information from multiple scales are aggregated and corrected

jointly. Other aggregations of the information from different subsets of scales are also possible, but they will similarly be conceived based on one's understanding or assumptions of the underlying specific processes driving SM dynamics. Investigations into suitable aggregations are beyond the scope of this work, hence we implemented the most elaborate decomposition. If joint linearity exists between two or more scales, their $\alpha_{Y,j}$ values will be similarly valued for use in Eq. (10).

For illustrations, we correct the biases in AMS and MER SM with respect to INS SM at Kyeamba using the above five schemes. Using the above notations, AMS and MER are treated as Y, the corrected AMS* and MER* as Y^*, and INS as X. MRA-TC was applied to observe their consequences in Fig. 5. In the upper panel, estimated $\hat{\alpha}_{Y,j}$ and $\hat{\alpha}_{Y^*,j}$ values provide diagnostics for detecting the presence of multiplicative biases before and after application of the correction schemes. The lower panel plots the SD of Y_j and Y_j^* and their associated noises $\epsilon_{Y,j}$ and $\epsilon_{Y^*,j}$. The values of the scaling coefficients $\alpha_{Y,j}$ (before correction) and $\alpha_{Y^*,j}$ (after), and the noise $\mathrm{std}(\epsilon_{Y,j})$ and $\mathrm{std}(\epsilon_{Y^*,j})$ were estimated using TC. But where TC estimates could not be retrieved (for $j = 1–2$) due to negative correlation amongst the data triplet (e.g. resulting from significant noise and weak instrument), OLS-derived (under) estimates serve as a guide for the above diagnostic purposes. Similarly, the total SD is a guide for noise SD in these cases.

Figure 5a shows the MRA of the biases and noise in the pre-corrected data Y. There is considerable variability in $\hat{\alpha}_{Y,j}$ across the scales, ranging from 0.5 to 1.8 for AMS, and from 0.5 to 1.4 for MER. In particular, their $\hat{\alpha}_Y'$ and $\hat{\alpha}_{Y,8}$ deviate significantly from 1, and are responsible for the larger SD (cf. INS) observed in Fig. 3c. Biases also exist on almost all other scales of AMS and MER. In the lower panel, the values of $\mathrm{std}(\epsilon_{Y,j})$ relative to $\mathrm{std}(Y_j)$ indicate the dominance of noise in the small scales $j = 1–3$. This explains the low R values between AMS (and MER) and INS in Fig. 3a. Furthermore, the signal-to-noise ratios are variable with scales and data sets, highlighting the importance of using a correction scheme that takes the signal-vs.-noise statistics into considerations. The TC-based scheme is limited to the linear case, and the CDF scheme ignores such a variability.

The MRA of the corrected data Y^* are shown in Fig. 5b–f. In addition we assess the level of agreement between corrected AMS* and INS time series in Table 1 using their root-mean-squared deviation (RMSD) and correlation R. The time series plots are shown in Fig. 6 to support interpretations. These additional results focus on the AMS-INS pair that best illustrates the influence of noise in AMS.

The results of bulk, A/S and MS linear rescaling can be readily interpreted. For bulk (Fig. 5b) and A/S linear (Fig. 5d) rescaling, the values of $\hat{\alpha}_Y$ and $\hat{\alpha}_{Y_i}$ used for their implementation (Eqs. 7 and 9) are listed in the figure. As these values are greater than unity for both AMS and MER, this leads to the

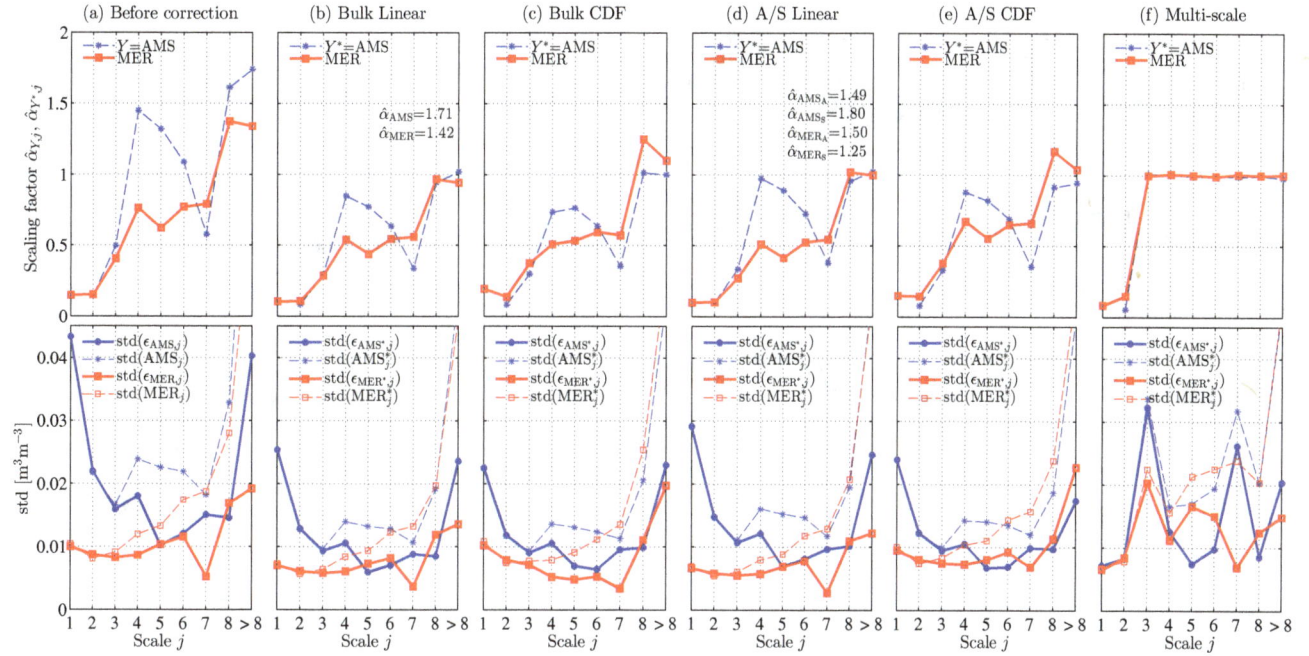

Figure 5. Bias correction of AMS and MER (as Y) with respect to INS (as X), showing the impact of 5 correction schemes on the scaling coefficients, noise and total SD on individual scales. Estimated $\hat{\alpha}_{Y,j} \neq 1$ or $\hat{\alpha}_{Y*,j} \neq 1$ suggests multiplicative bias in Y_j or Y_j^* as per Eq. (6). **(a)** is the diagnosis of Y before correction, and **(b–f)** are that of Y^* after correction. The estimated $\hat{\alpha}_{Y,j}$ and $\hat{\alpha}_{Y*,j}$ for the diagnoses are derived using OLS (for $j = 1, 2$) and TC ($j > 2$). The additional $\hat{\alpha}_Y$ values listed in **(b, d)** are the scaling coefficients used in the implementations of bulk and A/S linear rescaling. Scale $j > 8$ corresponds to $Y_8^{(a)}$.

Table 1. RMSD (in units of $m^3\,m^{-3}$) and correlation between INS and AMS SM at Kyeamba treated by various methods. The square brackets contain the 95 % confidence interval.

Methods	RMSD	Correlation
None	0.088	0.659 [14]
Bulk linear	0.055	0.659 [14]
Bulk CDF	0.053	0.679 [14]
A/S linear	0.059	0.635 [15]
A/S CDF	0.054	0.671 [14]
Multi-scale (MS)	0.062	0.650 [15]
Wavelet thresh. (WT)	0.069	0.709 [13]
WT + MS	0.048	0.711 [12]

suppression of the associated signal, as well as noise, components: $\text{std}(Y_j^*) < \text{std}(Y_j)$, and $\text{std}(\epsilon_{Y*,j}) < \text{std}(\epsilon_{Y,j})$. For AMS, the bulk linear scheme corrects the coarse-scale bias in $Y_8^{(a)}$ component and rescales the noise variance, reducing RMSD from 0.09 to $0.06\,m^3\,m^{-3}$. However, the fine-scale biases in Y_j^* are still present, and increased on some scales, e.g. at $j = 4, 7$ for AMS*. Additionally for A/S linear rescaling, $R(\text{AMS}^*,\text{INS})$ value does not change significantly and the noise are still clearly visible in Fig. 6b and d.

By construction, the MS rescaling uses the estimated $\hat{\alpha}_{Y,j}$ values from Fig. 5a to correct bias on all the scales. Fig. 5f

shows the analysis of MS-corrected Y^*. The equivalence $\hat{\alpha}_{Y*,j} = 1$ indicates that the multiplicative biases are eliminated at $j > 2$. At $j = 1$–2, as the scaling coefficients cannot be estimated by TC, CDF matching was applied to these scales such that the biases are still present on these scales. Amid the reduction of biases, we also observed noise amplification (i.e. $\text{std}(\epsilon_{Y*,j}) > \text{std}(\epsilon_{Y,j})$) in AMS* at $j = 3, 7$ and in MER* at $j = 3$–7, because of rescaling with less-than-unity $\hat{\alpha}_{Y,j}$ values in Eq. (10). Indeed it is evident from Eq. (6) that it is possible to increase the noise variance and MSE when reducing the bias component of the MSE. This in turn leads to larger disagreement between INS and AMS in terms of RMSD and R, and the increased amplitudes of the noise observed in AMS in Fig. 6f.

The bulk and A/S CDF methods produced very similar results with each other, and also with their linear counterparts. There is signal and noise suppression, but the scale-level biases are retained. The signal components of Y^* are negatively biased at $j = 3$–7 and positively biased at $j = 8$. The CDF-corrected AMS* shows slightly better RMSD and R with INS, owing to the reduced noise variance and a reduced bias at $\text{AMS}_8^{(a)*}$.

In summary, the MRA of the bulk and A/S schemes highlights the deficiency of using a correction scheme that does not take into account the scale variability of bias and the differences in noise statistics between the two data. The im-

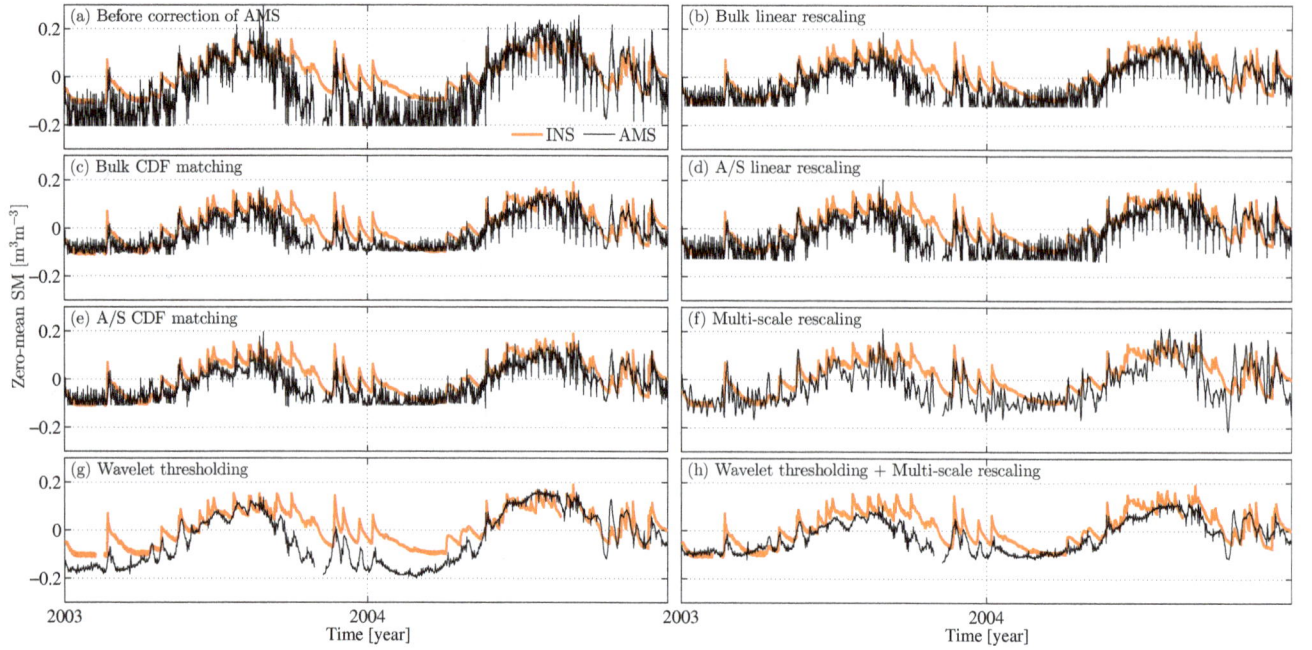

Figure 6. Time series of AMS SM at Kyeamba treated by various bias correction schemes. The use of WT-based de-noising has also been demonstrated in (**g, h**).

provements in RMSD and correlation between the corrected Y^* and the reference X are somewhat superficial, masking the fact that the bias correction is limited to the coarsest scales. On the other hand, the A/S-based and MS methods can modify the original noise profiles in the data across the scales, by amplifying (or suppressing) noise in individual components (either Y_j, Y_S, or Y_A) with less-than (greater-than) unity pre-correction α. This may be considered undesirable for an objective to produce more physically representative data with a simple error structure on the whole. Therefore, arguably, none of these methods is entirely satisfactory, in manners of not removing the multiplicative biases completely and/or changing error characteristics. From this viewpoint, the task of bias correction is seen as inseparable from that of noise reduction when considering MS (or A/S) bias correction, unless certain components in MRA were explicitly ignored.

7 Combining bias correction with wavelet de-noising

The last example presents an impetus to consider noise removal prior to bias correction and produce a simpler error structure in the bias-corrected data Y^*. Critically, TC provides noise and signal estimates that can be used for de-noising through thresholding of wavelet coefficients p_{jk}. The basic rationale for wavelet thresholding (WT) is that insignificant detail coefficients are likely due to noise, while significant ones are related to the signal component. Thus, a coefficient is eliminated if its magnitude is less than a given

threshold λ_p; otherwise, it is modified according to a transformation function $\Gamma(p_{jk})$ to remove the influence of the noise (Donoho and Johnstone, 1994).

One commonly used transformation is soft thresholding (Donoho, 1995), where the coefficients are modified according to

$$\Gamma_{\lambda_p}\left(p_{jk}\right) = \text{sign}\left(p_{jk}\right)\max\left(|p_{jk} - \lambda_p|, 0\right). \quad (11)$$

Such de-noising filters have near-optimal properties in the minmax sense. We follow the *BayesShrink* rule of Chang et al. (2000) to define a set of scale-dependent threshold values using

$$\lambda_{p,j} = \frac{\text{var}\left(\epsilon_{p,j}\right)}{\alpha_{p,j}\text{std}\left(f_j\right)}, \quad (12)$$

where the variances are provided by TC (Appendix B). This choice of threshold is near optimal under the assumption that the signal is generalised Gaussian distributed and the noise is Gaussian. When the threshold value for $j = 1\text{–}2$ could not be estimated using TC, CDF matching was applied. While TC is an ideal error estimator, alternative estimators for the threshold values are also available to make the de-noising a stand-alone process (Donoho and Johnstone, 1994; Donoho, 1995). After WT, the de-noised time series is constructed via inverse DWT of the modified coefficients, and can be subsequently corrected for biases. Combining with the MS bias correction scheme, biased-corrected, de-noised data are generated via

$$Y^* = \frac{Y'}{\hat{\alpha}'_Y} + \sum_{j=1}^{j_0} \sum_{k=1}^{n_j} \frac{\Gamma_{\lambda_{p,j}}(Y_{jk})}{\hat{\alpha}_{Y,j}} \psi_{jk}. \tag{13}$$

The prescription, which is essentially a two-stage operation, was applied to AMS for comparisons with the previous results. The first stage of de-noising leads to smoothing of the time series, improved R with INS by 0.05, and reduced RMSD by $0.02\,\mathrm{m}^3\,\mathrm{m}^{-3}$. The actual SM variability has become more apparent in Fig. 6g. Over-smoothing can occur due to our inability to properly distinguish signal from noise in AMS_1 and AMS_2 where the signal-to-noise ratio is very low. However, without the second stage of bias correction, the dynamic ranges of de-noised AMS and INS are visibly different, such that the improvement in RMSD with INS is limited. Combining WT and MS leads to improvement in both metrics of $RMSD = 0.048\,\mathrm{m}^3\,\mathrm{m}^{-3}$ and $R = 0.711$, with Fig. 6h confirming that the reduced noise was not amplified by the MS rescaling.

8 Conclusions

This work combines MRA and TC in a new analysis framework with increased capacity to provide a more comprehensive view of the inter-data relations on short and long timescales. TC (or CDF) rescaling can be exploited on individual scales to reduce scale-specific multiplicative biases, and provide "prior" knowledge of noise for calibrating a WT-based de-noising filter. As a demonstration of principle, these methods are applied to SM data from in situ and satellite sensors and a land surface model. Using MRA, we found that the three data exhibit significantly different wavelet spectra and variable degrees of agreement on different timescales. On fine scales, the contribution of noise is most prominent, undermining the correlation between the data sets. By contrast, the biases are most apparent on coarse scales. Furthermore, these biases are non-systematic across timescales in the study region and across spatial locations over Australia, and the signal-to-noise ratios vary with scales and between the various data, pointing to the need to use correction schemes that are capable of handling such complexities.

These observations raised concerns about the possible inadequate treatment of SM data in the linear regime, even with anomaly/seasonal decomposition. Scale-by-scale linear rescaling based on a MRA-TC analysis framework offers a more comprehensive treatment of different biases on different scales, but error characteristics are found to be modified by variable rescaling, and can lead to undesirable noise amplification. The method of removing biases and noise on individual scales offers a remedy, although a few caveats should be noted. First, TC analysis requires a strong instrument and large sample, and in cases where these prerequisites are not met, we resort to sub-optimal estimation and rescaling methods. Second, the issue of non-stationarity in errors and scaling has not been addressed so far, and this can lead to biased estimates of the correction parameters for rescaling and de-noising. Despite this, DWT offers additional degree of freedom in translation parameter k to accommodate non-stationarity. Third, given the theoretic viewpoint presented in this work, further evaluations based on assimilation of data treated by different schemes are still warranted to assess their practical impacts. Notwithstanding these factors, MRA-TC analysis can be an important tool to allow better characterisation of the inter-sensor differences and to develop more effective strategies in harmonising a broad range of observational data records in oceanography and hydrometeorology.

Appendix A: Wavelet statistical analysis

MRA enables the (bulk) variance var(p) of a time series p to be decomposed into wavelet variances var(p_j) on different scales j. Analogous to a Fourier spectrum, the expansion of var(p) yields a wavelet spectrum and is given by

$$\text{var}(p) = \sum_{j=1}^{J} \text{var}\left(p_j\right) \tag{A1}$$

$$= \text{var}\left(p_{j_0}^{(a)}\right) + \sum_{j=1}^{j_0} \text{var}\left(p_j\right) \tag{A2}$$

where the variance of the approximation time series $p_{j_0}^{(a)}$ can be expressed in terms of that of the detail time series p_j.

Similarly, wavelet covariance cov(X_j, Y_j) at a given j indicates the contribution of covariance between two time series (X, Y) on that scale. Specifically, the wavelet covariance on scale j can be expressed as

$$\text{cov}\left(X_j, Y_j\right) = \frac{1}{n_j} \sum_{k=1}^{n_j} X_{jk} Y_{jk}, \tag{A3}$$

noting that there is an equivalence of computing (co)variance in the wavelet and time domains. To exclude the boundary influence of a finite-length time series and missing values in the time series, an estimator of the wavelet covariance can be constructed by excluding the coefficients affected by the boundaries and gaps, followed by renormalisation. In the paper, we find it more intuitive to report the wavelet correlation, namely

$$R\left(X_j, Y_j\right) = \frac{\text{cov}\left(X_j, Y_j\right)}{\sqrt{\text{var}\left(X_j\right) \text{var}\left(Y_j\right)}}. \tag{A4}$$

Appendix B: Multi-scale triple collocation

Starting with the scale-level affine model of Eqs. (4) and (5), the associated scaling coefficients ($\alpha'_p, \alpha_{p,j}$) and error variances (var(ϵ'_p), var($\epsilon_{p,j}$)) for each scale can be estimated using TC. We use solutions of Su et al. (2014a) for the data triplet $p = \{X, Y, Z\}$ on each scale separately: with X as the reference by setting $\alpha_{X,j}, \alpha'_X = 1$,

$$\hat{\alpha}_{Y,j} = \frac{\text{cov}\left(Y_j, Z_j\right)}{\text{cov}\left(X_j, Z_j\right)}, \tag{B1}$$

$$\hat{\alpha}_{Z,j} = \frac{\text{cov}\left(Y_j, Z_j\right)}{\text{cov}\left(X_j, Y_j\right)}, \tag{B2}$$

$$\hat{\text{var}}\left(\epsilon_{p,j}\right) = \text{var}\left(p_j\right) - \frac{\text{cov}\left(p_j, q_j\right) \text{cov}\left(p_j, r_j\right)}{\text{cov}\left(q_j, r_j\right)}, \tag{B3}$$

$$\hat{\text{var}}\left(f_j\right) = \text{var}\left(X_j\right) - \hat{\text{var}}\left(\epsilon_{p,j}\right) \tag{B4}$$

where q and r are also data labels, but $p \neq q \neq r$. The hat notation is used throughout the paper to distinguish estimates from true values. It can be shown that, in probability, TC yields unbiased estimates whereby $\hat{\alpha}_{p,j} = \alpha_{p,j}$, $\hat{\text{var}}(\epsilon_{p,j}) = \text{var}(\epsilon_{p,j})$, and $\hat{\text{var}}(f_j) = \text{var}(f_j)$. These expressions were used to compute the results in Fig. 5 and the threshold values for wavelet de-noising. When TC does not produce physically meaningful estimates from negative or small covariance due to weak instruments and possible inadequacy of the considered signal and noise model, the OLS estimator was used,

$$\hat{\alpha}_{Y,j}^{\text{OLS}} = \frac{\text{cov}\left(X_j, Y_j\right)}{\text{var}\left(X_j\right)}, \tag{B5}$$

although its estimates are biased ($\hat{\alpha}_{Y,j}^{\text{OLS}} < \alpha_{Y,j}$) for our purpose, due to the extraneous contribution of noise variance in the denominator. Similarly the VAR estimator can be used, $\hat{\alpha}_{Y,j}^{\text{VAR}} = \sqrt{\text{var}(Y_j)/\text{var}(X_j)}$, but it is also biased.

Acknowledgements. We thank Wade Crow for valuable discussions and Clara Draper for her critiques of the early drafts. We acknowledge gratefully the feedback of Simon Zwieback, two anonymous reviewers, Wolfgang Wagner, and Editor Niko Verhoest in the refinement of our manuscript. We also thank all who contributed to the data sets used in this study. Kyeamba in situ data were produced by colleagues at Monash University and the University of Melbourne who have been involved in the OzNet programme. AMSR-E data were produced by Richard de Jeu and colleagues at Vrije University Amsterdam and NASA. The MERRA-Land data set was provided by NASA Goddard Earth Sciences Data and Information Services Center (GES DISC). The land cover/use map was produced by merging land cover (Lymburner et al., 2010) and land use (Australian Bureau of Rural Science, 2010) data sets. The recalibrated precipitation data of the Australian Water Availability Project (AWAP) (Jones et al., 2009) were obtained from the Australian Bureau of Meteorology. National soil data (McKenzie et al., 2005) were provided by the Australian Collaborative Land Evaluation Program ACLEP, endorsed through the National Committee on Soil and Terrain NCST (http://www.clw.csiro.au/aclep). The 9 s digital elevation map is obtained from Geoscience Australia (2008). This research was conducted with financial support from the Australian Research Council (ARC Linkage Project No. LP110200520) and the Bureau of Meteorology, Australia.

Edited by: N. Verhoest

References

Albergel, C., De Rosnay, P., Gruhier, C., Muñoz-Sabater, J., Hasenauer, S., Isaksen, L., Kerr, Y., and Wagner, W.: Evaluation of remotely sensed and modelled soil moisture products using global ground-based in situ observations, Remote Sens. Environ., 118, 215–226, doi:10.1016/j.rse.2011.11.017, 2012.

Australian Bureau of Rural Science: Land Use of Australia, version 4, 2005/2006, available at: http://data.daff.gov.au/anrdl/metadata_files/pa_luav4g9abl07811a00.xml (last access: 24 July 2014), 2010.

Brocca, L., Moramarco, T., Melone, F., Wagner, W., Hasenauer, S., and Hahn, S.: Assimilation of surface- and root-zone ASCAT soil moisture products into rainfall–runoff modeling, IEEE T. Geosci. Remote, 50, 2542–2555, doi:10.1109/TGRS.2011.2177468, 2012.

Caires, S. and Sterl, A.: Validation of ocean wind and wave data using triple collocation, J. Geophys. Res., 103, 3098, doi:10.1029/2002JC001491, 2003.

Chang, S. G., Yu, B., and Yeterli, M.: Adaptive wavelet thresholding for imaging denoising and compression, IEEE T. Image Process., 9, 1532–1546, doi:10.1109/83.862633, 2000.

Daubechies, I.: Ten Lectures on Wavelets, Society for Industrial and Applied Mathematics, doi:10.1137/1.9781611970104, 1992.

Donoho, D. L.: Denoising via soft thresholding, IEEE T. Inform. Theory, 41, 613–627, doi:10.1109/18.382009, 1995.

Donoho, D. L. and Johnstone, I. M.: Ideal spatial adaption via wavelet shrinkage, Biometrika, 81, 425–455, doi:10.1093/biomet/81.3.425, 1994.

Draper, C., Reichle, R., De Lannoy, G., and Liu, Q.: Assimilation of passive and active microwave soil moisture retrievals, Geophys. Res. Lett., 39, L04401, doi:10.1029/2011GL050655, 2012.

Draper, C. S., Reichle, R. R., de Jeu, R. A., Naeimi, V., Parinussa, R. M., and Wagner, W.: Estimating root mean square errors in remotely sensed soilmoisture over continental scale domains, Remote Sens. Environ., 137, 288–298, doi:10.1175/JHM-D-12-052.1, 2013.

Drusch, M., Wood, E. F., and Gao, H.: Observation operators for the direct assimilation of TRMM microwave imager retrieved soil moisture, Geophys. Res. Lett., 32, L15403, doi:10.1029/2005GL023623, 2005.

Entin, J. K., Robock, A., Vinnikov, K. Y., Hollinger, S. E., Liu, S., and Namkhai, A.: Temporal and spatial scales of observed soil moisture variations in the extratropics, J. Geophys. Res., 105, 11865–11877, doi:10.1029/2000JD900051, 2000.

Garcia, D.: Robust smoothing of gridded data in one and higher dimensions with missing values, Comput. Stat. Data Anal., 54, 1167–1178, doi:10.1016/j.csda.2009.09.020, 2010.

Geoscience Australia: GEODATA 9-Second DEM Version 3, available at: http://www.ga.gov.au/metadata-gateway/metadata/record/66006/ (last access: 24 July 2014), 2008.

GLACE Team, Koster, R. D., Dirmeyer, P. A., Guo, Z., Bonan, G., and Chan, E.: Regions of strong coupling between soil moisture and precipitation, Science, 305, 1138–1140, doi:10.1126/science.1100217, 2004.

Gruber, A., Dorigo, W. A., Zwieback, S., Xaver, A., and Wagner, W.: Characterizing Coarse-Scale Representativeness of in situ Soil Moisture Measurements from the International Soil Moisture Network, Vadose Zone J., 12, doi:10.2136/vzj2012.0170, 2013.

Janssen, P. A. E. M., Abdalla, S., Hersbach, H., and Bidlot, J.-R.: Error estimation of buoy, satellite, and model wave height data, J. Atmos. Ocean. Tech., 24, 1665–1677, doi:10.1175/JTECH2069.1, 2007.

Jones, D. A., Wang, W., and Fawcett, R.: High-quality spatial climate data-sets for Australia, Aust. Meteorol. Oceanogr., 58, 233–248, 2009.

Koster, R. D., Guo, Z., Yang, R., Dirmeyer, P. A., Mitchell, K., and Puma, M. J.: On the nature of soil moisture in land surface models, J. Climate, 22, 4322–4335, doi:10.1175/2009JCLI2832.1, 2009.

Kumar, S. V., Reichle, R. H., Harrison, K. W., Peters-Lidard, C. D., Yatheendradas, S., and Santanello, J. A.: A comparison of methods for a priori bias correction in soil moisture data assimilation, Water Resour. Res., 48, W03515, doi:10.1029/2010WR010261, 2012.

Legates, D. R., Mahmood, R., Levia, D. F., DeLiberty, T. L., Quiring, S. M., Houser, C., and Nelson, F. E.: Soil moisture: a central and unifying theme in physical geography, Prog. Phys. Geogr., 35, 65–86, doi:10.1177/0309133310386514, 2011.

Liu, Y. Y., Dorigo, W. A., Parinussa, R. M., De Jeu, R. A. M., Wagner, W., McCabe, M. F., Evans, J. P., and Van Dijk, A. I. J. M.: Trend-preserving blending of passive and active microwave soil moisture retrievals, Remote Sens. Environ., 123, 280–297, doi:10.1016/j.rse.2012.03.014, 2012.

Loew, A. and Schlenz, F.: A dynamic approach for evaluating coarse scale satellite soil moisture products, Hydrol. Earth Syst. Sci., 15, 75–90, doi:10.5194/hess-15-75-2011, 2011.

Lymburner, L., Tan, P., Mueller, N., Thackway, R., Lewis, A., Thankappan, M., Randall, L., Islam, A., and Senarath, U.: 250 metre Dynamic Land Cover Dataset of Australia, 1st Edn., Geoscience Australia, Canberra, 2010.

Mallat, S. G.: A theory for multiresolution signal decomposition: the wavelet representation, IEEE T. Pattern Anal., 11, 674–693, doi:10.1109/34.192463, 1989.

McKenzie, N. J., Jacquier, D. W., Maschmedt, D. J., Griffin, E. A., and Brough, D. M.: The Australian Soil Resource Information System: Technical Specifications, National Committee on Soil and Terrain Information/Australian Collaborative Land Evaluation Program, Canberra, 2005.

Miralles, D. G., De Jeu, R. A. M., Gash, J. H., Holmes, T. R. H., and Dolman, A. J.: Magnitude and variability of land evaporation and its components at the global scale, Hydrol. Earth Syst. Sci., 15, 967–981, doi:10.5194/hess-15-967-2011, 2011.

Mittelbach, H. and Seneviratne, S. I.: A new perspective on the spatio-temporal variability of soil moisture: temporal dynamics versus time-invariant contributions, Hydrol. Earth Syst. Sci., 16, 2169–2179, doi:10.5194/hess-16-2169-2012, 2012.

Owe, M., de Jeu, R., and Holmes, T.: Multisensor historical climatology of satellite-derived global land surface moisture, J. Geophys. Res., 113, F01002, doi:10.1029/2007JF000769, 2008.

Peel, M. C., Finlayson, B. L., and McMahon, T. A.: Updated world map of the Köppen–Geiger climate classification, Hydrol. Earth Syst. Sci., 11, 1633–1644, doi:10.5194/hess-11-1633-2007, 2007.

Reichle, R. H. and Koster, R. D.: Bias reduction in short records of satellite soil moisture, Geophys. Res. Lett., 31, L19501, doi:10.1029/2004GL020938, 2004.

Reichle, R. H., Koster, R. D., Dong, J., and Berg, A. A.: Global soil moisture from satellite observations, land surface models, and ground data: implications for data assimilation, J. Hydrometeorol., 5, 430–442, doi:10.1175/1525-7541(2004)005<0430:GSMFSO>2.0.CO;2, 2004.

Reichle, R. H., Koster, R. D., Liu, P., Mahanama, S. P. P., Njoku, E. G., and Owe, M.: Comparison and assimilation of global soil moisture retrievals from the Advanced Microwave Scanning Radiometer for the Earth Observing System (AMSR-E) and the Scanning Multichannel Microwave Radiometer (SMMR), J. Geophys. Res., 112, D09108, doi:10.1029/2006JD008033, 2007.

Reichle, R. H., Koster, R. D., De Lannoy, G. J. M., Forman, B. A., Liu, Q., Mahanama, S. P. P., and Touré, A.: Assessment and enhancement of MERRA land surface hydrology estimates, J. Climate, 24, 6322–6338, doi:10.1175/JCLI-D-10-05033.1, 2011.

Rienecker, M. M., Suarez, M. J., Gelaro, R., Todling, R., Bacmeister, J., Liu, E., Bosilovich, M. G., Schubert, S. D., Takacs, L., Kim, G.-K., Bloom, S., Chen, J., Collins, D., Conaty, A., da Silva, A., Gu, W., Joiner, J., Koster, R. D., Lucchesi, R., Molod, A., Owens, T., Pawson, S., Pegion, P., Redder, C. R., Reichle, R., Robertson, F. R., Ruddick, A. G., Sienkiewicz, M., and Woollen, J.: MERRA – NASA's Modern-Era Retrospective Analysis for Research and Applications, J. Climate, 24, 3624–3648, doi:10.1175/JCLI-D-11-00015.1, 2011.

Rodrigeuz-Iturbe, I.: Ecohydrology: a hydrologic perspective of climate-soil-vegetation dynamics, Water Resour. Res., 36, 3–9, doi:10.1029/1999WR900210, 2000.

Roebeling, R. A., Wolters, E. L. A., Meirink, J. F., and Leijnse, H.: Triple collocation of summer precipitation retrievals from SEVIRI over Europe with gridded rain gauge and weather radar data, J. Hydrometeorol., 13, 1552–1566, doi:10.1175/JHM-D-11-089.1, 2013.

Scipal, K., Drusch, M., and Wagner, W.: Assimilation of a ERS scatterometer derived soil moisture index in the ECMWF numerical weather prediction system, Adv. Water Resour., 31, 1101–1112, doi:10.1016/j.advwatres.2008.04.013, 2008a.

Scipal, K., Holmes, T., de Jeu, R., Naeimi, V., and Wagner, W.: A possible solution for the problem of estimating the error structure of global soil moisture data sets, Geophys. Res. Lett., 35, L24403, doi:10.1029/2008GL035599, 2008b.

Smith, A. B., Walker, J. P., Western, A. W., Young, R. I., Ellett, K. M., Pipunic, R. C., Grayson, R. B., Siriwardena, L., Chiew, F. H. S., and Richter, H.: The Murrumbidgee soil moisture network data set, Water Resour. Res., 48, W07701, doi:10.1029/2012WR011976, 2012.

Steelman, C. M., Endres, A. L., and Jones, J. P.: High-resolution ground penetrating radar monitoring of soil moisture dynamics: Field results, interpretation, and comparison with unsaturated flow model, Water Resour. Res., 48, W09538, doi:10.1029/2011WR011414, 2012.

Stoffelen, A.: Towards the true near-surface wind speed: Error modelling and calibration using triple collocation, J. Geophys. Res., 103, 7755–7766, doi:10.1029/97JC03180, 1998.

Su, C.-H., Ryu, D., Young, R. I., Western, A. W., and Wagner, W.: Inter-comparison of microwave satellite soil moisture retrivals over Murrumbidgee Basin, southeast Australia, Remote Sens. Environ., 134, 1–11, doi:10.1016/j.rse.2013.02.016, 2013a.

Su, C.-H., Ryu, D., Western, A. W., and W. Wagner, W.: De-noising of passive and active microwave satellite soil moisture time series, Geophys. Res. Lett., 40, 3624–3630, doi:10.1002/grl.50695, 2013b.

Su, C.-H., Ryu, D., Crow, W. T., and Western, A. W.: Beyond triple collocation: applications to satellite soil moisture, J. Geophys. Res.-Atmos., 119, 6419–6439, doi:10.1002/2013JD021043, 2014a.

Su, C.-H., Ryu, D., Crow, W. T., and Western, A. W.: Stand-alone error characterisation of microwave satellite soil moisture using a Fourier method, Remote Sens. Environ., 154, 115–126, doi:10.1016/j.rse.2014.08.014, 2014b.

Wang, G., Garcia, D., Liu, Y., de Jeu, R., and Dolman, A. J.: A three-dimensional gap filling method for large geophysical datasets: application to global satellite soil moisture observations, Environ. Modell. Softw., 30, 139–142, doi:10.1016/j.envsoft.2011.10.015, 2012.

Western, A. W., Grayson, R. B., and Bloschl, G.: Scaling of soil moisture: a hydrologic perspective, Annu. Rev. Earth Pl. Sc., 30, 149–180, doi:10.1146/annurev.earth.30.091201.140434, 2002.

Wright, P. G.: The Tariff on Animal and Vegetable Oils, Macmillan, New York, 1928.

Yilmaz, M. T. and Crow, W. T.: The optimality of potential rescaling approaches in land data assimilation, J. Hydrometeorol., 14, 650–660, doi:10.1175/JHM-D-12-052.1, 2013.

Zwieback, S., Scipal, K., Dorigo, W., and Wagner, W.: Structural and statistical properties of the collocation technique for error characterization, Nonlin. Processes Geophys., 19, 69–80, doi:10.5194/npg-19-69-2012, 2012.

Zwieback, S., Dorigo, W., and Wagner, W.: Estimation of the temporal autocorrelation structure by the collocation technique with emphasis on soil moisture studies, Hydrolog. Sci. J., 58, 1729–1747, doi:10.1080/02626667.2013.839876, 2013.

Hydrometeorological effects of historical land-conversion in an ecosystem-atmosphere model of Northern South America

R. G. Knox[1,*], M. Longo[2,**], A. L. S. Swann[3], K. Zhang[2,***], N. M. Levine[2,****], P. R. Moorcroft[2], and R. L. Bras[4]

[1] Massachusetts Institute of Technology, Cambridge, Massachusetts, USA
[2] Harvard University, Cambridge, Massachusetts, USA
[3] University of Washington, Seattle, Washington, USA
[4] Georgia Institute of Technology, Atlanta, Georgia, USA
[*] now at: Lawrence Berkeley National Laboratory, Berkeley, California, USA
[**] now at: EMBRAPA Satellite Monitoring, Campinas, São Paulo, Brazil
[***] now at: Cooperative Institute for Mesoscale Meteorological Studies, University of Oklahoma, Oklahoma, USA
[****] now at: University of Southern California, Los Angeles, California, USA

Correspondence to: R. G. Knox (rgknox@lbl.gov)

Abstract. This work investigates how the integrated land use of northern South America has affected the present day regional patterns of hydrology. A model of the terrestrial ecosystems (ecosystem demography model 2: ED2) is combined with an atmospheric model (Brazilian Regional Atmospheric Modeling System: BRAMS). Two realizations of the structure and composition of terrestrial vegetation are used as the sole differences in boundary conditions that drive two simulations. One realization captures the present day vegetation condition that includes deforestation and land conversion, the other is an estimate of the potential structure and composition of the region's vegetation without human influence. Model output is assessed for differences in resulting hydrometeorology.

The simulations suggest that the history of land conversion in northern South America is not associated with a significant precipitation bias in the northern part of the continent, but has shown evidence of a negative bias in mean regional evapotranspiration and a positive bias in mean regional runoff. Also, negative anomalies in evaporation rates showed pattern similarity with areas where deforestation has occurred. In the central eastern Amazon there was an area where deforestation and abandonment had lead to an overall reduction of above-ground biomass, but this was accompanied by a shift in forest composition towards early successional functional types and grid-average-patterned increases in annual transpiration.

Anomalies in annual precipitation showed mixed evidence of consistent patterning. Two focus areas were identified where more consistent precipitation anomalies formed, one in the Brazilian state of Pará where a dipole pattern formed, and one in the Bolivian Gran Chaco, where a negative anomaly was identified. These locations were scrutinized to understand the basis of their anomalous hydrometeorologic response. In both cases, deforestation led to increased total surface albedo, driving decreases in net radiation, boundary layer moist static energy and ultimately decreased convective precipitation. In the case of the Gran Chaco, decreased precipitation was also a result of decreased advective moisture transport, indicating that differences in local hydrometeorology may manifest via teleconnections with the greater region.

1 Introduction

It has been held that massive and widespread Amazonian deforestation would lead to regional reductions in precipitation, evaporation, and moisture convergence, with slight increases in surface temperature (Henderson-Sellers et al., 1993; Nobre et al., 1991; Lean and Warrilow, 1989; Dickinson and Henderson-Sellers, 1988). The Amazon Basin and its forest ecosystems are also an important component of the global circulation of energy (Gedney and Valdes, 2000), where changes (complete deforestation in either Amazonia or all tropical broadleaf forests) are thought to teleconnect beyond the continent (Avissar and Werth, 2005; Snyder, 2010). The literature documenting Amazonian land conversion and the surrounding areas is significant, the reader is referred to a small selection of non-exhaustive references for some background (Cardille and Foley, 2003; Skole and Tucker, 1993; INPE, 2003; Geist and Lambin, 2002; Laurance et al., 2001; Nepstad et al., 2001; Soares-Filho et al., 2006). The work presented here is motivated by a need to better understand how the history of land conversion has influenced the hydrology of the region. As will be outlined further, the mechanistic relationships between land conversion and hydrometeorological response is complex and has benefited from study with newer generations of land–atmosphere models with increased granularity and increased complexity in representing physical process.

There are several direct hydrologic mechanisms that connect changes in tropical forest structure (i.e., deforested versus intact canopies) to the regional climate system. Leaves, stems, and bare earth have variable light-scattering properties, such that intact forest canopies composed of dark vegetation typically have lower shortwave radiation albedo than areas with exposed soil (Chapin et al., 2002). This directly impacts the surface energy balance via net radiation. Forest canopies have a complex relationship with the surface moisture balance and mediate the transport of water in numerous ways. Model studies have shown that the representation of canopy interception can substantially impact the partitioning of evapotranspiration and surface runoff (Pitman et al., 1990; Wang et al., 2007; Crockford and Richardson, 2000). Some studies have found that forest canopies increase the interception of precipitation (Asdak et al., 1998; Dietz et al., 2006), and further that crown structure influences turbulent transport and evaporation of wet leaves (Dietz et al., 2006). Yet some have indicated that canopy interception increased in degraded forests (Chappell et al., 2001). Pastures and converted agricultural systems are generally associated with soil degradation such as decreased infiltration rates, nutrient loss and increased surface runoff, subject to variability and factors such as the soil texture and the existence of perennial under-story vegetation (Benegas et al., 2014). Practices such as grazing and agriculture promote soil compaction and decreased infiltration (Martinez and Zinck, 2004; Lal, 1996), and intense fires used for clearing lands may reduce soil organic matter that may favor infiltration (Kennard and Gholz, 2001). Forests with deep rooted trees draw from deeper soil moisture pools, which have different periodicity in available water and therefore alter the timing of latent heat flux via transpiration compared to grasslands (Kleidon and Heimann, 2000; Nepstad et al., 1994). Canopy structure also influences the turbulent exchange of heat, moisture and momentum with the atmosphere (Raupach et al., 1996).

The higher surface temperatures associated with widespread deforestation, as reported with the first generations of general circulation models and beyond, (Henderson-Sellers et al., 1993; Nobre et al., 1991; Lean and Warrilow, 1989; Dickinson and Henderson-Sellers, 1988) are thought to be the result of losses in evaporative cooling associated with cleared vegetation. The decrease in evaporative cooling is also thought to drive reduced precipitation, and subsequently reduces the heat released to the atmosphere through condensation (Eltahir and Bras, 1993). This has the potential to outcompete the effect of increased surface albedo of deforested lands (Eltahir, 1996), which suppresses net radiation thereby promoting surface cooling and divergence (Eltahir and Bras, 1993; Lean and Warrilow, 1989). Positive surface temperature anomalies induce convergent circulations coincident with a decrease in surface pressure. Decreased precipitation heating anomalies reduce the tendency towards convergence.

In the southwestern Amazonian dry season, statistical connections have been made between pastures and higher incidents of shallow cumulus clouds, compared to intact forests where shallow clouds are less frequent yet deep precipitating convective events are more frequent (Wang et al., 2009). The higher rate of deep precipitating convection over forests was associated with larger values of convective available potential energy (CAPE) (Williams and Renno, 1993), which in this case was driven by increased humidity and moisture flux from intact forest canopies. The increased frequency of shallow convection was attributed to more vigorous mesoscale circulations associated with deforestation induced land-surface heterogeneity (Wang et al., 2000; Souza et al., 2000).

Regional land–atmosphere simulations that can parameterize convective clouds indicate that structured land-conversion scenarios elicit shifts in mean basin precipitation, albeit less so than traditional coarse scale general circulation model studies (Silva et al., 2008). Coherent land surface patterns may strengthen convergence zones on the surface, creating vertical wind triggers to thunderstorms. For instance, Avissar and Werth (2005) determined that coherent land surface patterns transfer heat, moisture and wave energy to the higher latitudes through thunderstorm activity. Moreover, mesoscale simulations are found to capture key cloud feedback processes which fundamentally alter the atmospheric response to land-surface heterogeneities (Medvigy et al., 2011). The mesoscale simulation's ability to represent land-use scenarios at finer resolutions can impact spatial patterning of rainfall. For example, western propagating squall lines from the Atlantic are thought to dissipate over regions of wide-spread deforestation (Silva et al., 2008; d'Almeida et al., 2007). Evidence has also shown that convection can be driven by localized convergent air circulations triggered by land-surface heterogeneities, and that the likelihood and quality of resulting events are both dependent on the scale of heterogeneity and the position relative to disturbed and intact landscapes (Pielke, 2001; Dalu et al., 1996; Baldi et al., 2008; Anthes, 1984; Knox et al., 2011; Wang et al., 2009).

Regional scale-coupled land–atmosphere models can capture feed-backs resulting from land conversion at the scale of tens of kilometers and lower, particularly through improved resolution and the parameterization of atmospheric physics (such as convection and radiation scattering). At the land-surface, there is variability in canopy structure at the gap (size of a single large tree crown) scale and below. Processes that occur at these scales may be important to predicting ecosystem response and land–atmosphere exchange, as discussed above. Physics-based land surface models have non-linear representations of hydrologic and thermodynamic processes; therefore, using average canopy structure (such as in the "big-leaf" approach) to represent processes uniformly may provide different results compared to explicitly capturing these processes with sub-canopy and gap-scale structure.

This research uses the Brazilian Regional Atmospheric Modeling System (BRAMS, a variant of the Regional Atmospheric Modeling System (RAMS), Cotton et al., 2003) coupled with the ecosystem demography model 2 (ED2 or EDM2, Moorcroft et al., 2001; Medvigy et al., 2009), to explore the sensitivity of hydrologic climate of northern South America in response to present day land conversion. This modeling system can explicitly represent the processes of energy and mass transfer in the canopy and soil system, with sub-grid variability along ecosystem age-structured and vegetation size-structured axes. An experiment is conducted by comparing simulations that singularly differ in their representation of regional vegetation cover, one which captures the present day vegetation condition that includes deforestation and land-conversion, the other being an estimate of the potential structure and composition of the region's vegetation without human influence. Section 2 of this manuscript will detail experiment design of the coupled model experiment. The model system and experiment design is verified by comparison of model output with observations, see Appendices B–F.

In Sect. 3 we evaluate the hydrometeorological response to the changes in land-use history in a regional context. In Sect. 4 we evaluate the processes underlying the observed changes in hydrometeorology in two focus areas. A discussion and conclusion of the results follows.

2   Experiment design

The main task of this experiment is to conduct two regional simulations of the South American biosphere and atmosphere. The defining difference between the two simulations is how the land-surface model (ED2) represents the structure (the distribution of plant sizes) and composition (the distribution of plant types) of the region's terrestrial ecosystems, as a consequence of two different disturbance regimes. In one simulation, the vegetation reflects a structure and composition that has no effects of human land use, henceforth referred to as a *potential vegetation* (PV) condition. In the other simulation, the model will incorporate an estimate of modern (e.g., 2008) human land use, henceforth referred to as an *actual vegetation* (AV) condition. The procedure is broken down into steps and elaborated upon.

2.1   Description of the vegetation model – ED2

The ecosystem demography model 2 predicts the changes in the terrestrial vegetation structure, as modulated by the physically based conservation of water, carbon and enthalpy. Its central design philosophy assumes that the stochastic representation of plant communities integrated over a large sample can be portrayed deterministically as land fractions and plant groups, with explicit size (of the plants) and age (time since a patch of land housing the plants has experienced major disturbance) structure. By discretely representing the distribution of plant sizes and types, it can estimate vertical canopy structure, which directly impacts radiation scattering (throughfall interception), and in-canopy transport of scalars. By discretely representing variable disturbance history, the model can also explicitly simulate energy balance over a wide array of canopy types (closed canopies, recovering forests, grasslands, etc.) that exist within the footprint of driving meteorological data. In this experiment, the ED2 model resolves five different relevant tropical plant functional types (PFT): C4 grasses, early successional tropical evergreens, mid-successional tropical evergreens, late successional tropical evergreens and tropical C2 grasses. In the ED2 system, PFTs are used as sets of attributes that can be applied to numerous explicitly resolved plant groups of different size and in different parts of the disturbance strata.

2.2   Generation of surface boundary conditions

The creation of the initial vegetation conditions used a "spin-up" process. The spin-up process is an off-line dynamic ED2 simulation, where the driving atmospheric information comes from a pre-compiled forcing data set. The vegetation is initialized with an equal assortment of newly recruited (saplings) plant types. The off-line model is integrated over several centuries by sampling from the climate data set as the vegetation reaches an equilibrium. We identify equilibrium when the total biomass of each plant func-

Table 1. Simulation constraints describing the spin-up process creating the initial boundary conditions.

Specification	Value
Climate data	modified DS314*
Soil data	Quesada et al. (2011) + IGBP-DIS
Plant types	late succession tropical evergreens mid succession tropical evergreens early succession tropical evergreens subtropical grasses C4 grasses
Simulation period	508 years
Spatial resolution	gridded 1°
Bounding domain	30° S–15° N, 85° W–30° W
Tree allometry (DBH,height) (crown properties)	Chave et al. (2001), Baker et al. (2004b) Poorter et al. (2006), Dietze et al. (2008)
Turbulent transport	Beljaars and Holtslag (1991) atmospheric boundary Massman (1997) within canopy
Photosynthesis & leaf conductance	Collatz et al. (1991) Collatz et al. (1992) Leuning (1995)
Canopy radiation scattering	Zhao and Qualls (2005, 2006)
Soil hydrology	Walko et al. (2000), Tremback and Kessler (1985) Medvigy et al. (2009)

* Modified DS314 data is derived from Sheffield et al. (2006), precipitation down-scaling and radiation interpolation is applied, see footnote for data availability.

tional type in a grid-cell does not change more than 0.5 % over a period of 40 years. If equilibrium within this threshold was not achieved, the spin-up was allowed to continue to 508 years before stopping. For reference, ED2 simulations in old-growth central Amazonian forests take roughly 250 years to reach equilibrium biomass. The ED2 vegetation structure and composition at the end of the multi-century simulation was saved as the *potential vegetation* (PV) initial condition.

A summary of the simulation conditions in the spin-up is covered in Table 1. The model soil textures were derived from a combination of databases. Within the Amazon Basin, soil data were retrieved from Quesada et al. (2011); outside the basin soil data were retrieved from a combination of RADAMBRASIL and IGBP-DIS (Scholes et al., 1995; Rossato, 2001). The climate data used to drive the spin-up process was derived from the UCAR DS314 product (Sheffield et al., 2006)[1]. The DS314 is based on the National

Center for Environmental Prediction's Reanalysis Product (NCEP) and maintains the same global and temporal coverage period but has bias corrections and increased resolution based on the assimilation of composite data sets. The DS314 surface precipitation record was further processed such that grid cell average precipitation was downscaled to reflect the point-scale statistical qualities of local rain gauges. This technique used methods of Lammering and Dwyer (2000), and is explained in more detail in Knox (2012). The native NCEP reanalysis and European Center for Medium Range Weather Forecasting (ECMWF) 40 year Reanalysis (ERA-40) were also tested as driver data sets. The downscaled DS314 was ultimately chosen due to better agreement of estimated equilibrium biomass with observations (not shown).

The *actual vegetation* (AV) was created by continuing the simulation that produced the *potential vegetation*, and assumed that the starting year was 1900. This simulation was continued for another 108 years (until 2008) while incorporating human driven land-use change. Throughout the 508 year *potential vegetation* spin-up, as well as the 108 year continuation with human land use, the atmospheric carbon dioxide concentration was held constant at 378 ppm, which

[1] Original data sets used in the DS314 are from the Research Data Archive (RDA) which is maintained by the Computational and Information Systems Laboratory (CISL) at the National Center for Atmospheric Research (NCAR). The original Sheffield/DS314 data are available from the RDA (http://dss.ucar.edu) in data set number ds314.0.

approximates concentrations at the turn of the millennium (present day).

The model applies human land use by reading an externally compiled data set of land-use transition matrices (Albani et al., 2006) that defines the area fractions of which various land-cover types will change to another type over the course of the year. Two external data sets are used to create the land-use transition matrices, the global land-use data set (GLU) (Hurtt et al., 2006) and the SIMAMAZONIA-1 data set (Soares-Filho et al., 2006). The GLU data set incorporates the SAGE-HYDE 3.3.1 data set and provides land-use transitions in its native format globally, on a 1° grid from the years 1700 to 1999[2]. The SIMAMAZONIA 1 product provides a more intensive assessment of forest cover and deforestation focused in the Amazon Basin, starting in the year 2000. The data is formatted as yearly 1 km forest cover grids (forest, non-forest, and natural grasslands). The fraction of forest and non-forest cells that fall within each ED2 model simulation grid-cell are counted for each year. This enables the calculation of a rate of change equivalent to the transition matrix format of the GLU data set. The transitions from the GLU data set from 1990 to 1999 were linearly scaled to have continuity with the SIMAMAZONIA data set that is introduced in 2000. Land use reported in the GLU data prior to 1900 were lumped into a single combined transition and applied at the year 1900. A map of the fraction of the land surface containing human land use is provided (see Fig. 1).

Regional maps of above-ground biomass for the *potential vegetation* (PV) scenario and the differences between the two scenarios (*actual vegetation–potential vegetation*, or AV–PV) are provided in Fig. 2. The majority of above-ground biomass in the *potential* simulation is concentrated in the Amazon Basin and the Atlantic Forest of southern Brazil. Late successional broadleaf evergreens comprise most of the above-ground biomass in these regions. Early successional broadleaf evergreens are a prevalent but secondary contributor to biomass in the Amazon Basin. The early succession's contribute the majority of biomass in Cerrado (savanna like ecosystem, mixed open canopy forests with grasses) ecotones found roughly in central Brazil on the southern border of the Amazon rain-forest. This is consistent with their competition and resource niche which emphasizes fast growth and colonization of disturbed areas (such as fire and drought prone Cerrado). The model-estimated equilibrium above-ground biomass (AGB) and basal area (BA) that represent the initial condition are compared with a collection of census measurements in Baker et al. (2004a, b) (see Appendix B).

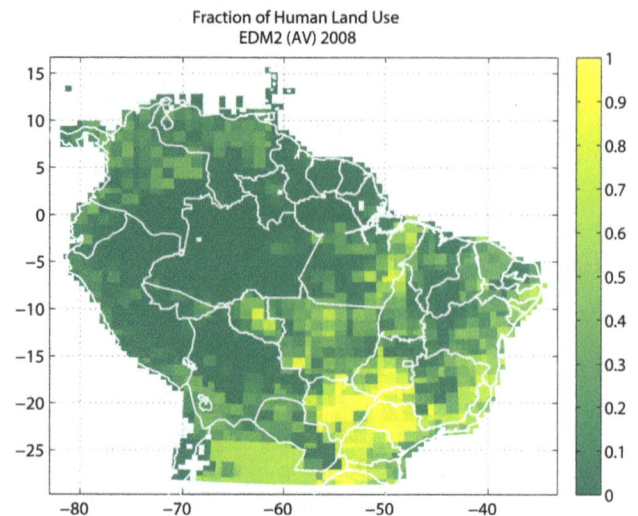

Figure 1. Fraction of the land surface with human land use. Output is taken from the ED2 the *actual vegetation* (AV) simulation, which was driven with global land-use (GLU) and SIMAMAZONIA-1-land-use transition data.

2.3 Land–atmosphere coupled simulations

The two coupled land–atmosphere simulations were conducted over 4 years, from January 2002 through December 2005. These 4 years were chosen because of the availability of lateral boundary conditions and validation data sets. With the exception of differing vegetation structure at the lower boundary, the lateral boundary conditions, model parameters, initialization of the atmospheric state, and timing are all identical between the two. The lateral boundary conditions (air temperature, specific humidity, geopotential, meridional wind speed, zonal wind speed) are taken from the European Centre for Medium Range Weather Forecasting's Interim Reanalysis (ERA-Interim) product (Dee et al., 2011). The data is interpolated from the ERA-Interim's model native Reduced Gaussian Grid (N128, which has an equatorial horizontal resolution of 0.75°).

The *actual* and *potential* boundary conditions utilized a dynamic model process to generate structure and composition. However, when applied to the coupled simulations, the land-surface dynamics including the processes of mortality, recruitment and growth are turned off and only phenology is left to vary in time. The motivation for this decision is to create a more simple comparison and efficient simulation. Further, the length of the coupled simulations are not long enough to generate large changes in above-ground biomass. As an example, Lewis (2006) and Baker et al. (2004a) estimate that in recent decades, the Amazon has sequestered approximately $0.6 \pm 0.2\,\mathrm{Mg\,C\,ha^{-1}\,yr^{-1}}$. Over a course of 4 years, this is less than $3\,\mathrm{Mg\,C\,ha^{-1}}$, which is on the order of 1–2 % of total forest biomass.

---

[2]The use of the SAGE-HYDE 3.3.1 global land-use data set acknowledges the University of New Hampshire, EOS-WEBSTER Earth Science Information Partner (ESIP) as the data distributor for this data set.

Table 2. Run time parameters and specifications in the ED2-BRAMS coupled simulations.

Specification	Value
Simulation period	January 2002–December 2005
Grid projection	polar stereographic
Grid dimensions	98° (E–W), 86° (N–S), 56° (vertical)
Horizontal grid resolution	64 km
Vertical grid resolution	110 m (lowest) stretching to 1500 m at 7 %
Atmospheric time step	30 s
Atmospheric acoustic time step	10 s
Land–surface model time step	120 s
Method of calculating Updraft base	level of maximum sum of mean and variance of vertical velocity
Number of prototype cloud scales	2
Mean radius of cloud 1	20 000 m
Minimum depth of cloud 1	4000 m
Mean radius of cloud 2	800 m
Minimum depth of cloud 2	80 m
Cumulus convective scheme	Grell and Dévényi (2002)
Cumulus convective trigger	pressure differential between updraft base and LFC* < 100 hpa
Cumulus dynamic control	Kain and Fritsch (1990), Kain (2004)
Condensate to precipitation conversion efficiency	3 %
Cloud # concentrations and distribution parameters	Medvigy et al. (2010)
Turbulent closure	Nakanishi and Niino (2006)
Shortwave radiation scattering	Harrington and Olsson (2001)
Longwave radiation scattering	Chen and Cotton (1983)
Advection	monotonic, Walcek and Aleksic (1998) & Freitas et al. (2012)
Cumulus feedback on radiation?	yes

* LFC = level of free convection.

A group of modeling parameters associated with convective parameterization and the radiation scattering of convective clouds were tuned using a manual binary search procedure. The parameters of mean cloud radius, mean cloud depth, cumulus convective trigger mechanism, dynamic control method and the condensate to precipitation conversion efficiency were calibrated against the Tropical Rainfall Measurement Mission 3B43 product and the surface radiation from the Global Energy and Water Cycle Experiment–Surface Radiation Budget (SRB) product version 2.5. Fitness metrics include monthly mean spatial bias, mean squared error and the variance ratios (i.e., the spatial variance of mean model output over the spatial variance of the observations). Manual binary search calibration was chosen because of the complexity of the parameter space, the need for human supervision and sanity checks, and the non-trivial computation requirements for each simulation. Fifty-four iterations were performed, utilizing a reduced domain in the first group of iterations to facilitate a more rapid calibration. A table of the finalized coupled model runtime conditions is provided

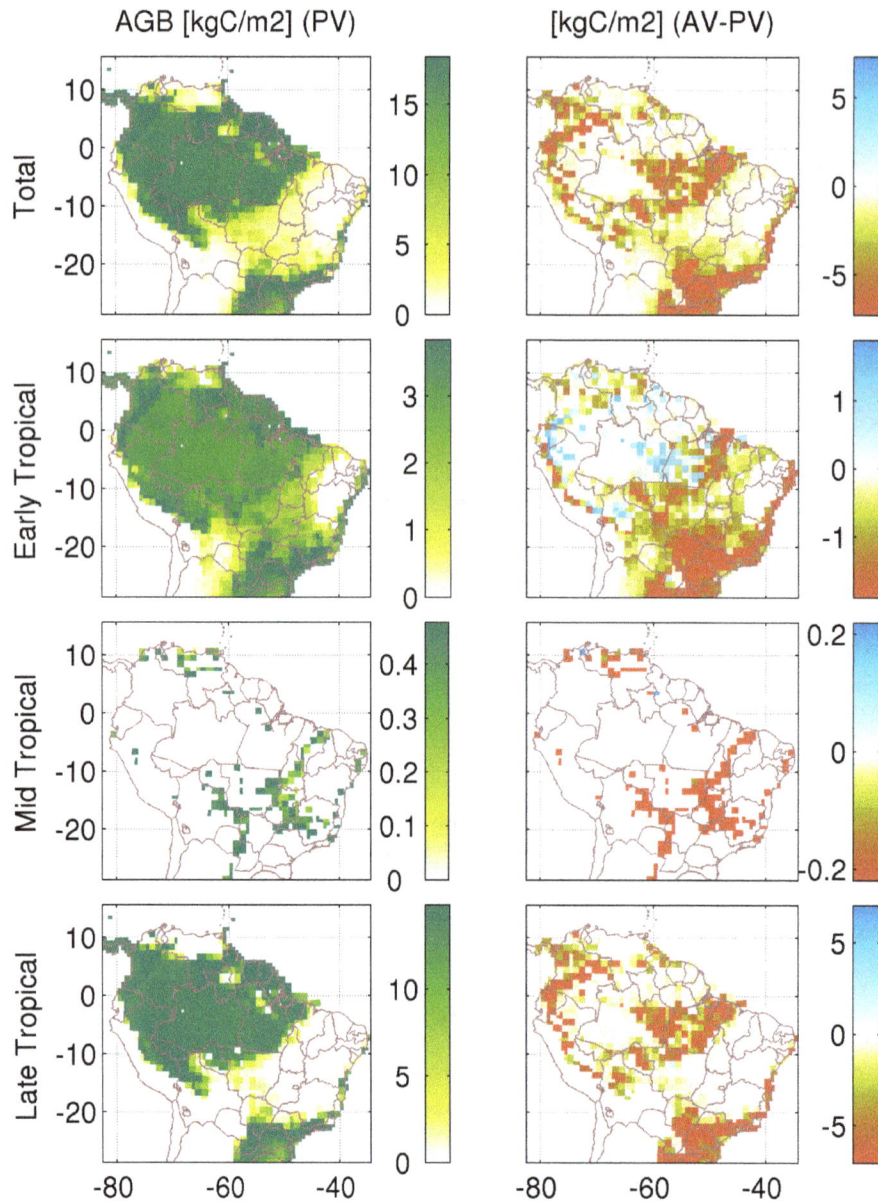

Figure 2. Regional maps of total above-ground biomass (AGB) (kg m^{-2}) from the ED2 initial condition. The left column indicates results are from the *potential vegetation* condition, the right column is the relative differences between the *actual* and *potential* scenarios, (AGB$_{AV}$− AGB$_{PV}$). Each row represents the partitioning of the above-ground biomass into respect plant functional types. "Early Tropical", "Mid Tropical" and "Late Tropical" refer to broadleaf tropical evergreen plant functional types.

in Table 2. Model output was then compared with observations of atmospheric thermodynamic variables, mean regional surface fluxes (precipitation, radiation, latent heat flux and sensible heat flux) and mean cloud cover profiles; these comparisons are provided in Appendices C–F, respectively.

3 Regional analysis of the actual and potential scenarios

The following analysis of results will repeatedly refer to anomalies, here defined as the subtracted differences of the *potential* vegetation scenario model output from the *actual* vegetation scenario model output (or alternatively, AV–PV).

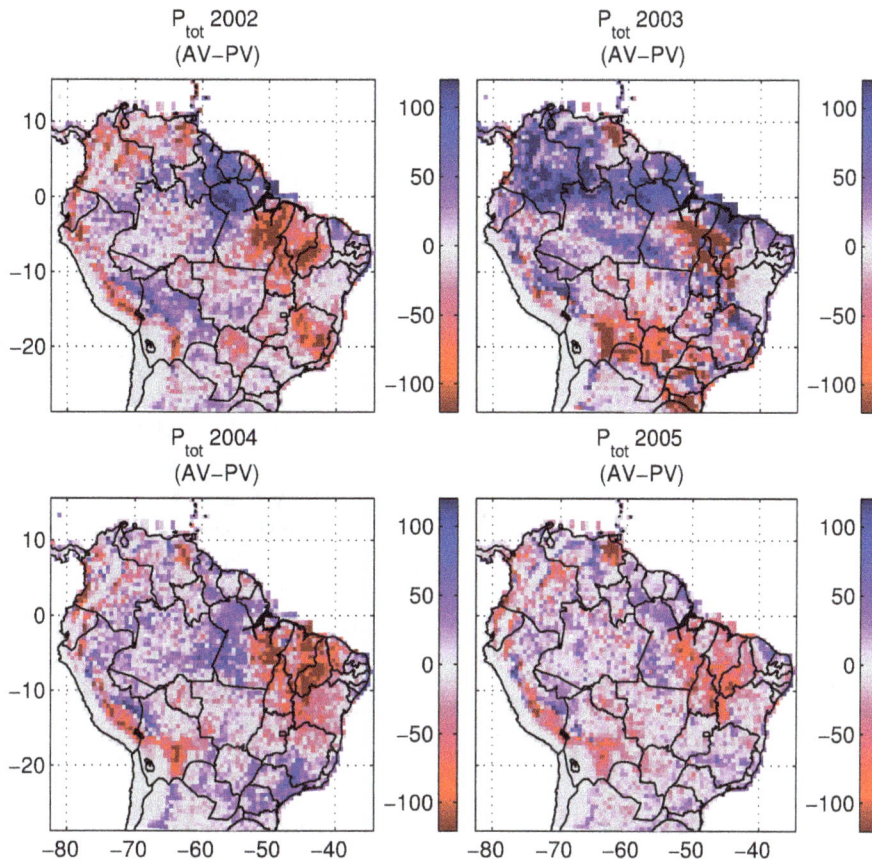

Figure 3. Differences in total annual surface precipitation (mm), 2002–2005. The *potential vegetation* (PV) condition is subtracted from the *actual vegetation* (AV) condition, (AV–PV).

3.1   Emergence of patterning and continental biases

The annual precipitation accumulations for each simulation were mapped in space and the anomalies between the two were compared for consistent patterning (see Fig. 3). Each anomaly appears to feature a dipole structure with a positive lobe (more *actual* scenario precipitation) in the north and west of the Amazon delta, and a negative lobe in the south and east of the Amazon delta region. Pattern differences also appear on the Peruvian–Bolivian border, although whether or not it can be considered a dipole is left to the reader. In each year, the precipitation anomaly shows increases on the foothills of the Andes Mountains in southern Peru and the northern tip of Bolivia. There is also a negative anomaly in precipitation in southern and central Bolivia. However, in each year there is also noise among the pattern. For instance in 2002, 2003 and 2005, there are locations in southern Bolivia that show increases in the precipitation anomaly adjacent to the area of decrease.

The patterning in downwelling shortwave surface radiation showed opposite behavior to that of precipitation (not shown), the response is strongly influenced by increased cloud optical depth where convective precipitation has in-

creased, and vice-versa. The atmospheric model did not incorporate dynamics of aerosols or atmospheric gases other than multi-phase water; therefore, variability in multi-phase water explains the variability and differences seen in optical depth. Maximum mean annual differences in surface irradiance peak at about $10\,\mathrm{W\,m^{-2}}$, and are strongest over the dipole associated with the precipitation anomaly, as well as over the eastern Brazilian dry lands (41° W).

The mean annual continental bias in accumulated precipitation, evapotranspiration and total runoff is presented in Fig. 4. There is little evidence of an overall continental bias in accumulated precipitation. However, the human land-use scenario generated a negative continental evapotranspiration anomaly and a positive runoff anomaly in each of the 4 years. Consistent patterning in evapotranspiration and transpiration anomaly were also evident (see Fig. 5). Generally speaking, a negative anomaly in transpiration and total evapotranspiration is evident over the "arc of Amazonian deforestation" (starting at 48° W 2° S going clockwise to 62° W 10° S, also see forest biomass differences in Fig. 2). The spatial correlation between the biomass and evapotranspiration anomalies ($R^2 = 0.4$), suggests that the variability in evapotranspiration cannot be explained purely by first order effects from

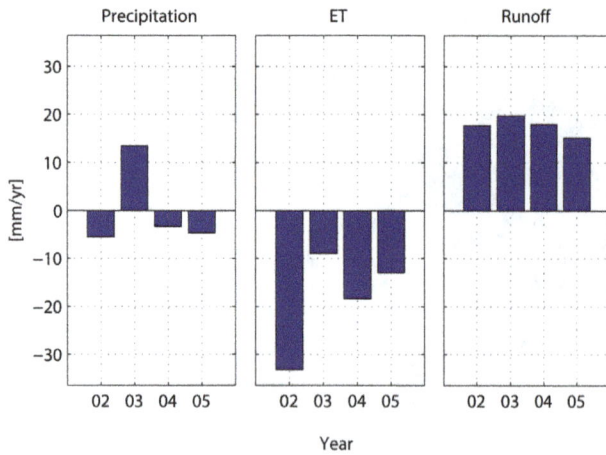

Figure 4. Difference in mean continental precipitation, evaporation and total runoff, between the *actual vegetation* case and *potential vegetation* case (AV–PV).

changes to forest structure. Second order effects and complex system feedbacks account for a portion of the variability. These effects include differences in precipitation, and potentially the effects of differences in surface heating and turbulent transport of scalars (heat and water).

3.2 Connecting hydrologic anomalies and ecosystem response

The availability of root-zone soil moisture, photosynthetically active radiation (PAR) and nutrients are examples of resource limitations that can potentially mediate the response of vegetation to changes in climate. Light and water are critical limiters of plant growth, disturbance (particularly through fire), and mortality (which can be functionally related to growth). However, the significance of these limiters in how they may drive ecosystem response is dependent on various factors other than the mean, such as the consistency of change (inter-annual variance), when the changes occur (seasonality) and how large the differences are relative to the total. A standard score "ζ" is one way to evaluate consistency, calculated as the inter-annual mean difference (denoted by brackets "$<>$") divided by the first standard deviation of the normalized difference η of variable x for year t (mean annual precipitation or downwelling shortwave radiation).

$$\eta_{(t)} = \frac{x_{\mathrm{AV},(t)} - x_{\mathrm{PV},(t)}}{0.5\left(x_{\mathrm{AV},(t)} + x_{\mathrm{PV},(t)}\right)} \tag{1}$$

$$\zeta = \frac{\langle \eta \rangle}{\sigma_\eta} \tag{2}$$

The spatial maps of the standard scores for precipitation and shortwave radiation anomalies are provided in the upper panels of Fig. 6. For reference, a standard difference of 1 suggests that the normalized difference is equal to its inter-annual standard deviation. The maps indicate that the dif-

ferences in precipitation and radiation from the two scenarios are relatively consistent at the two locations previously identified (Pará Brazil and northern Bolivia). They also indicate that the negative precipitation anomaly, and the positive radiation anomaly over the regions of intense deforestation (i.e., the arc of deforestation) are consistent.

The susceptibility of ecosystems to anomalous precipitation forcing may be derived from an ED2 model mechanic called the "moisture stress index" (MSI). This metric is simply the fraction of time that ED2 vegetation cohorts (plants) are actively keeping their stomata closed due to water limitations. For an ecosystem with N plant groups (also known as cohorts) indexed i, the mean land-surface moisture stress index is calculated by the leaf area index (LAI) weighting of the open-fraction f'_0 of stomata for each plant group in the community. Brackets "$<>$" denote an averaging in space and time. The stomatal open fraction f'_0, is based on the ratio of the plant's "demand" for root zone soil moisture, and the "supply" of water the roots are capable of extracting at that time. The demand requirement is driven by the maximum transpiration the plant would generate given the existing light, carbon and vapor pressure conditions with unlimited soil moisture.

$$\mathrm{msi} = 1 - \langle \frac{\sum_{i=1}^{N} \mathrm{LAI}_{(i)} f'_{\mathrm{o}(i)}}{\sum_{i=1}^{N} \mathrm{LAI}_{(i)}} \rangle \tag{3}$$

$$f'_{\mathrm{o}(i)} = \frac{1}{1 + \frac{\mathrm{Demand}}{\mathrm{Supply}}} \tag{4}$$

Vegetation communities that have experienced high moisture stress indices in the past are more likely to respond structurally to changes in precipitation, because subsequent changes in soil moisture availability will have immediate impacts on photosynthesis and the assimilation of carbon. The lower left panel of Fig. 6 shows the mean moisture stress index for the *actual* vegetation scenario. The lower right panel of Fig. 6 shows a map of above-ground biomass as a reference to the extents of the modeled Amazon tropical forests. Moisture stress is low in areas where there is copious precipitation (the supply term). Note that in the interior of the Amazon Basin, moisture stress has little to no influence on stomatal regulation (and subsequently photosynthesis). The open canopy dry forests in southern Brazil an Bolivia, as well as the Cerrado, have higher moisture stress.

4 Hydrometeorological focus areas – Pará Brazil and the Gran Chaco

Two locations that coincide with the pattern differences in precipitation are highlighted in Fig. 6. Each location shows decreases in normalized precipitation and increases in down-

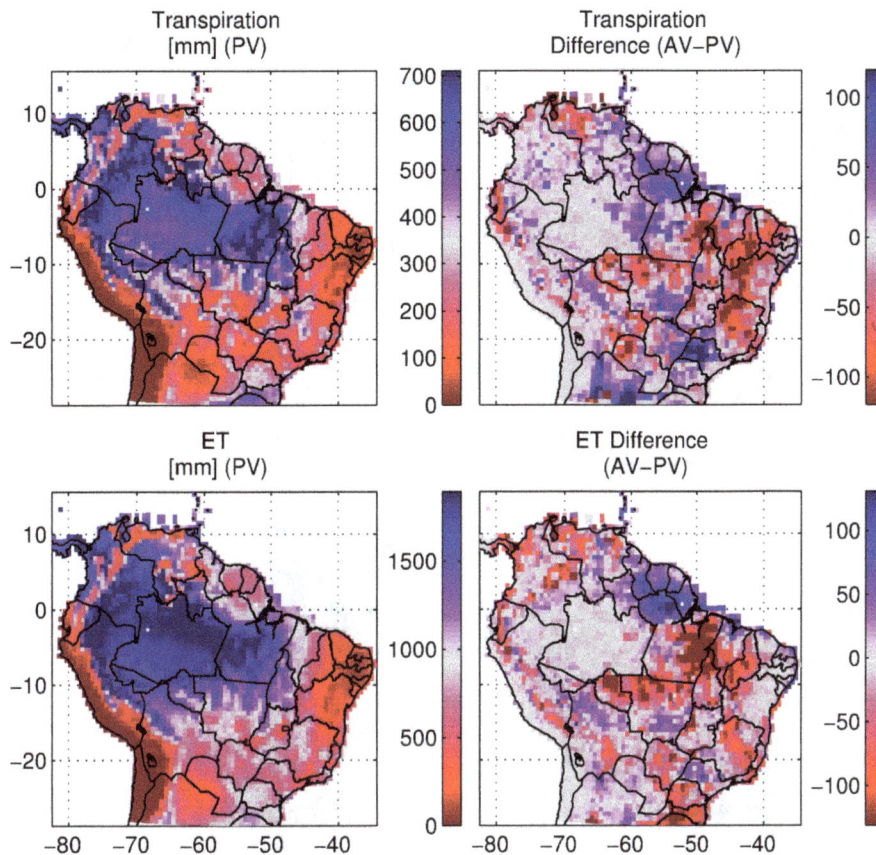

Figure 5. Left panels: mean annual transpiration and evapotranspiration in the *potential vegetation* (PV) scenario, from 2002 to 2005 (mm). Right panels: difference in mean annual transpiration and evapotranspiration between the *actual vegetation* case and *potential vegetation* case (AV–PV).

welling shortwave radiation associated with the *actual vegetation* scenario. The vegetation of these locations also show a degree of seasonal moisture stress according to the MSI metric presented in section 3.2. One site is centered on 4.5° S 50.5° W in the Brazilian state of Pará. The other site is centered on 19.5° S 63.5° W in the northwestern part of the Bolivian Gran Chaco where it meets the Andes mountains (sometimes referred to as the Montane Gran Chaco). These two locations are chosen as areas of focused evaluation of hydrology and hydrometeorology. For simplicity, these will be referred to as the *Pará* and Gran Chaco focus areas.

A representation of the vegetation demographics at the centroids of the two focus areas, as estimated by the ED2 model, are provided in Fig. 7. The natural landscapes at the Pará focus areas are dominated by tropical evergreen forests, and are close to (but not within) the ecotone transition between tropical forests and Cerrado. The offline model spin-up of the *actual* vegetation scenario imposed pastures on approximately one-third of the land-cover. Roughly 10 % of the landscape contains old-growth forests that have gone 200 years since the last disturbance. The focus area in the Gran Chaco is located in a region influenced by the outlet of

the South American Low Level Jet. The continental precipitation recycling ratio in this area is relatively high compared to the rest of the continent (Eltahir and Bras, 1994). This exact location in the Gran Chaco is a dry forest ecosystem that borders adjacent ecotones of tropical rainforests to the north, montane ecosystems to the west and grasslands to the south. The ED2 model estimated a *potential* vegetation demographic that is fairly consistent with the depiction of dry forests, a sparse cover of short trees with grasses in the understory. The *actual vegetation* simulation of the Gran Chaco, driven by the GLU data set (Hurtt et al., 2006), forced 25 % of the natural landscape to pasture (grasses), with an accompanying 20 % of abandoned and degraded lands. Human land use, as represented in the ED2 model at this specific location, led to a collapse of the estimated tree cover, which includes natural landscapes. This specific site is undoubtedly a more aggressive representation of the differences between the *actual* and *potential* scenario ecosystems in this region. As a whole, of course, human land-conversion has not lead to a collapse of the Gran Chaco's dry-forest ecosystems.

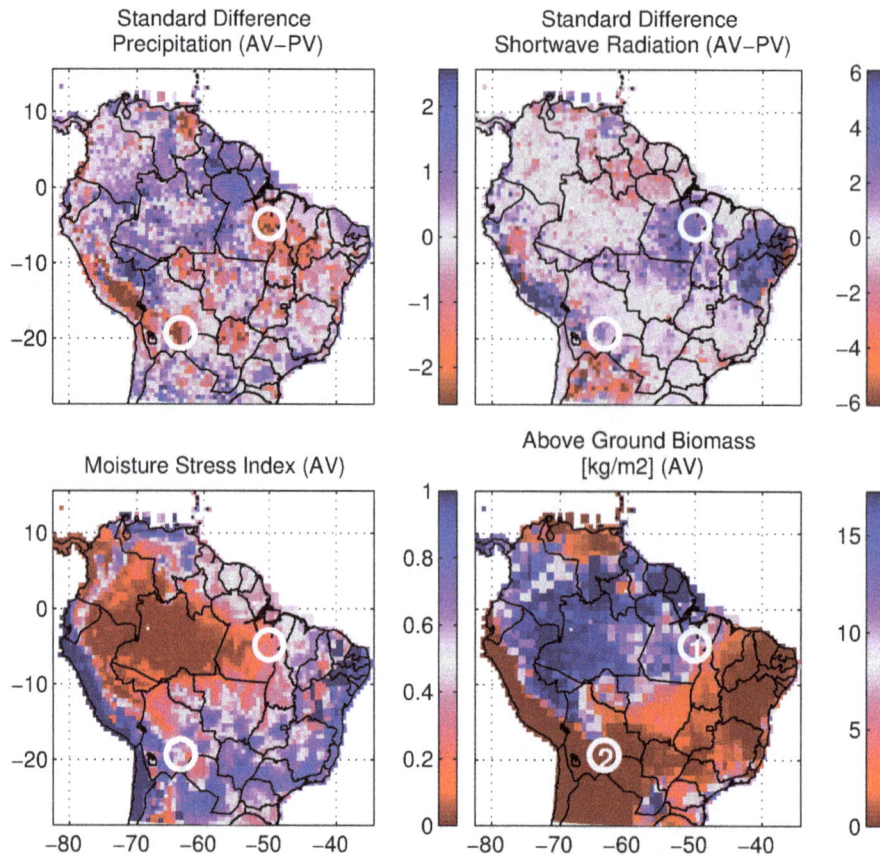

Figure 6. Combined assessment of the regional significance in differences between precipitation and radiation, and the susceptibility of the ecosystems. Upper panels show standard scores for consistency of differences *actual* and *potential* (AV–PV) surface precipitation and surface downwelling shortwave radiation. The lower left panel shows the moisture stress index for the *actual* (AV) scenario, see Eq. (3). For reference, *actual* (AV) scenario above-ground biomass is provided in the bottom right panel.

4.1 Canopy water and energy balance – Pará

Simulated annual precipitation at the Pará focus area was typically around $1500 \, \mathrm{mm \, yr^{-1}}$, the surface energy flux was dominated by leaf evaporation and transpiration. Transpiration dominated vapor flux in the dry season (May–November). Runoff in the form of drainage through the lower soil column occurred mostly during the wet season. The time series water and energy balance at the land-surface is summarized in Fig. 8. Accumulated water fluxes from the *potential* vegetation scenario are shown in the upper left panel a, anomalous accumulations are shown in the upper right panel b. The *actual* vegetation scenario experienced roughly 10 % less surface precipitation at the Pará focus area. However, the site experienced a small net increase $(30 \, \mathrm{mm \, yr^{-1}})$ in precipitation throughfall, due to a proportionally stronger decrease in leaf interception surfaces. There is also increased drainage in the *actual* vegetation scenario, which appears to be symptomatic of both increased throughfall and the decrease in the root-zone soil-moisture sink from transpiration.

The *actual* vegetation scenario receives more total shortwave and longwave radiation $(R_{\mathrm{SD}} + R_{\mathrm{LD}})$, which is directly attributable to the decrease in mean convective cloud albedo associated with the decrease in convective rainfall at the site. Although the site receives more total incoming radiation in the *actual* vegetation simulation, the surface albedo increases with the conversion of forests to pasture. This results in more reflected radiation and a decrease in combined sensible and latent heat flux $(H + L)$, see the bottom right panel d of Fig. 8.

4.2 Canopy water and energy balance – Gran Chaco

The annual precipitation at Gran Chaco in the *potential* simulation ranged from 500 to nearly 1000 mm. Annual precipitation was roughly 15 % lower in the *actual* vegetation simulation. A summary of the hydrologic response and the anomalies are shown in Fig. 9. Like the Pará site, land-conversion drove a decrease in leaf area, and therefore a significant decrease in leaf interception of precipitation in the *actual* (AV) simulation. However, in this case, the relative decrease in interception surfaces due to deforestation

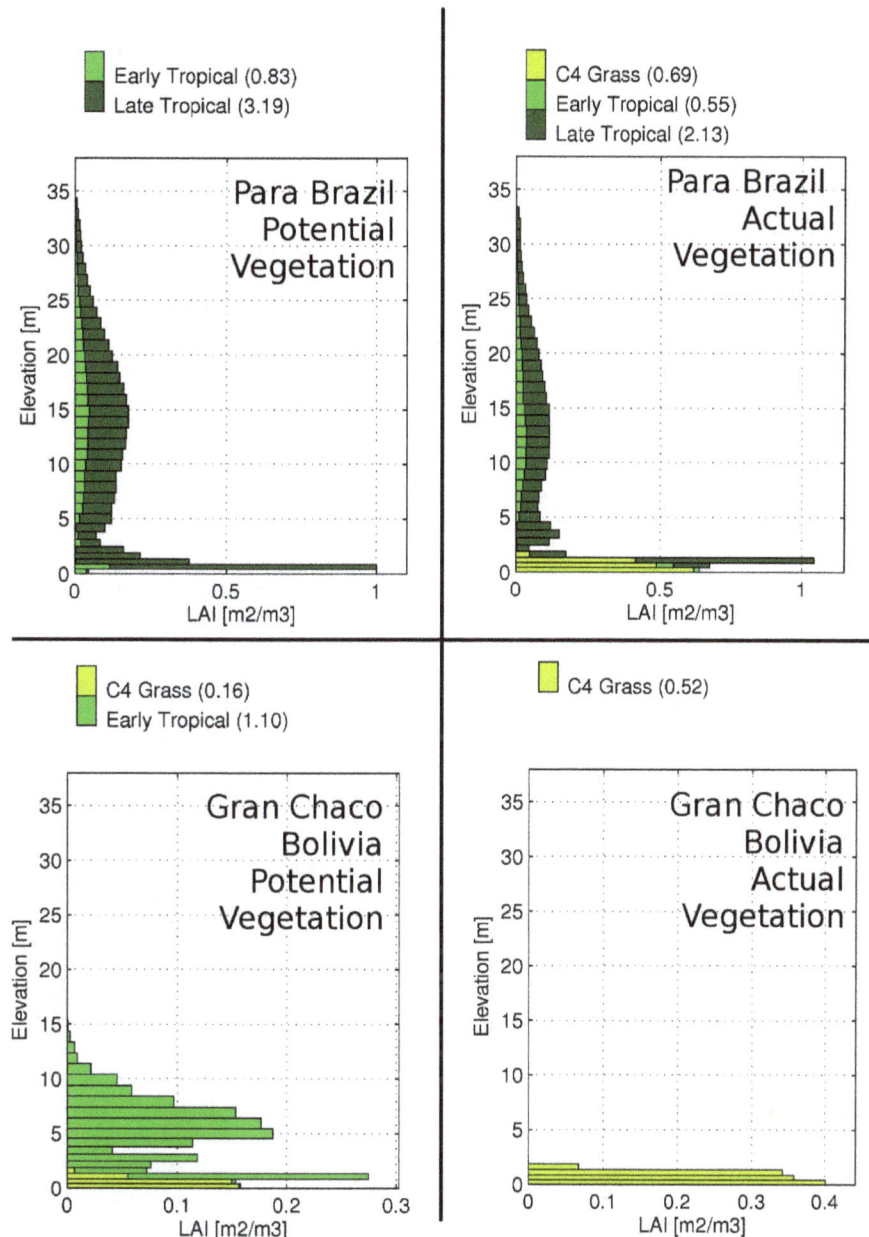

Figure 7. Mean vertical leaf area index profiles ($m^2 m^{-3}$) estimated by the ecosystem demography model 2 at the two focus areas. Vertically integrated leaf area index ($m^2 m^{-2}$) per each represented functional type of plant is shown in the key above each plot.

was less affecting than the decrease in total precipitation due to land–atmosphere feedbacks. Therefore, the *actual* vegetation simulation experienced a decrease in total precipitation throughfall. Soil evaporation accounted for half of the water losses, while leaf evaporation and transpiration equally combined to represent the other half. The relatively low precipitation rates promoted almost no detectable runoff. Transpiration decreased by 20 % in the *actual* simulation, which is a direct consequence of decreased stomatal density and precipitation throughfall.

Notwithstanding the decreased precipitation throughfall in the *actual* simulation, surface evaporation increased. Despite decreased precipitation throughfall, upper soil-column moisture from rain events has a longer residence time in the root zone, as shown in Fig. 10. This is an effect of decreased transpiration, and thus moisture in the grass root zone lasts comparatively longer into the dry season. Note that the relative reduction in precipitation throughfall nearly balanced the reduction in transpiration. There is little evidence to suggest that increased soil evaporation rates would maintain indef-

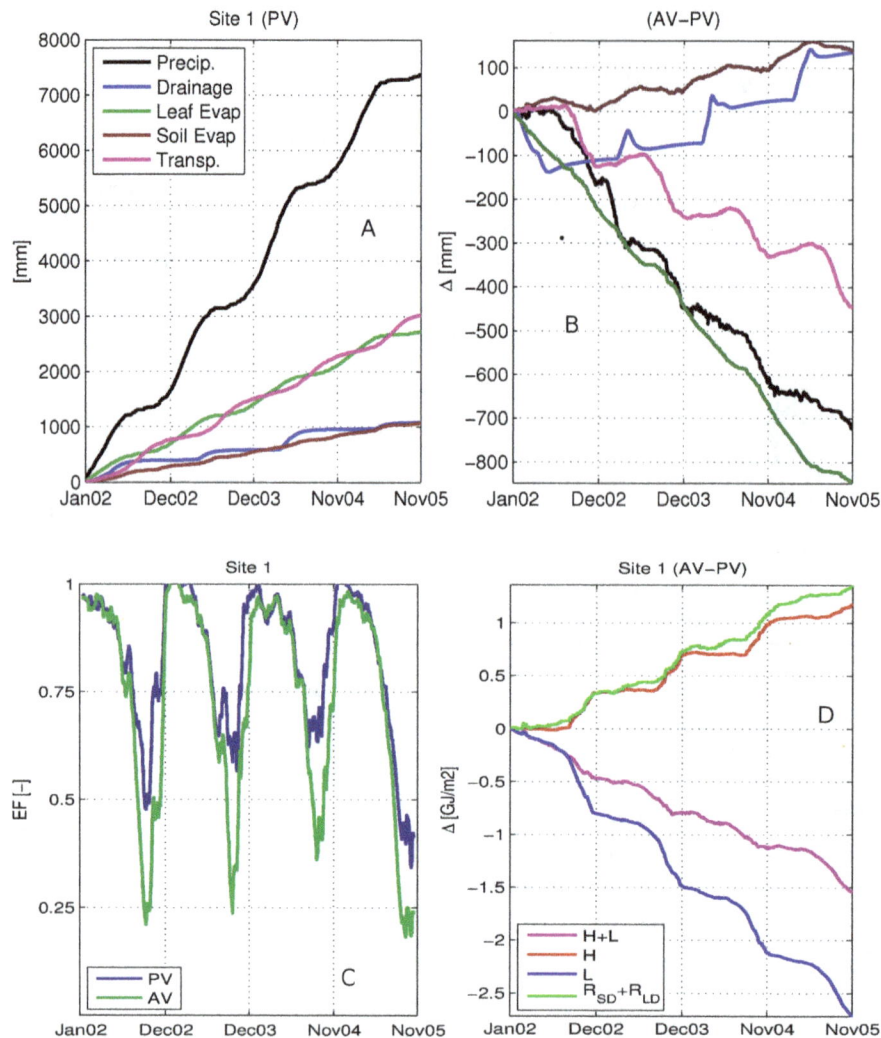

Figure 8. Time series analysis of the surface water and energy balance at the Pará focus site, 2002–2005. (**a**) Accumulated water flux for the *potential* scenario. (**b**) Accumulated differential water flux between the *actual* (AV) and *potential* (PV) scenarios. (**c**) Mean evaporative fraction, latent heat flux (L) divided by the sum of latent and sensible (H) heat flux ($L/(L + H)$). (**d**) Accumulated differential energy flux in gigajoules per square meter. R_{SD} is downwelling shortwave radiation incident on the surface, R_{LD} is downwelling longwave radiation incident on the surface.

initely in the *actual* scenario, which could alternatively be associated with transient changes in soil column storage.

Like the Pará site, land-conversion at Gran Chaco also drove an increase in total surface albedo, a direct effect due to the loss of dark foliage. More incident radiation is reflected, which reduces net radiation. Unlike the Pará focus area, the albedo effect is stronger than the increase in incident radiation, which leads to decreased sensible heat flux (see Fig. 9) and slightly cooler annual surface temperatures (not shown).

4.3 Land–atmosphere coupling – Pará

A box is constructed around this site for the month of September 2003 that contains the extents of a continuous space with negative precipitation anomaly (see Fig. 11). Ta-

ble 3 shows a selection of spatiotemporal mean indicators from the bounded domain. To summarize the differences in surface fluxes, the results are consistent with the single site time-series analysis, where the *actual* vegetation scenario experienced a decrease in net radiation (-10 W m^{-2}, despite increased incident shortwave radiation) and an increased mean surface albedo.

The decreased precipitation and surface energy flux of the *actual* scenario are accompanied by a boundary layer with lower equivalent potential temperature (see Fig. 12). The decrease in boundary layer equivalent potential temperature has a strong physical connection to explaining the decrease in precipitation, particularly since the vast majority of the precipitation was generated through the convective parameterization (data not shown). This was verified by recording a log of failures in deep convection (precipitating convective clouds) generated by the convective parameterization. All of the bias in these convective failures occurring in the *actual* simulation was attributed to the generation of convection that resulted in clouds that were too thin to be classified as precipitating deep convection clouds. This is indicative of the how much convective available potential energy (CAPE) can be released through convective buoyancy, which is controlled by the moist static energy in the surface parcels as well as the moist static energy of the mean atmosphere over the depth of the troposphere. Alternatively, there was no positive bias in the logs associated with the inability to trigger parcel buoyancy.

It is questioned if the driving force behind the reduced equivalent potential temperature profiles of the *actual* scenario is solely the result of local surface fluxes or caused by changes in the regional energy circulation. Both scenarios net a negative moisture convergence (divergent) budget for the month (total integrated water mass flux through the box boundaries, normalized by the box area). The *potential vegetation* scenario loses more water ($-51.32 \, \mathrm{kg \, m^{-2}}$) through its lateral boundaries than the AV scenario ($-37.14 \, \mathrm{kg \, m^{-2}}$), see Table 3. This can be visualized by flux vectors as well (see Fig. 13). In the *potential vegetation* case shown in the left panel, the flow vectors run east-to-west and up the gradient, which means the advecting air mass is gaining moisture and is consistent with the net water divergence described in Table 3.

4.4 Land–atmosphere coupling – Gran Chaco

Similar to the case study in Pará, a bounding box was constructed around the Gran Chaco site in April 2003 that captures a spatially continuous negative anomaly in precipitation. The boundaries are shown in black against the mean total and mean anomaly in monthly precipitation and evapotranspiration (see Fig. 14). Mean statistics are shown in Table 4. The *actual* scenario experienced less than half as much precipitation ($41 \, \mathrm{kg \, m^{-2}}$ compared to $85 \, \mathrm{kg \, m^{-2}}$). The evaporation anomaly between the two scenarios was not as strong ($111 \, \mathrm{kg \, m^{-2}}$ in the *potential* scenario compared to $83 \, \mathrm{kg \, m^{-2}}$ for the *actual* scenario).

The *actual vegetation* scenario experienced more downwelling shortwave radiation yet less net surface radiation, which was influenced by an increased surface albedo. Like the Pará case, the *actual* scenario experienced a lower equivalent potential temperature over the boundary layer (see Fig. 15). The convective parameterization logs accounted that the *actual* scenario experienced a great deal more failed convective events associated with an inability to generate deep clouds (the same reason as the Pará case). Note that the model's cloud depth parameterization is controlled by convective available potential energy. However, in this case, about 25 % of the bias in failures was also explained by an inability to trigger convection. This is interesting because this case was slightly different from the Pará in that the *actual* scenario did not experience higher levels of turbulent kinetic energy over the boundary layer.

An analysis of moisture convergence and advective flux was used to better understand the local versus regional controls that drive convective precipitation. Both scenarios showed negative moisture convergence, typical during the onset of the dry season in this region, refer to Table 4. The *potential* scenario showed less moisture divergence ($-37 \, \mathrm{kg \, m^{-2}}$) than the *actual* scenario ($-52 \, \mathrm{kg \, m^{-2}}$). The moisture advected into the Gran Chaco site comes via northerly winds from the moist Amazonian air mass, see the left panel of Fig. 16. Moisture transport from the north decreases in the *actual* scenario, see the right panel of Fig. 16.

5 Discussion

5.1 Secondary forests and evapotranspiration patterning

The maps of evapotranspiration and transpiration anomaly showed pattern similarity with the differences in aboveground biomass, as compared to a lack of pattern similarity between precipitation and above-ground biomass anomalies. The stronger correlation between the evapotranspiration and forest biomass anomalies can be rationalized by understanding how the ED2 model resolves canopy hydrologic process. In ED2, closed canopy tropical broadleaf evergreen forests have higher leaf area indices than grasses, with possible exceptions during drought deciduous leaf drop. Deforestation of closed canopy forests will therefore decrease total leaf area. Rainfall that is intercepted in ED2 has two outcomes: it can either re-evaporate or drip to the land surface (one shortcoming of this assumption is that epiphytes may store water directly from leaf interception). Throughfall precipitation has multiple outcomes: it can evaporate from the surface, become stored in the soil and vegetation indefinitely, or leave via transpiration or leave via runoff. The evaporation rates between the leaf and soil surfaces with canopy air space are regulated by two factors, the aerodynamic resistance and the effective vapor pressure deficit between the respective surfaces with the air space. In the ED2 model formulation, which scales in-canopy wind speeds following Massman (1997), the aerodynamic resistance in the forest canopy will attenuate from top to bottom as wind speeds monotonically decrease. Moreover, water that becomes bound in the soil matrix will have a decreased vapor pressure deficit with the adjacent air (compared to leaf water which is not bound, and assumed saturated) due to the effects of pore spaces at the soil surface (Lee and Pielke, 1992). Therefore in ED2, precipitation that is intercepted in the canopy has both an extra opportunity (simply considering the order of process) to evaporate back to the atmosphere, but also has a tendency towards both de-

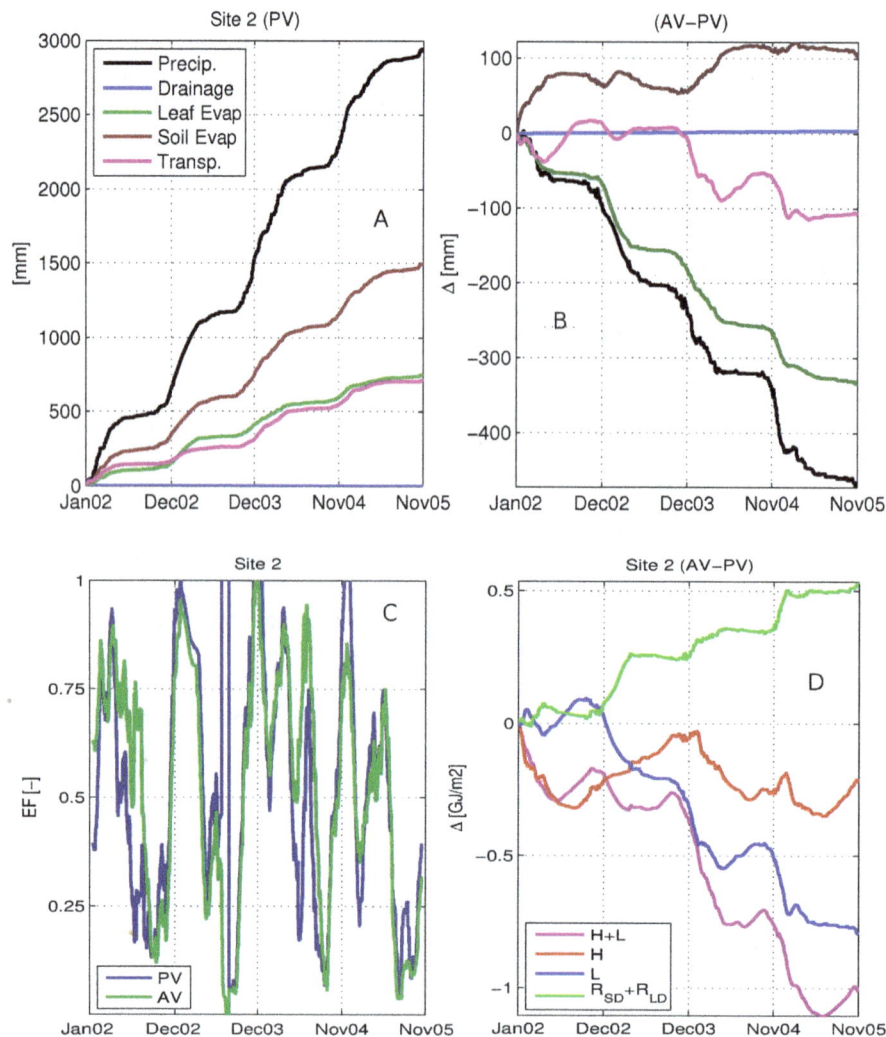

Figure 9. Time series analysis of the surface water and energy balance at the Gran Chaco site, 2002–2005. (**a**) Accumulated water flux for the *potential* scenario. (**b**) Accumulated differential water flux between *actual* (AV) and *potential* (PV) scenarios. (**c**) Mean evaporative fraction, latent heat flux (*L*) divided by the sum of latent and sensible (*H*) heat flux ($L/(L + H)$). (**d**) Accumulated differential energy flux in GigaJoules per square meter. R_{SD} is downwelling shortwave radiation incident on the surface, R_{LD} is downwelling longwave radiation incident on the surface.

creased aerodynamic resistance and increased vapor pressure deficit that drives evaporation rate.

However this explanation of process only considers the immediate structural effects of deforestation, which is not static but also has phases of recovery when left unmanaged. During the recovery cycle of tropical forests following *natural* disturbance, successful new growth in the canopy is typically dominated by pioneer species. Pioneer species have lower wood density, higher maximum photosynthetic capacity and quicker vertical growth than late successional species (Laurance et al., 2004; Poorter et al., 2006; Chave et al., 2006). The canopy leaf area may flush to previous levels within a century, yet it may take several centuries for total forest biomass to rebound. This type of behavior was observed in the model spin-up, where newly disturbed patches

of land in the central Amazon reached maximum leaf area over a span of a few decades, compared to the length of time (more than a century) it took for biomass to stabilize.

The results presented here support that secondary forests undergoing recovery from deforestation can drive detectable pattern increases in total transpired water across the region. It is rationalized that at these locations, photosynthetic capacity is scaled by leaf area, as well as a distribution of members skewed towards rapid growth and high photosynthetic capacity. If there is sufficient available soil water there would be an expected increase in total transpiration. In the *actual* model scenario containing deforestation effects, the model estimated an increase in early successional tropical evergreens (pioneers) in the recovering forests of northern Pará and eastern Amazonas (centered on 5° S 58° W) (see Fig. 2

Table 3. Hydrologic monthly means within the bounded area above the Pará case study, September 2003. Total change in column precipitable water for the month per square meter ΔM_{pw}, evapotranspiration ET, precipitation P and resolved moisture convergence Mc, 55 m air temperature T, mixing ration (55 m) r, equivalent potential temperature θ_e, surface albedo to shortwave radiation α, downwelling shortwave radiation R_{SD}, downwelling longwave radiation R_{LD}, upwelling longwave radiation R_{LU}, net surface radiation R_{net}, sensible heat flux SHF and latent heat flux LHF.

Case	ΔM_{pw}	ET	P	Mc	T	r
Units	$kg\,m^{-2}$	$kg\,m^{-2}$	$kg\,m^{-2}$	$kg\,m^{-2}$	°C	$g\,kg^{-1}$
AV	−3.457	63.1	29.8	−37.14	32.83	12.18
PV	−3.515	94.7	47.3	−51.32	32.35	12.93

Case	θ_e	α	R_{SD}	R_{LD}	R_{LU}	R_{net}	SHF	LHF
Units	K	–	$W\,m^{-2}$	$W\,m^{-2}$	$W\,m^{-2}$	$W\,m^{-2}$	$W\,m^{-2}$	$W\,m^{-2}$
AV	336.8	0.262	300.2	443.3	513.2	180.9	139.25	70.97
PV	338.2	0.257	285.6	443.9	498.0	187.8	114.50	106.45

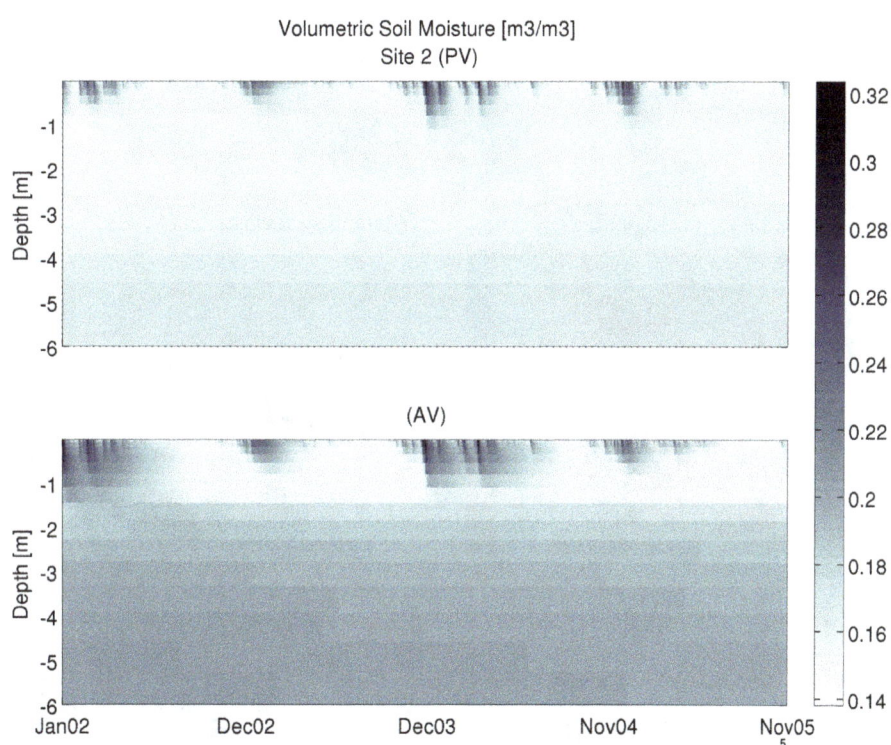

Figure 10. Time series profile of volumetric soil water at the Gran Chaco focus area. Both scenarios, *potential* (PV) and *actual* (AV), are shown separately.

second row right panel). There is an increase in regional transpiration here (Fig. 5) that has a strong pattern match with the increase in early successional biomass, moreover there is little evidence of influence from pattern precipitation here (see Fig. 3).

5.2 Regional surface water balance and runoff generation

There was a clear and consistent bias in the total annual evapotranspiration (negative) and runoff (positive) estimated

in the *actual* scenario when integrated over the entire domain of northern South America. There is some rudimentary explanation of the first order biases in evapotranspiration explained above. The regional runoff bias appears to have several potential explanations and remaining questions. It is clear that in the *actual* scenario, decreased canopy interception promotes a first order effect of increased canopy throughfall. However, we saw in comparing the Pará and Gran Chaco case studies that changes in canopy interception can be offset by changes in incident precipitation. Therefore increased canopy throughfall from deforestation is not ubiq-

Table 4. Hydrologic monthly means within the bounded area above the Gran Chaco case study, April 2003. Total change in column precipitable water for the month per square meter ΔM_{pw}, evapotranspiration ET, precipitation P and resolved moisture convergence Mc, 55 m air temperature T, mixing ration (55 m) r, equivalent potential temperature θ_e, surface albedo to shortwave radiation α, downwelling shortwave radiation R_{SD}, downwelling longwave radiation R_{LD}, upwelling longwave radiation R_{LU}, net surface radiation R_{net}, sensible heat flux SHF and latent heat flux LHF.

Case	ΔM_{pw}	ET	P	Mc	T	r
Units	$kg\,m^{-2}$	$kg\,m^{-2}$	$kg\,m^{-2}$	$kg\,m^{-2}$	°C	$g\,kg^{-1}$
AV	−11.42	82.95	41.89	−52.49	25.98	12.73
PV	−11.02	111.89	85.91	−36.99	27.36	15.15

Case	θ_e	α	R_{SD}	R_{LD}	R_{LU}	R_{net}	SHF	LHF
Units	K	–	$W\,m^{-2}$	$W\,m^{-2}$	$W\,m^{-2}$	$W\,m^{-2}$	$W\,m^{-2}$	$W\,m^{-2}$
AV	334.4	0.330	252.6	400.2	466.9	111.74	38.54	91.0
PV	342.0	0.297	218.7	424.9	462.6	130.2	28.15	122.5

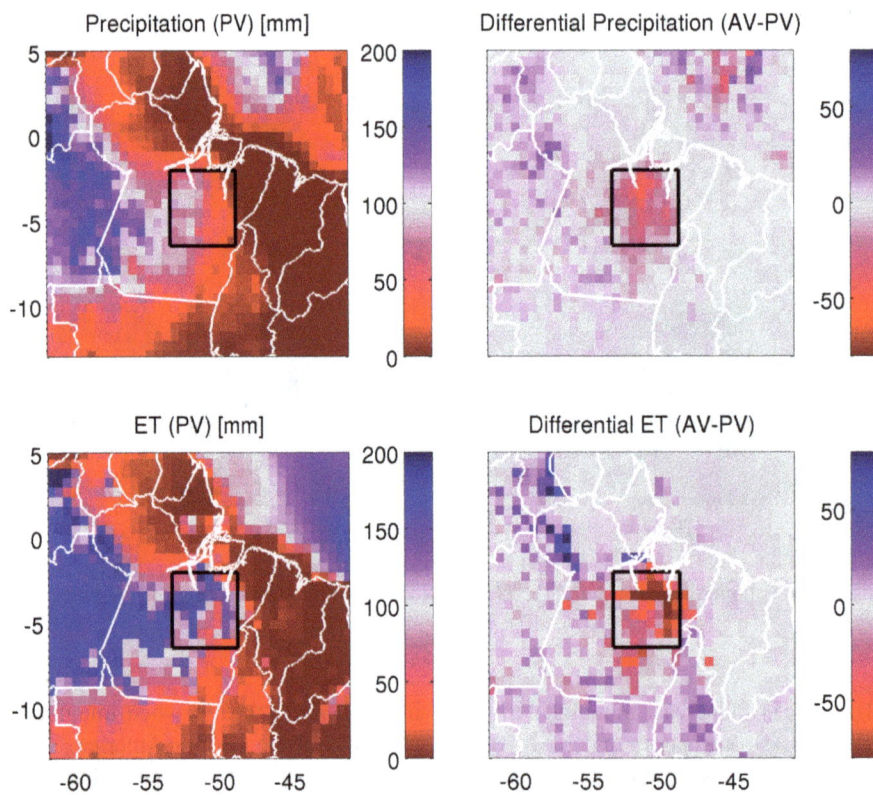

Figure 11. Monthly integrated surface water fluxes over the Pará focus region, September 2003. Upper left panel: map of integrated monthly precipitation for the *potential vegetation* simulation (PV). Upper right panel: map of the integrated difference in monthly precipitation, *actual vegetation* case minus the *potential vegetation* case (AV–PV). Lower left panel: map of integrated monthly evapotranspiration for the *potential vegetation* simulation. Lower right panel: map of the integrated difference in monthly evapotranspiration, *actual vegetation* case minus the *potential vegetation* case.

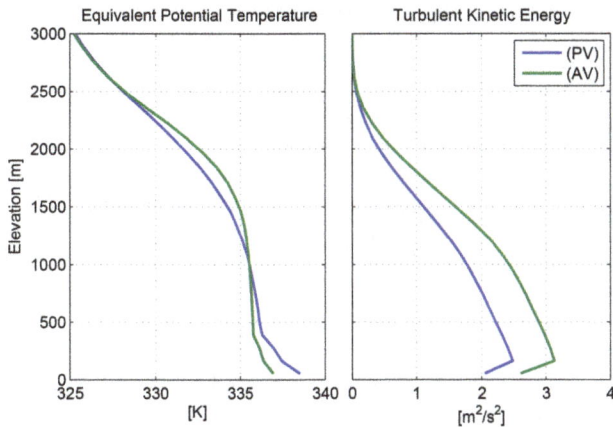

Figure 12. Mean profiles of Equivalent Potential Temperature and Turbulent Kinetic Energy at 15Z within the bounded domain at Pará, September 2003.

uitously associated with regional increases in runoff. In a regional water balance analysis, d'Almeida et al. (2007) also observed that wide-spread regional deforestation promoted decreased evapotranspiration and increased runoff. Similar to this study, they also found that precipitation feedback response to deforestation had the potential to impact the water cycling on par with direct effects of surface hydrologic parameters (although in their results, bi-directionally weakening or strengthening the water cycle depending on heterogeneity and land-cover fractionation).

In the simulations presented here, the increased continental runoff from the *actual* scenario is driven by higher mean annual soil moisture. The regional mean soil moisture depth simulated in the *actual* scenario oscillated around a inter-seasonal mean of 1.40 m (over an 8 m medium), averaging 5 cm greater than the *potential* case. As increased runoff has a negative feedback on increased soil water, and there was no consistent bias in precipitation, it is most likely that the positive shift in mean annual moisture in the *actual* scenario is driven by the decreases in regional evapotranspiration.

This experiment highlighted the use of relatively sophisticated vegetation biophysical processes, which incorporated variable vegetation structure, composition, rooting depth and uptake. However, the modeling framework did not incorporate lateral transport of any surface moisture. Therefore, these results must be interpreted with the understanding that lateral re-infiltration, lateral vadose-zone flow, interflow and water table dynamics could not influence soil moisture dynamics. In light of this, this experiment provides a gauge on the strength of the control that evaporation response to deforestation can have on regional water balance and runoff generation. There has been some evidence that soil hydrologic properties can be affected by land conversion in the tropics, Zimmermann et al. (2006) found that both infiltrability and upper root-zone saturated hydraulic conductivity was highest in intact rainforest compared to pasture and tree planta-

tions. Decreased infiltration in pastures has been related to increased runoff generation as well (Muñoz-Villers and Mc-Donnell, 2013). But there seems less certainty in the literature in quantifying the evaporative response from canopy and soil to regional Amazonian deforestation, degradation and recovery.

5.3 Intersection of seasonal hydrology and represented plant functional types on canopy process

The two case studies showed that the structure of the vegetation canopy can influence the seasonal cycle of moisture storage and land–atmosphere moisture flux. At the Gran Chaco site, transpiration was greater in the *potential vegetation* scenario during the wet season when the deeper roots and higher stomatal density of the open canopy forest could access available soil moisture. Alternatively, total evapotranspiration was greater in the *actual vegetation* scenario at the onset of the dry-season, mostly due to the fact that the grasslands had more available water stored in the upper root zone (recall Fig. 10).

The natural vegetation at the Gran Chaco site is represented "in model" with early successional broadleaf evergreen plant functional types, with accompanying C4 grasses. While the demographic size structure, composition, and the openness of the canopy shows some similarity with dry-forest structure, it must be realized that the wider range of water conservation strategies observed in nature could influence how the differences in surface-to-atmosphere energy fluxes play out at this site.

These findings suggest that the next generations of earth system models may benefit from improvements in representing plant diversity. The seasonal flux of surface-to-atmosphere water vapor is regulated by plants, and can potentially impact the hydrometeorological dynamic of the region. Total evapotranspiration during the transition from the late wet season to early dry season (April) at the Gran Chaco site was larger in the potential vegetation scenario. As shown in the hydrometeorological analysis in Sect. 4.4, this was a time in the seasonal cycle that exhibited relatively strong differences in the instability profiles in the atmosphere, albeit from competing local and regionally driven mechanisms.

5.4 Drivers of anomalous convective precipitation

The negative precipitation bias at the two focus areas in the *actual* scenario were both accompanied by reductions in net radiation, decreased annual latent heat flux (evapotranspiration) and increased albedo. They also experienced boundary layers with lower mean equivalent potential temperature, which was then related to fewer cumulus events that lacked sufficient convective available potential energy to generate deep clouds (diagnosed through convective logs). It is believed that differences in convective available potential energy underlies the convective precipitation anomaly.

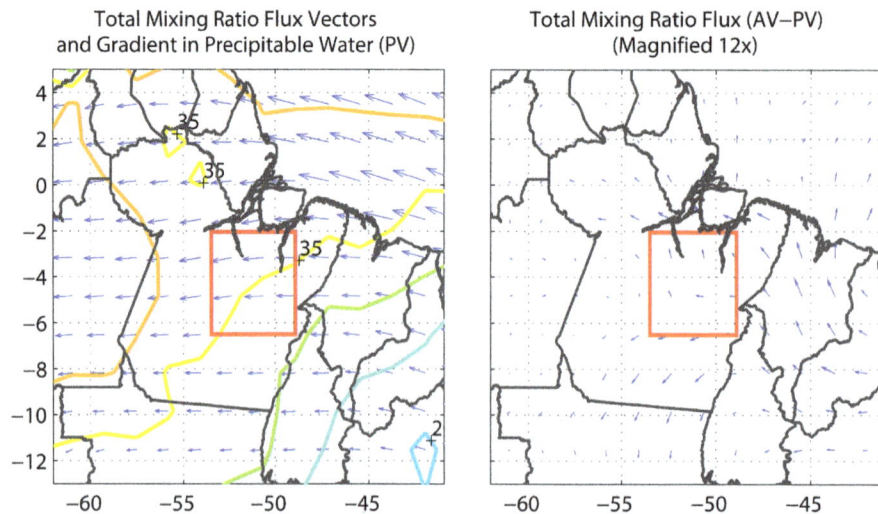

Figure 13. Left panel: map of vertically integrated total water advective flux vectors (quivers) and vertically integrated precipitable water (contours) for the *potential vegetation* case (PV), region near the Pará site, September 2003. Quivers are scaled and convey only directionality and relative magnitude. Contours of low precipitable water are shown by cool colors (blues) and high precipitable water with warm colors (reds). Right panel: the differential in vertically integrated advection of total precipitable water, *actual vegetation* minus *potential vegetation* (AV–PV). Quivers are scaled to 12 times relative to the left panel. In both panels the sub-domain bounding the Pará focus region is shown with a red box.

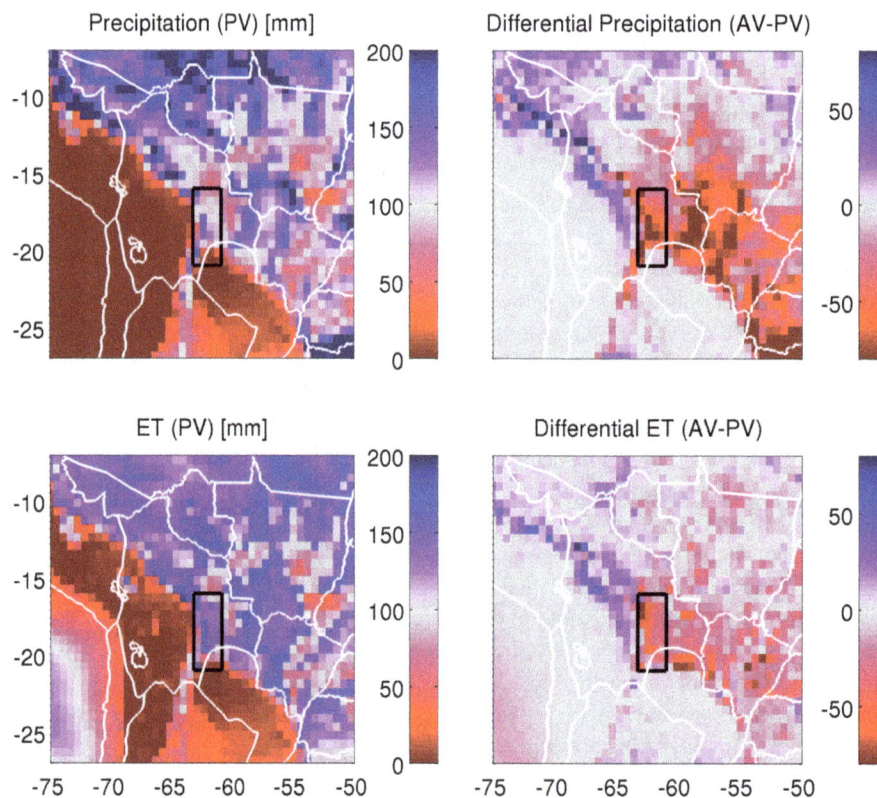

Figure 14. Monthly integrated surface water fluxes over the Gran Chaco focus region, April 2003. Upper left panel: map of integrated monthly precipitation for the *potential vegetation* simulation (PV). Upper right panel: map of the integrated difference in monthly precipitation, *Actual* vegetation case minus the *potential vegetation* case (AV–PV). Lower left panel: map of integrated monthly evapotranspiration for the *potential vegetation* simulation. Lower right panel: map of the integrated difference in monthly evapotranspiration, *actual vegetation* case minus the *potential vegetation* case.

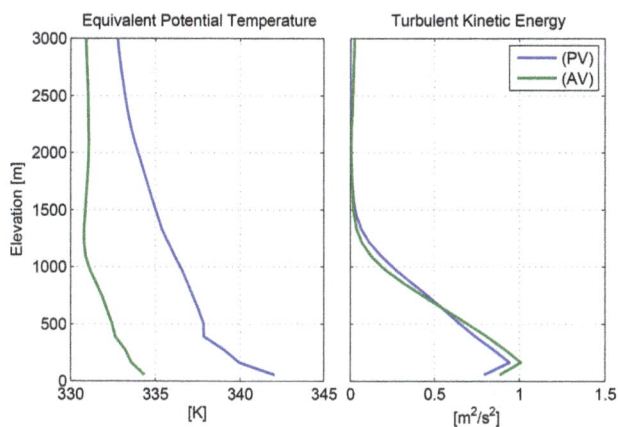

Figure 15. Mean profiles of equivalent potential temperature and turbulent kinetic energy at 15Z within the bounded domain at Gran Chaco, April 2003.

Positive anomalies in sensible heat flux for the *actual* scenario are concurrent with increased turbulent kinetic energy, which is thought to promote the circulations and boundary layer development that lifts air parcels to trigger convection (Wang et al., 2009; Fisch et al., 2004). The case study at Pará did show increased boundary layer turbulent kinetic energy and sensible heat flux, and it is possible that increased boundary layer turbulence may have helped recoup some losses in precipitation. This is deductive reasoning; however, following that the Gran Chaco case study did not have increased turbulent kinetic energy associated with the *actual* scenario and it did experience more failed convective events associated with the inability to overcome convective inhibition (where there was no such trend at Pará).

The two case studies offer evidence that the negative precipitation anomaly, can be mediated by by primarily local effects and also by a combination of local effects and regional circulation effects. The Pará case study suggests that the negative precipitation anomaly is driven primarily by changes in the local surface energy flux. The dry season prevailing winds at Pará flowed up the moisture gradient (i.e., gaining moisture, not losing moisture). And according to the flux vectors in Fig. 13, despite the decreased precipitation experienced in the *actual* scenario, the prevailing winds fluxed more moisture into the domain when compared to the *potential* case. In contrast, the Gran Chaco case study showed evidence that the precipitation anomaly was responding to a change in the regional circulation as well as changes in the local surface energy fluxes. Here, the prevailing winds came out of the north and flowed down the moisture gradient (i.e., losing moisture). While the domain was a net source of moisture (divergent) the net flux into the domain via advection was positive. The *actual* scenario experienced a relative decrease in advected moisture flux from the prevailing winds (see Fig. 16).

5.5 Uncertainty in model estimates

Estimation uncertainty in coupled regional models exist in many sources, including the initial condition, the boundary conditions, the scale limitations of the resolved processes, the mechanics of the model processes and the parameters that govern the processes. In limited area models, there is also variability in the lateral boundary conditions. For the current research, this variability can affect the differences detected between the two scenarios, potentially impacting results. Ideally, this variability space can be explored in depth, perhaps using ensembles over multiple decades, including differing starting years and perturbations. The range of the variability space that can be sampled is limited by computational expense of the simulations. The 4–7 year simulations presented here took approximately 2 months each using 96 parallel computational cores with high-speed interconnects, the computational time being an result of the highly memory intensive ecosystem model and atmospheric time stepping that is relatively short (30 s) compared to general circulation models. However, we maintain that the the pattern differences in precipitation and evaporation between the two scenarios showed consistency enough to merit commentary.

The simulations used a number of different external data sets, each of which contained information at different spatial scales. This includes the soils information (variable scales), the human disturbance transitions (1°), the scale at which the forest structure was "spun-up" from climate driver data (1°) versus the scale at which that data was re-sampled (using nearest neighbor) in the coupled simulation (64 km). The uncertainty associated with the dynamics of the coupled simulations are subject to the scale of the information provided by these external data sets, as well as any biases that might be inherent in those data sets. This may be particularly true in the case of the lateral boundary condition data.

It is also necessary to acknowledge that there is uncertainty inherent in model process. For instance, in the absence of explicitly resolving buoyant updrafts (which is not possible in mesoscale simulations), the successful triggering of convection was based on the negative energy between the level of updraft and the level of free convection. There are alternative methods for estimating where updrafts start and how a parcel may or may not overcome inhibition to reach free buoyancy. We chose a straight forward parameterization that compares the negative buoyancy to a threshold. There are other trigger mechanisms that can be used, such as estimating the statistical distribution of vertical kinetic energy present in eddy motions at the level of updraft, and using that to estimating the likelihood that eddies will overcome negative buoyancy through force balance computations. This is simply an example of how the simulations were undoubtedly influenced by these choices, and this is something that must be considered when interpreting the output from any complex numerical model.

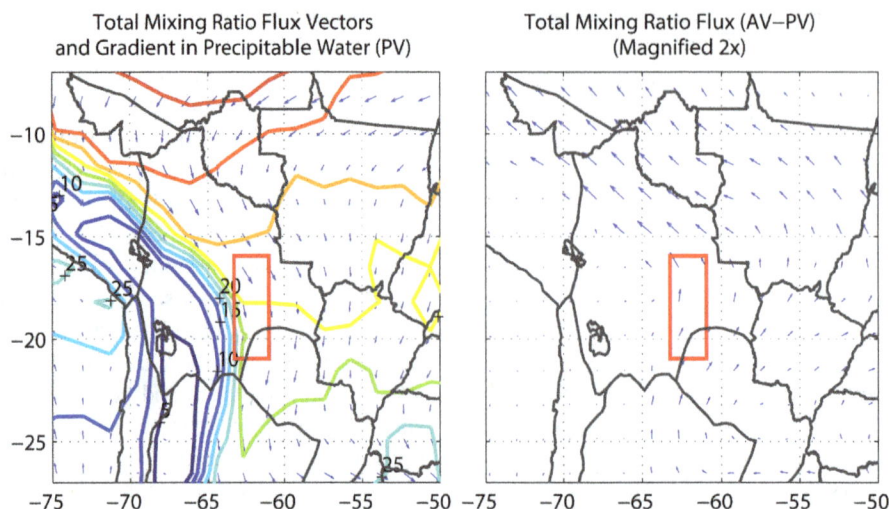

Figure 16. Left panel: map of vertically integrated total water advective flux vectors (quivers) and vertically integrated precipitable water (contours) for the *potential vegetation* case (PV), region near the Gran Chaco site, April 2003. Quivers are scaled and convey only directionality and relative magnitude. Contours of low precipitable water are shown by cool colors (blues) and high precipitable water with warm colors (reds). Right panel: the differential in vertically integrated advection of total precipitable water, *actual vegetation* minus *potential vegetation* (AV–PV). Quivers are scaled to 2 times relative to the left panel. In both panels the sub-domain bounding the Gran Chaco focus region is shown with a red box.

In the acknowledging uncertainty inherent in the simulations presented here, we have also tried to verify if these simulations can show agreement with observations. The *actual* scenario's simulations were compared to a group of different observations with the intention of verifying the modeling system's ability to represent key processes. The model's regional demographic of vegetation biomass was compared to field inventory data, atmospheric thermodynamic profiles were compared to soundings, mean all-sky profiles of cloud water were compared to satellite estimates and the seasonality of precipitation, net radiation, latent heat flux and sensible heat flux were compared with multi-data composite data products (see Appendices B–F). The comparisons with observations (which harbor their own uncertainty) suggest the model system is adequate to make a meaningful comparison between the two scenarios, yet not without room for improvement. In particular, this modeling study (and regional coupled simulations in general) would stand to benefit from improvements in how large-scale precipitation (weak spatiotemporal variability compared to point scale measurements) and canopy interception is handled.

A small selection of published research has conducted similar simulations to those reported here, with results that offer limited comparison. Perhaps most similar, Bagley et al. (2014) compared different land-cover scenarios with some similarities to our *potential* and *actual vegetation*, finding a generally weaker patterning of precipitation anomaly. However, evidence of the strongest pattern differences also occurred south of the Amazon delta, which may be viewed as correlated with the depression in the precipitation dipole pre-

sented in this work. The simulations conducted by Bagley et al. (2014) also used a different modeling system (The Weather Research and Forecasting System WRF) and land-surface model (Noah LSM), suggesting that model intercomparison would be a useful endeavor in rectifying the regional hydrometeorological effects of South American land-cover change.

Other studies have used similar simulation approaches to the work described here, but targeted land–atmosphere effects under future conditions. Global simulations conducted by Medvigy et al. (2011) found that future business as usual (aggressive) deforestation would generate dipole precipitation differential with respect to conservative deforestation scenarios. The length scales of the patterning in the dipoles (100s of kilometers) showed similarity with results presented here. However, the locations of differences they presented were different, as of course were the driving land-use scenarios as well. Walker et al. (2009) also evaluated the effects of future deforestation scenarios on the regional hydrometeorology, finding that massive deforestation outside of protected areas will lead to basin-wide changes in rainfall, both positive and negative. In summary, there are some commonalities between the various regional simulations, albeit from differing land-use scenarios; impacts on hydrometeorological climate are expressed most strongly as changes in dry-season precipitation, and that patterning exists to varying degrees at large length scales (> 100 km) (Walker et al., 2009; Bagley et al., 2014; Medvigy et al., 2011).

6   Conclusions

The simulations presented here produced negative anomalies in evapotranspiration rates that showed pattern similarity with areas where deforestation had driven noticeable differences in aboveground biomass. The results showed mixed support that historical land conversion has had influence on the patterning of South American precipitation. Over 4 years of simulation most of the region showed little consistency in annual anomalies. However, patterns in the precipitation anomaly emerged at specific locations, where mean annual precipitation showed moderate yet consistent differences as evaluated through a standard score. One pattern showed a dipole structure that occurred near eastern Pará Brazil. The other pattern, showed a negative anomaly in the Gran Chaco region of central and southern Bolivia. In this area, their was positive precipitation anomaly at the Peruvian–Bolivian border, yet the general patterning may not-necessarily be considered a dipole.

The simulations suggest land conversion in South America has not had a large impact on mean precipitation in the region as a whole. Mean regional precipitation was lower in the *actual* scenario in only 3 of the 4 simulation years. In contrast, differences in mean continental runoff (increased with human land conversion) and total evapotranspiration (decreased with human land conversion) were both consistent from year to year and showed greater differences compared to precipitation.

It was also observed that a large-scale shift in forest composition to early successional tropical evergreens following deforestation and abandonment in central eastern Amazon produced a noticeable increase in grid-average transpiration. This increase was striking not only because the pattern difference matched the pattern shift in composition, but particularly because the overall biomass in this area had decreased compared to the natural (*potential*) condition, suggesting that the physiological differences of the vegetation had influenced regional fluxes and not necessarily the quantity.

Beyond characterizing mean and pattern differences in the regional water balance, it was identified that changes (be they moderate) in precipitation may have occurred in locations where terrestrial vegetation actively regulates photosynthesis due to water limitations and that the changes in precipitation can be attributed to combinations of changes in local surface fluxes and changes in the regional atmospheric circulation. An assessment of the regional vegetation's response to moisture stress has indicated that both these locations show some susceptibility to changes in root-zone soil moisture (plant stomata were actively regulated to conserve water over a broad range depending on exact location, varying from 20–80 % of the time); however, the ecosystems of Gran Chaco are generally dryer and show greater susceptibility.

In both focus cases, deforestation led to increases in total surface albedo, driving decreases in net radiation, boundary layer moist static energy and ultimately convective precipitation. However, the differences in precipitation at Pará Brazil are more strongly connected with these localized differences in land-surface energy flux. The hydrometeorological analysis near Gran Chaco suggests that human land-conversion has had some impact on the strength of the South American moisture circulation in the southwestern portion of the Amazon, which claims partial responsibility along with differences in surface fluxes for an estimated decrease in annual precipitation in the Gran Chaco.

Appendix A: Land–atmosphere model coupling

The atmospheric model (BRAMS) and the land surface model (ED2) are loosely and asynchronously coupled. This assumes that the two models pass each others' boundary conditions at a frequency that captures the natural variability of the flux, yet the fluxes between models are not dynamically changing as prognostic variables within the numerical integrator sub-stepping. The atmosphere (BRAMS) provides information at the beginning of the ED2 forward step, while ED2 assumes this information is constant over the duration of its forward step and makes a time average of the fluxes during the step to return.

Mesoscale atmospheric models at spatial resolutions of tens of kilometers typically perform integrations on the order of tens of seconds to maintain numerical stability and convergence. The simulations in this work used an atmospheric time step of 30 s (for non-acoustic dynamics, acoustics were 10 s). Ideally, the land surface and atmospheric models operate at time steps that approach the infinitesimally small. The ED2 has a non-trivial computational overhead due to the large number of vegetation cohorts experiencing energy balancing within each grid cell. Ultimately, the ED2.1 model used a 120 s time step while coupled to BRAMS.

Atmospheric variables such as air temperature, humidity and wind-speed are provided to ED2 at a reference height of 70 m. This is required because ED2 internally calculates turbulent transport of heat, moisture and momentum through the canopy and into the inertial sub-layer of the lower atmosphere. The turbulent transport of scalars through the sub-layer above the canopy were calculated following Beljaars and Holtslag (1991). These calculations relied on gradients that were based on the enthalpy, density and specific humidity of air at reference temperature and within the canopy's interstitial air-space. A list of the variables required to drive ED2 is provided in Table A1.

The ED2 model passes boundary fluxes at the grid scale to the atmospheric surface layer as an area-weighted average across patches. There are three groups of information the land surface must provide the atmosphere model: (1) the topography which governs the geometry of the atmosphere's coordinate system and drag, (2) a lower boundary condition for turbulent closure, i.e., the vertical velocity perturbations of momentum, heat, moisture and carbon, and (3) a surface albedo for the atmospheric radiative transfer calculations. These variables are listed in Table A2. Because the ED2 model prognoses spatial variables at the patch sale (such as canopy temperature, humidity, etc), spatial averaging is required for flux variables. For any generic variable S, at a grid cell with M patches of area A_{patch}, the area averaged quantity is straight forward.

$$S_{grid} = \sum_{j}^{M} S_{patch, j} A_{patch, j} \qquad (A1)$$

$$\sum_{j}^{M} A_{patch, j} = 1 \qquad (A2)$$

More detailed explanation of how turbulent fluxes are calculated in the ED2 model can be found in Medvigy et al. (2009) and Knox (2012).

Table A1. Atmospheric boundary conditions provided by BRAMS, that drive the ED2 model.

Symbol	Units	Description
u_x	$[\mathrm{m\,s^{-1}}]$	Zonal wind speed
u_y	$[\mathrm{m\,s^{-1}}]$	Meridional wind speed
T_a	$[\mathrm{K}]$	Air temperature
q_a	$[\mathrm{kg\,kg^{-1}}]$	Air specific humidity
$\dot{m}_{pcp}$	$[\mathrm{kg\,s^{-1}}]$	Precipitation mass rate
z_{ref}	$[\mathrm{m}]$	Height of the reference point
R_{ld}	$[\mathrm{w\,m^{-2}}]$	Downward longwave radiation
R_{vb}	$[\mathrm{w\,m^{-2}}]$	Downward shortwave radiation, visible beam
R_{vd}	$[\mathrm{w\,m^{-2}}]$	Downward shortwave radiation, visible diffuse
R_{nb}	$[\mathrm{w\,m^{-2}}]$	Downward shortwave radiation, near infrared beam
R_{nd}	$[\mathrm{w\,m^{-2}}]$	Downward shortwave radiation, near infrared diffuse

Table A2. ED2 flux variables providing the lower boundary condition for the BRAMS atmospheric model.

Symbol	Units	Description
$\overline{(u'w')}$	$[\mathrm{m^2\,s^{-2}}]$	Average vertical flux of horizontal wind velocity perturbations
$\overline{(w'w')}$	$[\mathrm{m^2\,s^{-2}}]$	Average vertical flux of vertical wind velocity perturbations
$\overline{(t'w')}$	$[\mathrm{mK\,s^{-2}}]$	Average vertical flux of temperature perturbations
$\overline{(q'w')}$	$[\mathrm{kg\,m^{-2}\,s^{-2}}]$	Average vertical flux of moisture perturbations
$\overline{(c'w')}$	$[\mathrm{\mu mol\,m^{-2}\,s^{-2}}]$	Average vertical flux of carbon perturbations
χ_s	$[-]$	Average total shortwave albedo
χ_l	$[-]$	Average total longwave albedo
R_{lu}	$[\mathrm{w\,m^{-2}}]$	Average upwelling longwave radiation

Appendix B: Regional above-ground biomass

The model estimated live above-ground biomass (AGB) and basal area (BA) that represent the initial condition is compared with a collection of census measurements in Baker et al. (2004a, b) (see Fig. B1). A map is provided showing the locations of the plot experiments (see Fig. B2). The coordinates from the measurement stations were matched with ED2 nearest neighbor grid cells. Consistent with observations, only ED2 plants greater than 10 cm in primary forests were included in the comparison. The published measurements were taken at different times over the previous decades, the lag between these sites and the time of the simulation initial condition (January 2008) varies and is reported. Tree diameters in ED2 is diagnosed allometrically from structural carbon, similar to allometric equations in Chave et al. (2001) and Baker et al. (2004b), with differences in parameterization to reflect functional groups.

Please note that the forest inventory data makes no assumption that aboveground biomass is in equilibrium. This comparison is only intended to evaluate the present day static representation of forest structure. As applied to the coupled simulation, the were treated as static. As stated earlier, the length of the simulation did not merit the need to incorporate dynamics.

The majority of sites show fair agreement with model estimated above-ground biomass and basal area. The exception to this is the cluster of sites located in eastern Bolivia at Huanchaca Dos (HCC), Chore 1 (CHO), Los Fierros Bosque (LFB) and Cerro Pelao (CRP). There are several potential reasons for this discrepancy attributed to the model such as: variability in (1) climate forcing data (most notably precipitation and vapor pressure deficit as these are water limited growth conditions), (2) edaphic conditions and (3) plant functional parameters. These plots are close to the southern Amazonian transition between tropical rain-forests and Cerrado type open canopy forests where gradients in vegetation types are large and uncertainty is expected to be greater. Large spatial gradients in biomass are also reflected in the differences among the cluster of plots (HCC, CHO, LFB and CRP), (124.8 Mg ha^{-1} above-ground biomass at CHO and 260.0 Mg ha^{-1} AGB at HCC). The sharp gradient in forest biomass suggests that the omission of sub-pixel variability in the modified DS314 climate forcing could explain a portion of this difference, along with any persistent biases.

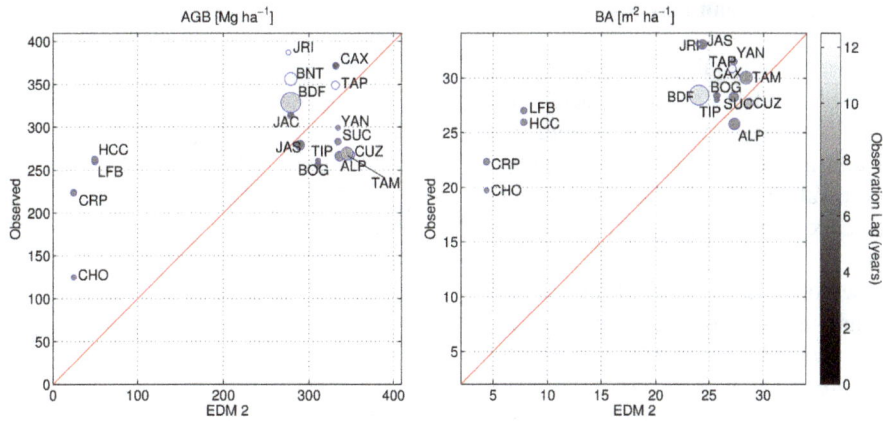

Figure B1. Comparison of model estimated mean above-ground biomass (AGB) and basal area (BA) with measurements presented in Baker et al. (2004a, b). Circle size shows relative approximation of the number of census sites used in the field measurements reported in Baker – maximum = 11 separate plots at BDF (code for the Biological Dynamics Forest Fragments Project). Darker circles indicate that measurements were taken more recently and therefore have less time lag in the comparison with the ED2 initial condition (January 2008). In accord with methods of Baker et al. (2004a, b), model estimates were filtered to include only primary forests and ignored vegetation less than 10 cm diameter. Coarse woody debris was excluded from comparison, only live stems were accounted for.

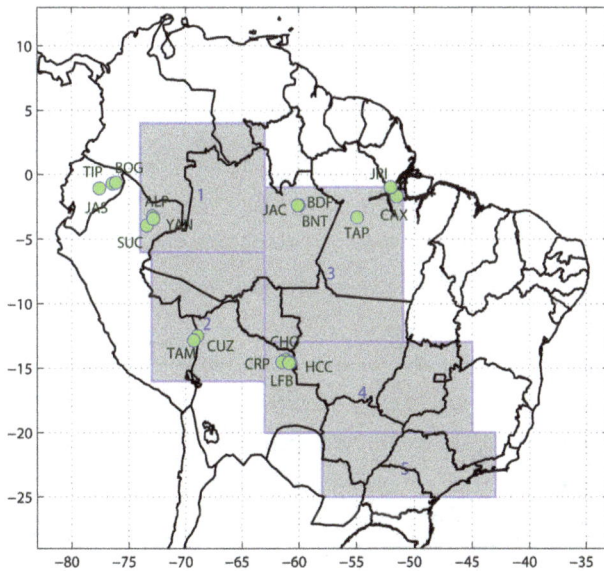

Figure B2. Locations of zones and sites of analysis. Zones are numbered 1 through 5, and reflect geographic areas where model and observation spatial means are compared for validation (see Appendix D and E). Forest census plot sites from Baker et al. (2004a, b) are referenced with green markers and their station code. Station codes designate the following site names: Allpahuayo (ALP), BDFFP (BDF), Bionte (BNT), Bogi (BOG), Caxiuana (CAX), Chore (CHO), Cerro Pelao (CRP), Cuzco Amazonico (CUZ), Huanchaca Dos (HCC), Jacaranda (JAC), Jatun Sacha (JAS), Jari (JRI), Los Fierros Bosque (LFB), Sucusari (SUC), Tambopata (TAM), Tapajos (TAP), Tiputini (TIP) and Yanamono (YAM).

Appendix C: Thermodynamic mean profiles

The objectives of this experiment require that modeling system output match mean observations to such a degree that there is trust in the model's ability to represent physical processes. It is believed that the relative differences between the two simulations have validity if model processes reproduce mean observed tendencies in the land surface and atmosphere.

Mean monthly profiles of model estimated air temperature, specific humidity and moist static energy are compared with mean radiosonde data over Manaus Brazil (see Figs. C1 and C2). Comparisons are made at 00:00 UTC (20:00 LT) and 12:00 UTC (20:00 LT). The model estimates a consistently warmer atmosphere, in the range of about two degrees both morning and evening. Model estimated specific humidity in the lower troposphere is lower than the radiosondes (see Fig. C1). Moist static energy is slightly underestimated by the model in the lower troposphere and then overestimated in the mid to upper troposphere. This may suggest that the model is convecting relatively large quantities of warm, moist air at the surface and entraining it to the upper atmosphere.

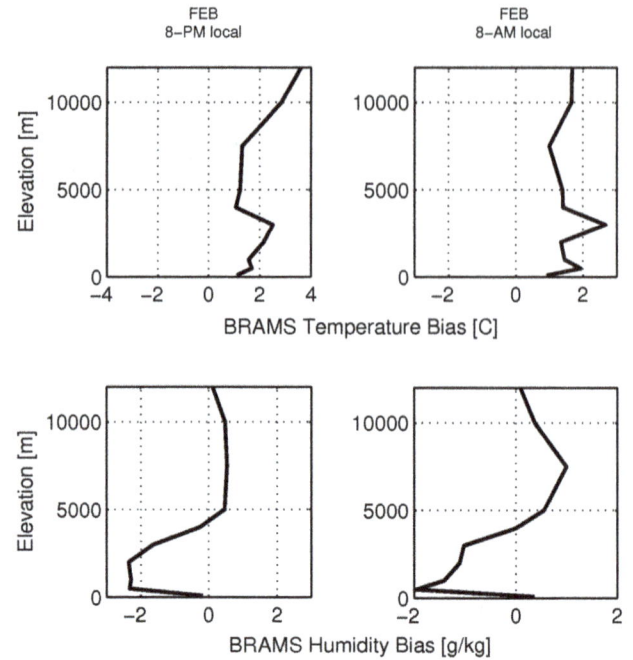

Figure C1. Comparison of model estimates with radiosonde data, differences in mean air temperature and specific humidity. Manaus, February 2003.

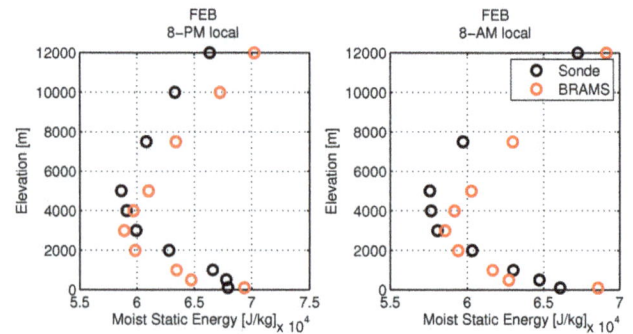

Figure C2. Comparison of model estimated mean moist static energy with rawinsonde measurements. Manaus, February 2003.

Appendix D: Inter-seasonal precipitation and surface radiation

Monthly precipitation and downwelling shortwave radiation in the model is evaluated as spatial means within five separate zones of analysis. The boundaries of the zones of analysis are shown in Fig. B2. Monthly mean model estimates are again compared to precipitation estimates from the Tropical Rainfall Measurement Mission (TRMM) 3B43 product and surface radiation from the Global Energy and Water Exchanges Project (GEWEX) Surface Radiation Budget (SRB) product version 2.5.[3] There is generally acceptable agreement between the model and TRMM estimated precipitation. The seasonal variability in both data sets is greater than their differences (see Fig. D1). The largest differences are reflected in the strength of of the wet-season peak precipitation in Zone 3 (central eastern Amazon) and the severity of the dry season precipitation in Zone 5 (southern Brazil). The timing of peak and minimum rainfall show generally good agreement, particularly in Zones 2–5. The lower estimate of mean precipitation in southern Brazil is consistent with decreased cloud albedo and increased downwelling shortwave radiation at the surface (see Fig. D2). Surface shortwave radiation is modestly over-estimated compared with the SRB estimates for most other cases.

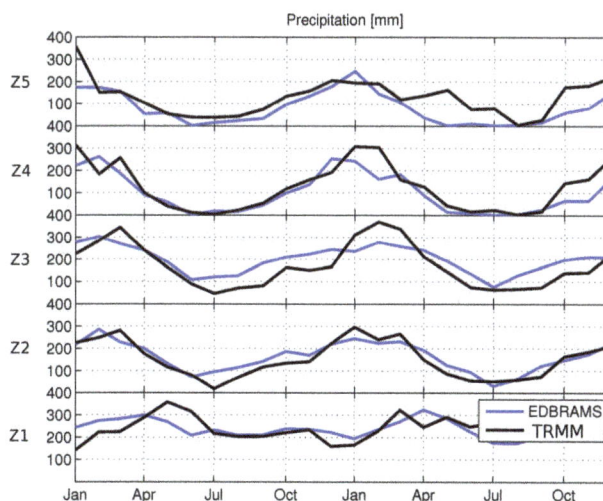

Figure D1. Mean monthly precipitation from ED2-BRAMS and the Tropical Rainfall Measurement Mission (TRMM) 3B43 product, years 2002–2003. Spatial means are taken within zones according to Fig. B2.

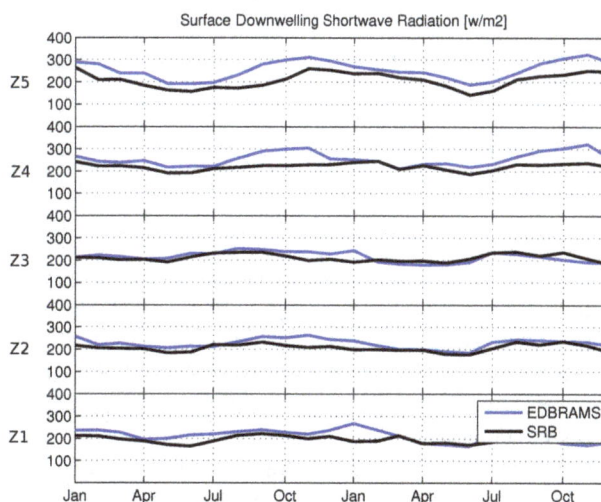

Figure D2. Mean monthly surface radiation from ED2-BRAMS and the Surface Radiation Budget (SRB) product version 2.5, years 2002–2003. Spatial means are taken within zones according to Fig. B2.

---

[3]These data were obtained from the NASA Langley Research Center Atmospheric Sciences Data Center NASA/GEWEX SRB Project.

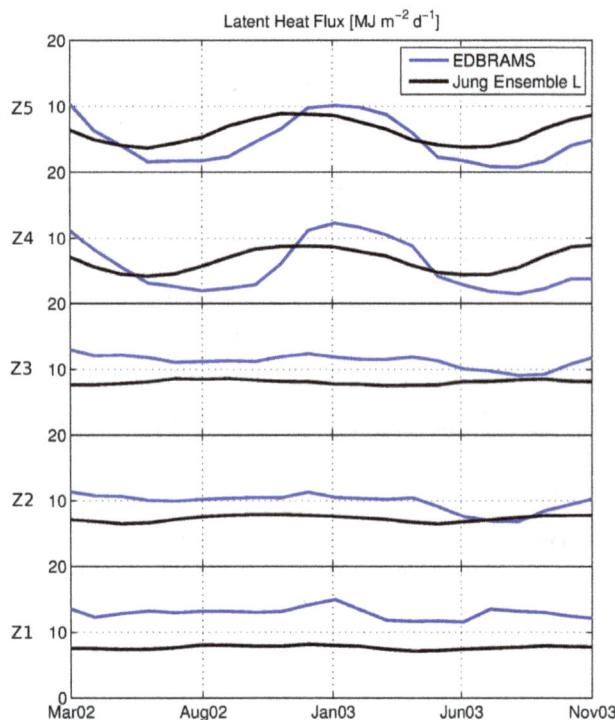

Figure E1. Mean monthly surface-to-atmosphere latent heat flux from ED2-BRAMS and the synthesis product from Jung et al. (2011, 2009). Spatial means are calculated over the zones shown in Fig. B2.

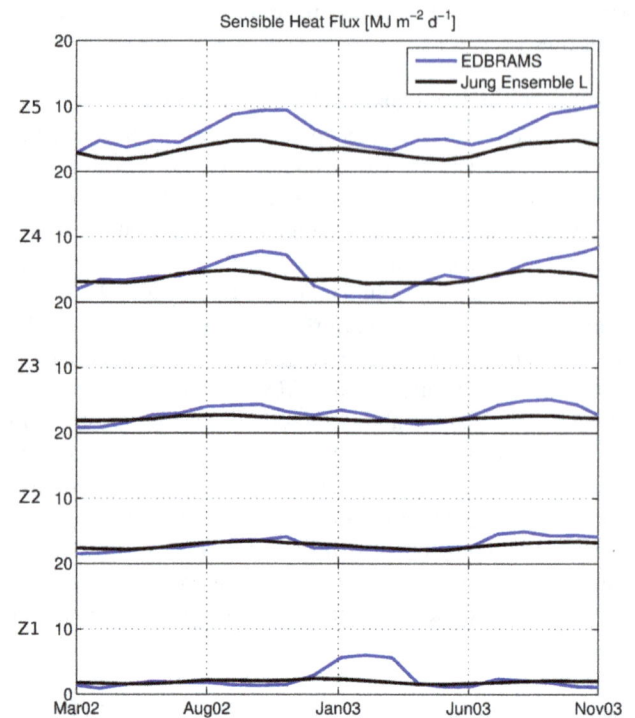

Figure E2. Mean monthly surface-to-atmosphere sensible heat flux from ED2-BRAMS and the synthesis product from Jung et al. (2011, 2009). Spatial means are calculated over the zones shown in Fig. B2.

Appendix E: Inter-seasonal latent and sensible heat flux

Similar to the comparison in Appendix D, monthly mean model estimated latent (see Fig. E1) and sensible heat flux (see Fig. E2) are compared with means from a benchmark. In this case, we compare output with the products of Jung et al. (2011). These products are based off of surface observations, which are upscaled using gridded explanatory variables from various sources including the Climate Research Unit (CRU), Global Precipitation Climatology Centre (GPCC) as well as the ERA-Interim product used in this study. The inter-seasonal means also incorporate spatial means within the domains shown in Fig. B2.

The purpose of this comparison is to give some benchmark of the ecosystem model's ability to partition energy flux at the land surface. The patterned flux of energy is dependant on the atmospheric model and its lateral boundary conditions as well. Compared to the synthesis product, the model system estimated a stronger inter-seasonal variability of latent and sensible heat flux over southern Brazil and the South American Convergence zone. Both model and observations showed relatively lower seasonal variability over the Amazon. In these regions, the model estimated small biases in sensible heat flux.

The greatest discrepancies between benchmark and model output was a latent heat flux bias (higher in the EDBRAMS

model) among the three zones covering the Amazon Basin. There are several possible explanations for this, outside of uncertainty inherent in the benchmark product. Latent and sensible heat flux contribute a portion of the total energy flux balance through the land surface, which also includes contributions from change in storage, diffusive ground heat flux, net radiation and the enthalpy contained in the mass flux of precipitation and runoff (enthalpy flux from density and pressure changes can be assumed near zero). Latent heat flux also contributes to a portion of the water mass balance at the land surface, which also includes precipitation mass flux, change in storage and total runoff. During previous experiments with the ED2 model used in offline simulations, we found that the surface water balance was sensitive to the scale of the precipitation input. When driving the land-surface model with precipitation resolved at coarse scales (such as native NCEP and ECMWF products), the leaf evaporation rates were disproportionately high. It was found that low but continuous precipitation rates from these products promoted a slow wetting of canopy leaves, and as a result the canopy leaves overflowed to the point of generating throughfall with less frequency and magnitude. Point scale precipitation rates from rain gauges in the Amazon showed a much larger variability in precipitation intensity. After using downscaling routines based on Lammering and Dwyer (2000) to preserve the monthly volume of grid-cell precipitation and creating point-scale precipitation intensity, these products (specifically the DS314 from UCAR) elicited a shift in canopy throughfall rates thereby decreasing latent heat flux and increasing surface runoff. Precipitation scale and how it affects the distribution of intensity, storm duration and the time between storms is a challenge in couple model simulation, that cannot be overcome using the same techniques as offline simulations. The spatial and temporal resolution of the simulations used in this work (40 km with 15 min time step between convective precipitation calls) are smaller than the reanalysis models (larger than 1°), yet they cannot generate point-scale precipitation. There are various approaches to ameliorating precipitation scale effects, such as using multi-atmosphere multi-land (MAML) sub-grid methods and by employing creative ways at the land-surface to generate throughfall volumes that match observations even when driven with precipitation rates that cannot match those that are observed. Regional and mesoscale couple simulations such as the work presented here, could benefit greatly from advances in this area.

Appendix F: All-sky cloud water content profiles

Cloud profile validation data sets were constructed from CloudSat cloud water content (2B-CWC-RO) and cloud classification (2B-CLDCLASS-LIDAR) data sets.[4] Overpasses during February 2007–2011 that intersected the geographic subset between 3° N–12° S and 70–55° W were collected and interpolated to a constant vertical datum above the surface. Overpasses typically occurred near 17:00 UTC.

Making a rigorous comparison of the model estimated cloud water and observation is challenging. Consider that the simulation time frame does not overlap with the CloudSat mission time frame, so these comparisons are treated as proxies to climatology and not weather validation. CloudSat measurements are known to have signal loss, attenuation and clutter during moderate to intense rainfall; events such as these could not be filtered from the comparison. It must also be assumed that the cloud classification algorithm is not without error. Nonetheless, the purpose of the comparison was to get a sense of whether the simulations estimated reasonable mean ranges of water contents and cloud fractions, and also if the phase transitions (liquid to ice) were occurring at reasonable elevations.

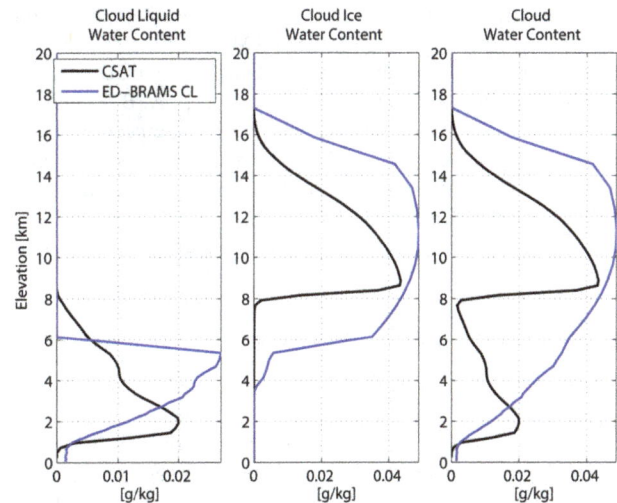

Figure F1. CloudSat climatological water content profiles and model estimated water content profiles for February 2003, 17:00 UTC, 3° N–12° S and 70–55° W.

The all-sky cloud water content profiles for both cumulus and non-cumulus clouds are provided in Fig. F1. The peaks in model estimated mean cloud water content showed reasonable agreement across liquid and ice cloud types. The model estimated generally more water content in both phases, skewed towards higher altitudes and showed a unimodal shape in the vertical distribution. It is possible that CloudSat relative underestimation could be explained by the omission of precipitating clouds.

---

[4]CloudSat data sets were provided on-line by CloudSat Data Processing Center, courtesy of NASA, Colorado State University and their partners.

Acknowledgements. This work was made possible through both the National Science Foundation Grant ATM-0449793 and National Aeronautics and Space Administration Grant NNG06GD63G. The authors would like to thank D. Entekhabi and E. A. B. Eltahir for their generous discussion and council. Further, the peer reviewers provided us with thorough, challenging and fair recommendations. We believe their efforts have contributed to significant improvements to this manuscript and we thank them.

Edited by: P. Regnier

References

Albani, M., Medvigy, D., Hurtt, G. C., and Moorcroft, P. R.: The contributions of land-use change, CO_2 fertilization, and climate variability to the Eastern US carbon sink, Global Change Biol., 12, 2370–2390, 2006.

Anthes, R. A.: Enhancement of convective precipitation by mesoscale variations in vegetative covering in semiarid regions, J. Clim. Appl. Meteorol., 23, 541–554, 1984.

Asdak, C., Jarvis, P. G., van Gardingen, P., and Fraser, A.: Rainfall interception loss in unlogged and logged forest areas of Central Kalimantan, Indonesia, J. Hydrol., 206, 237–244, 1998.

Avissar, R. and Werth, D.: Global hydroclimatological teleconnections resulting from tropical deforestation, J. Hydrometeorol., 6, 134–146, 2005.

Bagley, J., Desai, A., Harding, K., Synder, P., and Foley, J.: Drought and deforestation: Has land cover change influenced recent precipitation extremes in the Amazon?, J. Climate, 27, 345–361, 2014.

Baker, T., Phillips, O., Malhi, Y., Almeida, S., Arroyo, L., Fiore, A. D., Erwin, T., amd T.J. Killeen, N. H., Laurance, S., Laurance, W., Lewis, S., Monteagudo, A., Neill, D., Vargas, P., Pitman, N., Silva, N., and Vasquez-Martinez, R.: Increasing biomass in Amazonian forest plots, Philos. T. Roy. Soc. Lond. B, 359, 353–365, 2004a.

Baker, T., Phillips, O., Malhi, Y., Almeida, S., Arroyo, L., Fiore, A. D., Erwin, T., Killeen, S., Laurance, S., Laurance, W., Lewis, S., Lloyd, J., Monteagudo, A., Neill, D., Patino, S., Pitman, N., Silva, N., and Martinez, R. V.: Variation in wood density determines spatial patterns in Amazonian forest biomass, Global Change Biol., 10, 545–562, 2004b.

Baldi, M., Dalu, G. A., and Pielke, R. A.: Vertical velocities and available potential energy generated by landscape variability – theory, J. Appl. Meteorol. Clim., 47, 397–410, 2008.

Beljaars, A. C. M. and Holtslag, A. A. M.: Flux Parameterization over Land Surfaces for Atmospheric Models, J. Appl. Meteorol., 30, 327–341, 1991.

Benegas, L., Ilstedt, U., Roupsard, O., Jones, J., and Malmer, A.: Effects of trees on infiltrability and preferential flow in two contrasting agroecosystems in Central America, Agr. Ecosyst. Environ., 183, 185–196, 2014.

Cardille, J. and Foley, J.: Agricultural Land-use Change in Brazilian Amazonia Between 1980 and 1995: Evidence from Integrated Satellite and Census Data, Remote Sens. Environ., 87, 551–562, 2003.

Chapin, F., Matson, P., and Mooney, H.: Principles of Terrestrial Ecosystem Ecology, Springer-Verlag, New York, 2002.

Chappell, N., Bidin, K., and Tych, W.: Modelling rainfall and canopy controls on net-precipitation beneath selectively-logged tropical forest, Plant Ecology, 153, 215–229, 2001.

Chave, J., Riera, B., and Dubois, M.: Estimation of biomass in a neotropical forest of French Guiana: spatial and temporal variability, J. Trop. Ecol., 17, 79–96, 2001.

Chave, J., Muller-Landau, H. C., Baker, T. R., Easdale, T. A., Ter Steege, H., and Webb, C. O.: Regional and phylogenetic variation of wood density across 2456 neotropical tree species, Ecol. Appl., 16, 2356–2367, 2006.

Chen, C. and Cotton, W.: A One-Dimensional Simulations of the Stratocumulus-Capped Mixed Layer, Bound.-Lay. Meteorol., 25, 289–321, 1983.

Collatz, G. J., Ball, J., Grivet, C., and Berry, J. A.: Physiological and environmental regulation of stomatal conductance, photosynthesis and transpiration: a model that includes a laminar boundary layer, Agr. Forest Meteorol., 54, 107–136, 1991.

Collatz, G. J., Ribas-Carbo, M., and Berry, J.: Coupled Photosynthesis-Stomatal Conductance Model for Leaves of C_4 Plants, Aust. J. Plant Physiol., 19, 519–538, 1992.

Cotton, W., Pielke, R., Walko, R., Liston, G., Tremback, C., Jiang, H., McAnelly, R., Harrington, J., Nicholls, M., Carrio, G., and McFadden, J.: RAMS 2001: Current status and future directions, Meteorol. Atmos. Phys., 82, 5–29, 2003.

Crockford, R. and Richardson, D.: Partitioning of rainfall into throughfall, stemflow and interception: effect of forest type, ground cover and climate, Hydrol. Process., 14, 2903–2920, 2000.

d'Almeida, C., Vörösmarty, C. J., Hurtt, G. C., Marengo, J. A., Dingman, S. L., and Keim, B. D.: The effects of deforestation on the hydrological cycle in Amazonia: a review on scale and resolution, Int. J. Climatol., 27, 633–647, 2007.

Dalu, G. A., Pielke, R. A., Baldi, M., and Zeng, X.: Heat and momentum fluxes induced by thermal inhomogeneities, J. Atmos. Sci., 53, 3286–3302, 1996.

Dee, D. P., Uppala, S. M., Simmons, A. J., Berrisford, P., Poli, P., Kobayashi, S., Andrae, U., Balmaseda, M. A., Balsamo, G., Bauer, P., Bechtold, P., Beljaars, A. C. M., van de Berg, L., Bidlot, J., Bormann, N., Delsol, C., Dragani, R., Fuentes, M., Geer, A. J., Haimberger, L., Healy, S. B., Hersbach, H., Hólm, E. V., Isaksen, L., Kållberg, P., Köhler, M., Matricardi, M., McNally, A. P., Monge-Sanz, B. M., Morcrette, J.-J., Park, B.-K., Peubey, C., de Rosnay, P., Tavolato, C., Thépaut, J.-N., and Vitart, F.: The ERA-Interim reanalysis: configuration and performance of the data assimilation system, Q. J. Roy. Meteorol. Soc., 137, 553–597, 2011.

Dickinson, R. and Henderson-Sellers, A.: Modeling Tropical Deforestation: A Study of GCM Land-Surface Parameterization, Q. J. Roy. Meteorol. Soc., 114, 439–462, 1988.

Dietz, J., Hoelscher, D., Leuschner, C., and Hendrayanto: Rainfall partitioning in relation to forest structure in differently managed montane forest stands in Central Sulawesi, Indonesia, Forest Ecol. Manage., 237, 170–178, 2006.

Dietze, M., Wolosin, M., and Clark, J.: Capturing diversity and inerspecific variability in allometries: A hierarchical approach, Forest Ecol. Manage., 256, 1939–1948, 2008.

Eltahir, E.: Role of Vegetation in Sustaining Large-Scale Atmospheric Circulations in the Tropics, J. Geophys. Res.-Atmos., 101, 4255–4268, 1996.

Eltahir, E. and Bras, R.: On the Response of the Tropical Atmosphere to Large-Scale Deforestation, Q. J. Roy. Meteorol. Soc., 119, 779–793, 1993.

Eltahir, E. A. B. and Bras, R.: Precipitation recyling in the Amazon basin, Q. J. Roy. Meteorol. Soc., 120, 861–880, 1994.

Fisch, G., Tota, J., Machado, L., Dias, M., Lyra, R., Nobre, C., Dolman, A., and Gash, J.: The convective boundary layer over pasture and forest in Amazonia, Theor. Appl. Climatol., 78, 47–59, 2004.

Freitas, S. R., Rodrigues, L. F., Longo, K. M., and Panetta, J.: Impact of a monotonic advection scheme with low numerical diffusion on transport modeling of emissions from biomass burning, J. Adv. Model. Earth Syst., 4, Q1, doi:10.1029/2011MS000084, 2012.

Gedney, N. and Valdes, P.: The effect of Amazonian deforestation on the northern hemisphere circulation and climate, Geophys. Res. Lett., 27, 3053–3056, 2000.

Geist, H. and Lambin, E.: Proximate Causes of Underlying Driving Forces of Tropical Deforestation, Bioscience, 52, 143–150, 2002.

Grell, G. A. and Dévényi, D.: A generalized approach to parameterizing convection combining ensemble and data assimilation techniques, Geophys. Res. Lett., 29, 38-1–38-4, 2002.

Harrington, J. and Olsson, P.: A Method for the Parameterization of Cloud Optical Properties in Bulk and Bin Microphysical Models. Implications for Arctic Cloud Boundary Layers, Atmos. Res., 57, 51–80, 2001.

Henderson-Sellers, A., Dickinson, R., Durbridge, T., Kennedy, P., McGufie, K., and Pitman, A.: Tropical Deforestation: Modeling Local to Regional Scale Climate Change, J. Geophys. Res., 98, 7289–7315, 1993.

Hurtt, G., Frolking, S., Fearon III, M. B. M., Shevialokova, E., Malyshew, S., Pacala, S., and Houghton, R.: The underpinnings of land-use history: three centuries of global gridded land-use transitions, wood harvest activity and resulting secondary lands, Global Change Biol., 12, 1–22, 2006.

INPE: Monitoring of the Amazon forest by satellite 2001–2002, Instituto Nacional de Pesquisas Espaciais, Technical Paper, Sao Jose Dos Campos, Brazil, 2003.

Jung, M., Reichstein, M., and Bondeau, A.: Towards global empirical upscaling of FLUXNET eddy covariance observations: validation of a model tree ensemble approach using a biosphere model, Biogeosciences, 6, 2001–2013, doi:10.5194/bg-6-2001-2009, 2009.

Jung, M., Reighstein, M., Margolis, H., Cescatti, A., Richardson, A., Arain, M. A., Arneth, A., Bernhofer, C., Bonal, D., Chen, J., Gianelle, D., Gobron, N., Kiely, G., Kutsch, W., Lasslop, G., Law, B., Lindroth, A., Merbold, L., Montagnani, L., Moors, E. J., Papale, D., Sottoconrola, M., Vaccari, F., and Williams, C.: Global Patterns of Land-Atmosphere Fluxes of Carbon Dioxide, Latent Heat, and Sensible Heat Derived from Eddy Covariance, Satellite and Meteorological Observations, J. Geophys. Res., 116, 1–16, 2011.

Kain, J.: The Kain-Fritsch Convective Parameterization: An Update, J. Appl. Meteorol., 43, 170–181, 2004.

Kain, J. and Fritsch, J.: A One-Dimensional Entraining-Detraining Plume Model and Its Application in Convective Parameterization, J. Atmos. Sci., 47, 2784–2802, 1990.

Kennard, D. and Gholz, H.: Effects of high- and low-intensity fires on soil properties and plant growth in a Bolivian dry forest, Plant Soil, 234, 119–129, 2001.

Kleidon, A. and Heimann, M.: Assessing the Role of Deep Rooted Vegetation in the Climate System with Model Simulations: Mechanism, Comparison to Observations and Implications for Amazonian Deforestation, Clim. Dynam., 16, 183–199, 2000.

Knox, R.: Land Conversion in Amazonia and Northern South America; Influences on Regional Hydrology and Ecosystem Response, PhD thesis, Massachusetts Institute of Technology, Cambridge, Massachusetts, USA, 2012.

Knox, R., Bisht, G., Wang, J., and Bras, R.: Precipitation Variability over the Forest-to-Nonforest Transition in Southwestern Amazonia, J. Climate, 24, 2368–2377, 2011.

Lal, R.: Deforestation and land-use effects on soil degradation and rehabilitation in western Nigeria, 1. Soil physical and hydrological properties, Land Degred. Develop., 7, 19–45, 1996.

Lammering, B. and Dwyer, I.: Improvement of Water Balance in Land Surface Schemes by Random Cascade Disaggregation of Rainfall, Int. J. Climatol., 20, 681–695, 2000.

Laurance, W., Cochrane, M., Bergen, S., Fearnside, P., Delamonica, P., Barber, C., D'Angelo, S., and Fernandes, T.: The future of the Brazilian Amazon, Science, 291, 438–439, doi:10.1126/science.291.5503.438 2001.

Laurance, W., Nascimento, H., Laurance, S., Condit, R., D'Angelo, S., and Andrade, A.: Inferred longevity of Amazonian rainforest trees based on a long-term demographic study, Forest Ecol. Manage., 190, 131–143, 2004.

Lean, J. and Warrilow, D.: Simulation of the Regional Climatic Impact of Amazon Deforestation, Nature, 342, 411–413, 1989.

Lee, T. and Pielke, R.: Estimating the Soil Surface Specific-Humidity, J. Appl. Meteorol., 31, 480–484, 1992.

Leuning, R.: A critical appraisal of a combined stomatal-photosynthesis model for C3 plants, Plant Cell Environ., 18, 339–355, 1995.

Lewis, S.: Tropical forests and the changing earth system, Philos. T. Roy. Soc. B, 361, 195–210, 2006.

Martinez, L. and Zinck, J.: Temporal variation of soil compaction and deterioration of soil quality in pasture areas of Colombian Amazonia, Soil Till. Res., 75, 3–17, 2004.

Massman, W.: An Analytical One-Dimensional Model of Momentum Transfer by Vegetation of Arbitrary Structure, Bound.-Lay. Meteorol., 83, 407–421, 1997.

Medvigy, D., Wofsy, S., Munger, J., Hollinger, D., and Moorcroft, P.: Mechanistic scaling of ecosystem function and dynamics in space and time: Ecosystem Demography model version 2, J. Geophys. Res., 114, 1–21, 2009.

Medvigy, D., Walko, R., Otte, M., and Avissar, R.: The Ocean-Land-Atmosphere-Model: Optimization and Evaluation of Simulated Radiative Fluxes and Precipitation, Mon. Weather Rev., 138, 1923–1939, 2010.

Medvigy, D., Walko, R., and Avissar, R.: Effects of Deforestation on Spatiotemporal Distributions of Precipitation in South America, J. Climate, 24, 2147–2163, 2011.

Moorcroft, P., Hurtt, G., and Pacala, S.: A Method for Scaling Vegetation Dynamics: The Ecosystem Demography Model, Ecol. Monogr., 71, 557–586, 2001.

Muñoz-Villers, L. E. and McDonnell, J. J.: Land use change effects on runoff generation in a humid tropical montane cloud forest region, Hydrol. Earth Syst. Sci., 17, 3543–3560, doi:10.5194/hess-17-3543-2013, 2013.

Nakanishi, M. and Niino, H.: An improved Mellor-Yamada level-3 model with condensation physics: its numerical stability and application to a regional prediction of advection fog, Bound.-Lay. Meteorol., 119, 397–407, 2006.

Nepstad, D., de Carvalho, C., Davidson, E., Jipp, P., Lefebvre, P., Negreiros, H., dal Silva, E., Stone, T., Trubore, S., and Vieira, S.: The Role of Deep Roots in the Hydrological and Carbon Cycles of Amazonian Forests and Pastures, Nature, 372, 666–669, 1994.

Nepstad, D., Carvalho, G., Barros, A., Alencar, A., Capobianco, J., amd P. Moutinho, J. B., Lefebvre, P., Silva, U. L., and Prins, E.: Road paving, Fire Regime Feedbacks and the Future of Amazon Forests, Forest Ecol. Manage., 154, 395–407, 2001.

Nobre, C., Sellers, P., and Shukla, J.: Amazonian Deforestation and Regional Climate Change, J. Climate, 4, 957–988, 1991.

Pielke, R.: Influence of the spatial distribution of vegetation and soils on the prediction of cumulus convective rainfall, Rev. Geophys., 39, 151–171, 2001.

Pitman, A., Henderson-Sellers, A., and Yang, Z.: Sensitivity of Regional Climates to Localized Precipitation in Global-Models, Nature, 346, 734–737, 1990.

Poorter, L., Bongers, L., and Bongers, F.: Architecture of 54 moist-forest tree species: traits, trade-offs and functional groups, Ecology, 87, 1289–1301, 2006.

Quesada, C. A., Lloyd, J., Anderson, L. O., Fyllas, N. M., Schwarz, M., and Czimczik, C. I.: Soils of Amazonia with particular reference to the RAINFOR sites, Biogeosciences, 8, 1415–1440, doi:10.5194/bg-8-1415-2011, 2011.

Raupach, M., Finnigan, J., and Brunet, Y.: Coherent eddies and turbulence in vegetation canopies: The mixing-layer analogy, Bound.-Lay. Meteorol., 78, 351–382, 1996.

Rossato, L.: Estimativa da capacidade de armazenamento de água no solo do Brasil, Msc. thesis, Instituto Nacional de Pesquisas Espaciais (INPE), São José dos Campos, Brazil, 2001.

Scholes, R., Skole, D., and (eds.), J. I.: A Global Database of Soil Properties: Proposal for Implementation, Report of the Global Soils Task Group, Tech. Rep. IGBP-DIS Working Paper 10a, International Geosphere-Biosphere Programme – Data and Information System (IGBP-DIS), University of Paris, Paris, France, 1995.

Sheffield, J., Goteti, G., and Wood, E.: Development of a 50-Year High-Resolution Global Dataset of Meteorological Forcings for Land Surface Modeling, J. Climate, 19, 3088–3111, 2006.

Silva, R. R. D., Werth, D., and Avissar, R.: Regional impacts of future land-cover changes on the Amazon basin wet-season climate, J. Climate, 21, 1153–1170, 2008.

Skole, D. and Tucker, C.: Tropical Deforestation and Habitat Fragmentation in the Amazon: Satellite Data from 1978 to 1988, Science, 260, 1905–1910, 1993.

Snyder, P. K.: The Influence of Tropical Deforestation on the Northern Hemisphere Climate by Atmospheric Teleconnections, Earth Interact., 14, 1–34, doi:10.1175/2010EI280.1, 2010.

Soares-Filho, B. S., Nepstad, D. C., Curran, L. M., Cerqueira, G. C., Garcia, R. A., Ramos, C. A., Voll, E., McDonald, A., Lefebvre, P., and Schlesinger, P.: Modelling conservation in the Amazon basin, Nature, 440, 520–523, 2006.

Souza, E. P., Renno, N. O., and Silva-Dias, M. A. F.: Convective circulations induced by surface heterogeneities, J. Atmos. Sci., 57, 2915–2922, 2000.

Tremback, C. and Kessler, R.: A surface temperature and moisture parameterization for use in mesoscale models, Preprints, Seventh Conf. on Numerical Weather Prediction, Montreal, PQ, Canada, Amer. Meteor. Soc., 355–358, 1985.

Walcek, C. and Aleksic, N.: A simple but accurate mass conservative, peak-preserving, mixing ratio bounded advection algorithm with Fortran code, Atmos. Environ., 32, 3863–3880, 1998.

Walker, R., Moore, N., Arima, E., Perz, S., Simmons, C., Caldas, M., Vergara, D., and Bohrer, C.: Protecting the Amazon with protected areas, P. Natl. Acad. Sci. USA, 106, 10582–10586, 2009.

Walko, R., Band, L., Baron, J., Kittel, T., Lammers, R., Lee, T., Ojima, D., Pielke, R., Taylor, C., Tague, C., Tremback, C., and Vidale, P.: Coupled atmosphere-biophysics-hydrology models for environmental modeling, J. Appl. Meteorol., 39, 931–944, 2000.

Wang, D., Wang, G., and Anagnostou, E. N.: Evaluation of canopy interception schemes in band surface models, J. Hydrol., 347, 308–318, 2007.

Wang, J., Bras, R., and Eltahir, E.: The Impact of Observed Deforestation on the Mesoscale Distribution of Rainfall and Clouds in Amazonia, J. Hydrometeorol., 1, 267–286, 2000.

Wang, J., Chagnon, F., Williams, E., Betts, A., Renno, N., Machado, L., Bisht, G., Knox, R., and Bras, R.: The impact of deforestation in the Amazon basin on cloud climatology, P. Natl. Acad. Sci., 106, 3670–3674, 2009.

Williams, E. and Renno, N.: An Analysis of the Conditional Stability of the Tropical Atmosphere, Mon. Weather Rev., 121, 21–36, 1993.

Zhao, W. and Qualls, R. J.: A multiple-layer canopy scattering model to simulate shortwave radiation distribution within a homogeneous plant canopy, Water Resour. Res., 41, W08409, doi:10.1029/2005WR004016, 2005.

Zhao, W. and Qualls, R. J.: Modeling of long-wave and net radiation energy distribution within a homogeneous plant canopy via multiple scattering processes, Water Resour. Res., 42, W08436, doi:10.1029/2005WR004581, 2006.

Zimmermann, B., Elsenbeer, H., and De Moraes, J.: The influence of land-use changes on soil hydraulic properties: Implications for runoff generation, Forest Ecol. Manage., 222, 29–38, 2006.

ERA-Interim/Land: a global land surface reanalysis data set

G. Balsamo[1], C. Albergel[1], A. Beljaars[1], S. Boussetta[1], E. Brun[2], H. Cloke[3], D. Dee[1], E. Dutra[1], J. Muñoz-Sabater[1], F. Pappenberger[1], P. de Rosnay[1], T. Stockdale[1], and F. Vitart[1]

[1] European Centre for Medium-Range Weather Forecasts (ECMWF), Reading, UK
[2] Météo-France, Toulouse, France
[3] University of Reading, Reading, UK

Correspondence to: G. Balsamo (gianpaolo.balsamo@ecmwf.int)

Abstract. ERA-Interim/Land is a global land surface reanalysis data set covering the period 1979–2010. It describes the evolution of soil moisture, soil temperature and snowpack. ERA-Interim/Land is the result of a single 32-year simulation with the latest ECMWF (European Centre for Medium-Range Weather Forecasts) land surface model driven by meteorological forcing from the ERA-Interim atmospheric reanalysis and precipitation adjustments based on monthly GPCP v2.1 (Global Precipitation Climatology Project). The horizontal resolution is about 80 km and the time frequency is 3-hourly. ERA-Interim/Land includes a number of parameterization improvements in the land surface scheme with respect to the original ERA-Interim data set, which makes it more suitable for climate studies involving land water resources. The quality of ERA-Interim/Land is assessed by comparing with ground-based and remote sensing observations. In particular, estimates of soil moisture, snow depth, surface albedo, turbulent latent and sensible fluxes, and river discharges are verified against a large number of site measurements. ERA-Interim/Land provides a global integrated and coherent estimate of soil moisture and snow water equivalent, which can also be used for the initialization of numerical weather prediction and climate models.

1 Introduction

Multimodel land surface simulations, such as those performed within the Global Soil Wetness Project (Dirmeyer, 2011; Dirmeyer et al., 2002, 2006), combined with seasonal forecasting systems have been crucial in triggering advances in land-related predictability as documented in the Global Land–Atmosphere Coupling Experiments (Koster et al., 2006, 2009, 2011). The land surface state estimates used in those studies were generally obtained with offline model simulations, forced by 3-hourly meteorological fields from atmospheric reanalyses, and combined with simple schemes to address climatic biases. Bias corrections of the precipitation fields are particularly important to maintain consistency of the land hydrology. The resulting land surface data sets have been of paramount importance for hydrological studies addressing global water resources (e.g. Oki and Kanae, 2006). A state-of-the-art land surface reanalysis covering the most recent decades is highly relevant to foster research into intraseasonal forecasting in a changing climate, as it can provide consistent land initial conditions to weather and seasonal forecast models.

In recent years several improved global atmospheric reanalyses of the satellite era from 1979 onwards have been produced that enable new applications of offline land surface simulations. These include ECMWF's (European Centre for Medium-Range Weather Forecasts) Interim reanalysis (ERA-Interim; Dee et al., 2011) and NASA's Modern Era Retrospective-analysis for Research and Applications (MERRA; Rienecker et al., 2011). Simmons et al. (2010) have demonstrated the quality of ERA-Interim near-surface fields by comparing with observations-only climatic data records. Balsamo et al. (2010a) evaluated the suitability of ERA-Interim precipitation estimates for land applications at various timescales from daily to annual over the conterminous US. They proposed a scale-selective rescaling method to address remaining biases based on the Global Precipitation Climatology Project monthly precipitation data (GPCP; Huffman et al., 2009). This bias correction method addresses issues related to systematic model errors and non-conservation typical of data assimilation systems (Berrisford et al., 2011). Szczypta et al. (2011) have evaluated the incoming solar radiation provided by the ERA-Interim reanalysis with ground-based measurements over France. They showed a slight positive bias, with a modest impact on land surface simulations. Decker et al. (2012) confirmed these findings using flux tower observations and showed that the land surface evaporation of ERA-Interim compared favourably with the observations and with other reanalyses.

Offline land surface only simulations forced by meteorological fields from reanalyses are not only useful for land-model development but can also offer an affordable mean to improve the land surface component of reanalysis itself. Reichle et al. (2011) have used this approach to generate an improved MERRA-based land surface product (MERRA-Land; http://gmao.gsfc.nasa.gov/research/merra/merra-land. php). Similarly, we have produced ERA-Interim/Land, a new global land surface data set associated with the ERA-Interim reanalysis, by incorporating recent land model developments at ECMWF combined with precipitation bias corrections based on GPCP v2.1. Albergel et al. (2013) have already shown the value of an ERA-Interim/Land variant (with no precipitation readjustment) together with other model-based and remote sensing data sets for the detection of soil moisture climate trends in the past 30 years.

To produce ERA-Interim/Land, near-surface meteorological fields from ERA-Interim were used to force the latest version of the HTESSEL land surface model (Hydrology-Tiled ECMWF Scheme for Surface Exchanges over Land). This scheme is an extension of the TESSEL scheme (van den Hurk et al., 2000) used in ERA-Interim, which was based on the 2006 version of ECMWF's operational Integrated Forecasting System (IFS). HTESSEL includes an improved soil hydrology (Balsamo et al., 2009), a new snow scheme (Dutra et al., 2010), a multiyear satellite-based vegetation climatology (Boussetta et al., 2013a), and a revised bare soil evaporation (Balsamo et al., 2011; Albergel et al., 2012a). The majority of improvements in ERA-Interim/Land in the Northern Hemisphere can be attributed to land parameterization revisions, while the precipitation correction is important in the tropics and the Southern Hemisphere.

The purpose of this paper is to document ERA-Interim/Land and its added value from ECMWF's perspective. This will be done by providing some limited verification and diagnostics comparing ERA-Interim/Land and ERA-Interim with the purpose of explaining what is the origin of the differences. A very basic question is how can offline assimilation have added value, because in its current form it does not include data assimilation of soil moisture and snow? Alternatively one could ask: would it have been beneficial to have no soil moisture and snow assimilation in ERA-Interim? The answer is non-trivial, but it is known that in a coupled system, data assimilation for soil moisture is a necessity; otherwise precipitation can "run away" through a

positive precipitation/evaporation feedback at the continental scale (Viterbo and Betts, 1999; Beljaars et al., 1996). The soil moisture increments keep precipitation under control and tend to be beneficial for fluxes, but not always for soil moisture (Drusch and Viterbo, 2007). An offline land simulation produced after the coupled reanalysis has the advantage that there is no positive feedback because precipitation is prescribed and the surface water budget is closed as there are no soil moisture increments. The problems with snow reanalysis are mainly related to observations; snow gauges can have large biases, and the simple analysis scheme used in ERA-Interim occasionally results in a negative impact of the assimilated observations.

The next section describes the various data sets used for production and verification of ERA-Interim/Land. Section 3 describes the offline land surface model integrations. Section 4 presents the main results on verification of land surface fluxes, soil moisture, snow, and surface albedo. The land surface estimates from ERA-Interim/Land are a preferred choice for initializing ECMWF's seasonal forecasting system (System-4; Molteni et al., 2011), as well as the monthly forecasting system (Vitart et al., 2008), since both systems make use of the ERA-Interim/Land scheme. A summary and recommendations for the usage of the ERA-Interim/Land product are reported in the conclusions.

2 Data set and methods

The experimental set-up makes use of offline (or stand-alone) land simulations, which represent a convenient framework for isolating benefits and deficiencies of different land surface parameterizations (Polcher et al., 1998). In addition, given the complexity of the coupling with the atmosphere, offline simulations are much more cost-effective (faster) to run than a coupled atmosphere–land assimilation system.

In this study, offline runs are performed both at the global and point scales. All the 3-hourly meteorological forcing parameters were linearly interpolated in time to the land surface model integration time step of 30 min. The land-use information has been derived from the United States Geophysical Survey–Global Land Cover Classification (USGS-GLCC) and the United Nations–Food and Agriculture Organization (UN-FAO) data set at the same resolution as the forcing data. A comprehensive description of the land surface model and the ancillary data sets is given in the IFS documentation (2012, Part IV, Chapters 8 and 11; http://www. ecmwf.int/research/ifsdocs/CY37r2/index.html).

2.1 Validation and supporting data sets

The quality of ERA-Interim/Land relies on (i) the accuracy of the ERA-Interim forcing, (ii) bias correction of precipitation with the GPCP v2.1 data, and (iii) the realism of the land surface model. Its accuracy can be documented by veri-

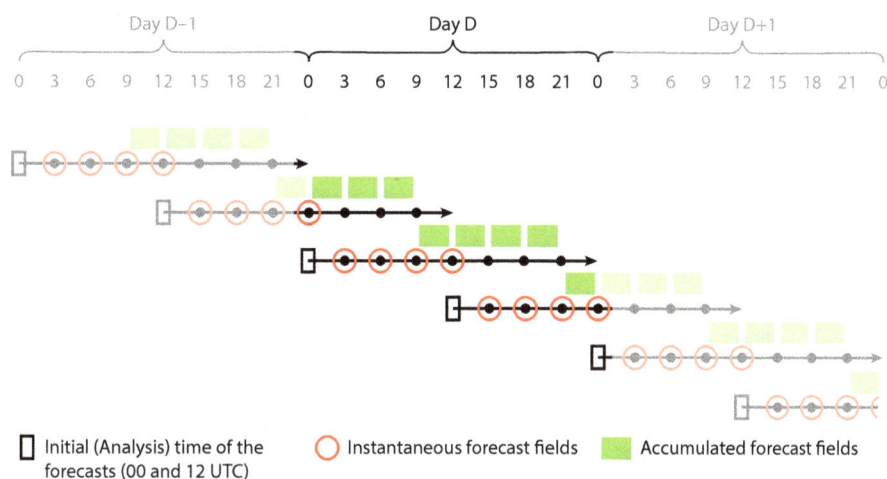

Figure 1. Schematic representation of the ERA-Interim meteorological forecasts concatenation for the creation of the 3-hourly forcing time series used in ERA-Interim/Land for a given day. Orange circles indicate instantaneous variables valid at their time stamp: 10 m temperature, humidity, wind speed, and surface pressure. Green boxes indicate fluxes valid on the accumulation period: surface incoming short-wave and long-wave radiation, rainfall, and snowfall.

fication with independent data e.g. surface fluxes, runoff, and soil temperature/moisture. In the following, the data sets entering the ERA-Interim/Land generation and its verification are briefly presented.

2.1.1 ERA-Interim meteorological reanalysis

ERA-Interim (Dee et al., 2011) is produced at T255 spectral resolution (about 80 km) and covers the period from January 1979 to present, with product updates at approximately 1 month delay from real-time. The ERA-Interim atmospheric reanalysis is built upon a consistent assimilation of an extensive set of observations (typically tens of millions daily) distributed worldwide (from satellite remote sensing, in situ, radio sounding, profilers, etc.). The analysis step combines the observations with a prior estimate of the atmospheric state produced with a global forecast model in a statistically optimal manner. In ERA-Interim two analyses per day are performed at 00:00 and 12:00 UTC (universal time coordinated), which serve as initial conditions for the subsequent forecasts. As a result of the data assimilation, the short-range forecasts (first-guess fields) stay close to the real atmosphere and the 12-hourly adjustments due to observations remain small. This justifies the use of a concatenation of short-range forecasts for forcing the offline land surface reanalysis. The forecasts have the advantage of being available every 3 h and they also provide estimates of precipitation and radiation. Experience with ERA-Interim has shown that the estimates of wind, temperature and moisture (at the lowest model level), which are well-constrained by observations, are generally of high quality in the 0–12 h forecast range and show only very small jumps from one 12 h cycle to the next (see Simmons et al., 2010, for a comparison of reanalysis temperature estimates with observations). Estimates of precipitation and ra-

diation, however, although indirectly constrained by temperature and humidity observations, are generated by the forecast model and are therefore subject to a small but systematic spin-up during the first few hours of the forecasts (Kållberg, 2011). Therefore, the 9–21 h forecast range is used for the fluxes co-located in time with the other fields as illustrated in Fig. 1.

2.1.2 GPCP v2.1 precipitation

The monthly GPCP data set merges satellite and rain gauge data from a number of satellite sources including the global precipitation index, the outgoing long-wave radiation precipitation index (OPI), the Special Sensor Microwave/Imager (SSM/I) emission, the SSM/I scattering, and the TIROS Operational Vertical Sounder (TOVS). In addition, rain gauge data from the combination of the Global Historical Climate Network (GHCN) and the Climate Anomaly Monitoring System (CAMS), as well as the Global Precipitation Climatology Centre (GPCC) data set, which consists of approximately 6700 quality controlled stations around the globe interpolated into monthly area averages, are used over land. Adler et al. (2003) detail the data sets and methods used to merge these data.

Compared to earlier releases, version 2.1 of GPCP used in this study takes advantage of the improved GPCC gauge analysis and the usage of the OPI estimates for the new SSM/I era. Thus, the main differences between the two versions are the result of the use of the new GPCC full data reanalysis (version 4) for 1997–2007, the new GPCC monitoring Product (version 2) thereafter, and the recalibration of the OPI data to a longer 20-year record of the new SSM/I-era GPCP data. Further details on the new version can be found in Huffman et al. (2009).

Table 1. List of flux tower sites used for the verification. The listed biome types are deciduous broadleaf forest (DBF), evergreen broadleaf forest (EBF), deciduous needle-leaf forest (DNF), evergreen needle-leaf forest (ENF), mixed forest (MF), woody savannahs (WSA), grasslands (GRA), crops (CRO), and wetlands (WET).

N	Site	Lat.	Long.	Veg. type	N	Site	Lat.	Long.	Veg. type
1	sk-oa	53.63	−106.20	DBF	18	it-ro2	42.39	11.92	DBF
2	sk-obs	53.99	−105.12	ENF/WET	19	nl-ca1	51.97	4.93	GRA
3	brasilia	−15.93	−47.92	WSA/GRA/SH	20	nl-haa	52.00	4.81	GRA
4	at-neu	47.12	11.32	GRA	21	nl-hor	52.03	5.07	GRA
5	ca-mer	45.41	−75.52	WET	22	nl-loo	52.17	5.74	ENF
6	ca-qfo	49.69	−74.34	ENF	23	ru-fyo	56.46	32.92	ENF
7	ca-sf1	54.49	−105.82	ENF	24	ru-ha1	54.73	90.00	GRA
8	ca-sf2	54.25	−105.88	ENF	25	ru-ha3	54.70	89.08	GRA
9	ch-oe1	47.29	7.73	GRA	26	se-sk2	60.13	17.84	ENF
10	fi-hyy	61.85	24.29	ENF	27	us-arm	36.61	−97.49	CRO
11	fr-hes	48.67	7.06	DBF	28	us-bar	44.06	−71.29	DBF
12	fr-lbr	44.72	−0.77	ENF	29	us-ha1	42.54	-72.17	DBF
13	il-yat	31.34	35.05	ENF	30	us-mms	39.32	−86.41	DBF
14	it-amp	41.90	13.61	GRA	31	us-syv	46.24	−89.35	MF
15	it-cpz	41.71	12.38	EBF	32	us-ton	38.43	−120.97	MF/WSA
16	it-mbo	46.02	11.05	GRA	33	us-var	38.41	−120.95	GRA
17	it-ro1	42.41	11.93	DBF	34	us-wtr	45.81	−90.08	DBF

The motivation for rescaling ERA-Interim precipitation estimates using GPCP data is to combine the best aspects of both data sets. ERA-Interim precipitation shows excellent synoptic variability but can be biased. Bias adjustments based on GPCP add the constraint of observations on a monthly timescale e.g. through the calibration of GPCP with SYNOP (synoptic) gauges. Balsamo et al. (2010a) evaluate ERA-Interim precipitation before and after rescaling with independent high-resolution data over the US. They conclude that in the extratropics, ERA-Interim is already close to GPCP in terms of performance, but that the monthly bias correction with GPCP gives an improvement. Much less is known about the tropics and areas with snow. Errors in ERA-Interim precipitation are much larger in the tropics (Betts et al., 2009; Agustì-Panareda et al., 2010) than in the extratropics and benefit from bias correction with GPCP is expected to be substantial. Runoff verification results shown below provide indirect evidence for this conclusion. For snowfall, Brun et al. (2013) conclude, on the basis of snow accumulation verification, that the quality of ERA-Interim is excellent and exceeds those based on gauge observations, which tend to suffer from substantial undercatch. The impact of GPCP bias correction on snowfall is fairly small.

2.1.3 FLUXNET land energy fluxes

FLUXNET is a global surface energy, water, and CO_2 flux observation network and consists of a collection of regional networks (Baldocchi et al., 2001; http://fluxnet.ornl.gov). Additionally, observational data for the year 2006 from the Boreal Ecosystem Research and Monitoring Sites (BERMS;

Betts et al., 2006), and the Coordinated Energy and water cycle Observations Project (CEOP) were used in this study.

The FLUXNET observations are part of the La Thuile data set, which provides flux tower measurements of latent heat flux (LE), sensible heat flux (H) and net ecosystem exchange (NEE) at high temporal resolution (30–60 min). For verification purposes, hourly observations from the year 2004 were selected from the original observational archive (excluding gap filled values) with a high-quality flag only (see Table 1).

As part of the CEOP program, reference site observations from the Amazonian region also belonging to the LBA experiments (the Large Scale Biosphere–Atmosphere Experiment in Amazonia) are available for scientific use. In this study, observations are taken from flux towers located within a woody savannah region (Brasilia).

2.1.4 ISMN soil moisture observing network

In situ soil moisture observations are extremely useful for the evaluation of modelled soil moisture. In recent years, huge efforts were made to collect observations representing contrasting biomes and climate conditions. Some of them are now freely available such as data from The International Soil Moisture Network (ISMN; Dorigo et al., 2011, 2013, http://ismn.geo.tuwien.ac.at/). The ISMN is a new data-hosting centre where globally available ground-based soil moisture measurements are collected, harmonized and made available to users. This includes a collection of nearly 1000 stations (with data from 2007 up to present) gathered and quality controlled at ECMWF. Albergel et al. (2012a, b, c) have used these data to validate various soil moisture estimates

Table 2. Comparison of surface soil moisture with in situ observations for ERA-Interim/Land (bold) and ERA-Interim (normal font) in 2010: mean correlation (R), bias (observation minus ERA), root mean square error (RMSE), and normalized standard deviation (NSD = SD$_{model}$/SD$_{obs}$). Scores are given for significant correlations with p values < 0.05. For each R estimate a 95 % confidence interval (CI) was calculated using a Fisher Z transform.

Network (N stations with significant R)	R (95 % CI)	Bias (m³ m⁻³)	RMSE (m³ m⁻³)	NSD = σ_model/σ_obs
AMMA, W. Africa (3)	**0.63** (**±0.06**)	**−0.060**	**0.082**	**2.67**
Pellarin et al. (2009)	0.61 (±0.07)	−0.153	0.154	0.69
OZNET, Australia (36)	**0.79** (**±0.05**)	**−0.112**	**0.131**	**1.01**
Smith et al. (2012)	0.78 (±0.05)	−0.078	0.106	0.55
SMOSMANIA, France (12)	**0.83** (**±0.04**)	**−0.080**	**0.108**	**0.83**
Albergel et al. (2008)	0.82 (±0.05)	−0.037	0.099	0.41
REMEDHUS, Spain (17)	**0.76** (**±0.04**)	**−0.152**	**0.175**	**1.57**
Ceballos et al. (2005)	0.79 (±0.04)	−0.110	0.135	0.84
SCAN, US (119)	**0.64** (**±0.07**)	**−0.078**	**0.130**	**0.95**
Schaefer and Paetzold (2001)	0.62 (±0.07)	−0.063	0.110	0.54
SNOTEL, US (193)	**0.62** (**±0.10**)	**−0.045**	**0.115**	**0.78**
Schaefer and Paetzold (2001)	0.69 (±0.08)	−0.088	0.123	0.44

produced at ECMWF, including from ERA-Interim as well as from offline land simulations. Data from six networks are considered for 2010: NRCS-SCAN (Natural Resources Conservation Service–Soil Climate Analysis Network) and SNOTEL (SNOwpack TELemetry) over the United States, with 177 and 348 stations, respectively; SMOSMANIA (Soil Moisture Observing System–Meteorological Automatic Network Integrated Application) with 12 stations in France; REMEDHUS (REd de MEDición de la HUmedad del Suelo) in Spain with 20 stations, the Australian hydrological observing network labelled OZNET with 38 stations; and AMMA (African Monsoon Multidisciplinary Analyses) in western Africa with 3 stations. Data at 5 cm and the year 2010 are used for the comparison because it is the depth and year for which most of the stations have observations (Table 2 includes references for different networks).

2.1.5 The GTS-SYNOP observing network

The GTS-SYNOP (Global Telecommunications System–surface SYNOPtic observation) is an operationally maintained data set under coordination of the World Meteorological Organization (WMO), which provides daily ground-based observations of the main weather parameters and selected land surface quantities, such as snow depth, at a large number of sites worldwide. The snow data are acquired at a minimum frequency of once a day and represent the only quantitative snow-depth measurements in contrast to remote sensing observations, which have limited information on snow depth. These data are operationally used at ECMWF

for the daily global snow analysis as described in Drusch et al. (2004) and de Rosnay et al. (2014).

2.1.6 Satellite surface albedo

The Moderate Resolution Imaging Spectroradiometer (MODIS) albedo product MCD43C3 provided data describing both directional hemispheric reflectance (black-sky albedo) and bihemispherical reflectance (white-sky albedo) in seven different bands and aggregated bands. Data from the Terra and Aqua platforms are merged in the generation of the product that is produced every 8 days on a 0.05° global grid. The accuracy and quality of the product has been studied by several authors (e.g. Román et al., 2009; Salomon et al., 2006). The MODIS product has served as a reference for model validation (e.g. Dutra et al., 2010, 2012; Wang and Zeng, 2010; Zhou et al., 2003). In this study, we compare the white-sky broadband short-wave albedo (2000–2010) with ERA-Interim and ERA-Interim/Land. MODIS albedo was averaged for each month and spatially aggregated to the model grid.

2.1.7 The GRDC river discharge data set

The Global Runoff Data Centre (GRDC) operates under the auspice of the World Meteorological Organization and provides data for verification of atmospheric and hydrologic models. The GRDC database is updated continuously and contains daily and monthly discharge data information for over 3000 hydrologic stations in river basins located in 143 countries. Over the GSWP-2 period, the runoff data of

1352 discharge gauging stations was available and used for verification of the soil hydrology (Balsamo et al., 2009). Pappenberger et al. (2009) and Balsamo et al. (2010b) used the GRDC discharge to evaluate a coupled land surface–river discharge scheme for river flood prediction.

2.2 Land modelling component

ERA-Interim/Land differs from the land component of ERA-Interim in a number of parameterization improvements introduced in the operational ECMWF forecast model since 2006, when the ERA-Interim reanalysis started. The meteorological forcing described in Sect. 2.1.1 is used to drive an 11 year spin-up run (1979–1989). The average of the 11 "1 Januaries" is taken as a plausible initial condition for 1 January 1979.

A single continuous 32-year simulation starting on 1 January 1979 is then realized with the latest ECMWF land surface scheme. The modelling components that were updated with respect to ERA-Interim are briefly described in the following subsections with emphasis on those changes that have an impact on ERA-Interim/Land performance.

2.2.1 Soil hydrology

A revised soil hydrology in TESSEL was proposed by van den Hurk and Viterbo (2003) for the Baltic Basin. These model developments were in response to known weaknesses of the TESSEL hydrology: specifically, the choice of a single global soil texture, which does not characterize different soil moisture regimes, and a Hortonian runoff scheme which produces hardly any surface runoff. Therefore, a revised formulation of the soil hydrological conductivity and diffusivity (spatially variable according to a global soil texture map) and surface runoff (based on the variable infiltration capacity approach) were operationally introduced in IFS in November 2007. Balsamo et al. (2009) verified the impact of the soil hydrological revisions from field site to global, atmospheric coupled experiments and in data assimilation.

2.2.2 Snow hydrology

A fully revised snow scheme was introduced in 2009 to replace the existing scheme based on Douville et al. (1995). The snow density formulation was changed and liquid water storage in the snowpack was introduced, which also allows for the interception of rainfall. On the radiative side, the snow albedo and the snow cover fraction have been revised and the forest albedo in presence of snow has been retuned based on MODIS satellite estimates. A detailed description of the new snow scheme and verification from field site experiments to global offline simulations are presented in Dutra et al. (2010). The results showed an improved evolution of the simulated snowpack with positive effects on the timing of runoff and terrestrial water storage variation and a better match of the albedo to satellite products.

2.2.3 Vegetation seasonality

The leaf area index (LAI), which expresses the phenological phase of vegetation (growing, mature, senescent, dormant), was kept constant in ERA-Interim and assigned by a lookup table depending on the vegetation type; thus, vegetation appeared to be fully developed throughout the year. To allow for seasonality, a LAI monthly climatology based on a MODIS satellite product was implemented in IFS in November 2010. The detailed description of the LAI monthly climatology and its evaluation is provided in Boussetta et al. (2013a).

2.2.4 Bare soil evaporation

In ERA-Interim, the bare ground evaporation is based on the same stress function as for vegetation. The result is that evaporation is not possible for soil moisture contents below the permanent wilting point. This has been improved by adopting a lower stress threshold for bare soil (Balsamo et al., 2011) which is in agreement with previous experimental findings (e.g. Mahfouf and Noilhan, 1991) and results in more realistic soil moisture for dry lands. The new bare soil evaporation in conjunction with the LAI update as reported in Balsamo et al. (2011) has been extensively evaluated by Albergel et al. (2012a) over the US. The evaluation was based on data from the Soil Climate Analysis Network (SCAN) as well as Soil Moisture and Ocean Salinity (SMOS) satellite data.

3 Results

The quality of ERA-Interim/Land builds upon reduced errors in the meteorological forcing and improved land surface modelling. In the following, selected verification results are illustrating the skill of ERA-Interim/Land in reproducing the main land water reservoirs and fluxes towards the atmosphere and river outlets. The two most active water reservoirs are the root-zone soil moisture (here the top 1 m of soil is considered) and the snow accumulated on the ground. These global reservoirs in its median of the distribution calculated over the period 1979–2010 are shown in Figs. 2 and 3 for ERA-Interim and ERA-Interim/Land, respectively. The median of soil moisture (SM) and snow water equivalent (SWE) are both expressed in millimetres of water or equivalently in kilograms per square metre. The medians over the 32-year SM and SWE are based on 11 daily values centred around 15 January and 15 July for 32 years, resulting in 352 samples. The median is of particular interest to illustrate the snow and soil moisture global climatology maps because it indicates "typical" values and a single exceptional year would leave the median invariant. The same argument is valid for mid-July SM in which a single exceptional flood will not affect the median.

Clear differences can be seen between ERA-Interim and ERA-Interim/Land in both January snow amount and July soil moisture (compare Figs. 2 and 3). The differences in

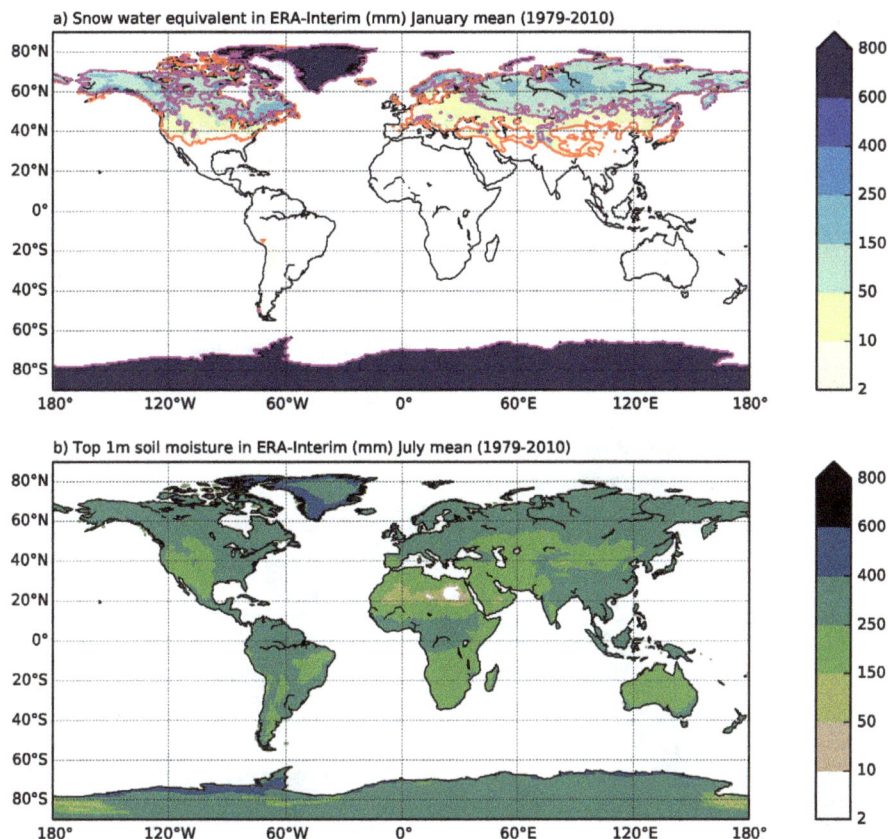

Figure 2. Median of the land water reservoirs in the 1979–2010 period for ERA-Interim: (**a**) snow water equivalent (SWE, kg m^{-2}) for the 10–20 January period, and (**b**) top 1 m soil moisture (TCSM, kg m^{-2}) for the 10–20 July period. The red and magenta contours in (**a**) indicate the 5th and 95th percentiles, respectively, of 10 kg m^{-2} snow water equivalent as an indication of the year to year variability of snow cover.

snow amount are due to (i) the GPCP bias correction of precipitation forcing, (ii) improved snowmelt, density and albedo in the land surface model, and (iii) the lack of data assimilation of snow depth in ERA-Interim/Land. The GPCP correction results in a slightly reduced snowfall, and the changes in the snow model lead predominantly to differences in the marginal snow areas and seasonal differences. The main difference comes from the data assimilation method used in ERA-Interim. It uses a Cressman (1959) scheme for the assimilation of SYNOP observations, which has documented deficiencies in areas with sparse observations, and strong relaxation to climatology before 2003. After 2003, qualitative information from a snow cover product is used instead of climatology (Drusch et al., 2004). In particular, the use in ERA-Interim of climatology before 2003 and the poor handling of sparse observations with the Cressman scheme make ERA-Interim/Land (which relies on forcing and the model only) more suitable for studies of interannual variability and extremes. From Figs. 2a and 3a, it can be seen that snow mass has more variability in ERA-Interim than in ERA-Interim/Land. This is the result of the Cressman analysis of SYNOP data, particularly in areas with low-density observations. To illustrate the dynamical range of the distri-

bution and the capability of reanalysis to reproduce anomalies, the 5th and 95th percentiles of the 10 kg m^{-2} contour is also plotted in Figs. 2a and 3a. As expected there is a large distance between the 5th and 95th percentiles, indicating a lot of interannual variability in the snow line.

The summer soil moisture also shows large differences between ERA-Interim and ERA-Interim/Land. As can be seen from Figs. 2b and 3b, soil moisture tends to be lower in ERA-Interim/Land. This is mainly the result of the modified soil hydrology properties which increases the effective size of the soil moisture reservoir, permits a larger amplitude of the seasonal cycle, and allows soil moisture to go lower in summer. Data assimilation in ERA-Interim also tends to reduce the seasonal cycle by adding water in summer (Drusch and Viterbo, 2007). ERA-Interim/Land shows more spatial variability than ERA-Interim. This is the result of the spatial variability of soil properties, which ERA-Interim does not have, and the reformulation of the bare soil evaporation.

The evolution of ERA-Interim/Land along a 10-year period of this data set and its differences with respect to ERA-Interim are illustrated in Figs. 4 and 5 for both soil moisture and snow water equivalent. The stability and the differences with respect to ERA-Interim can be appreciated in Figs. 4a

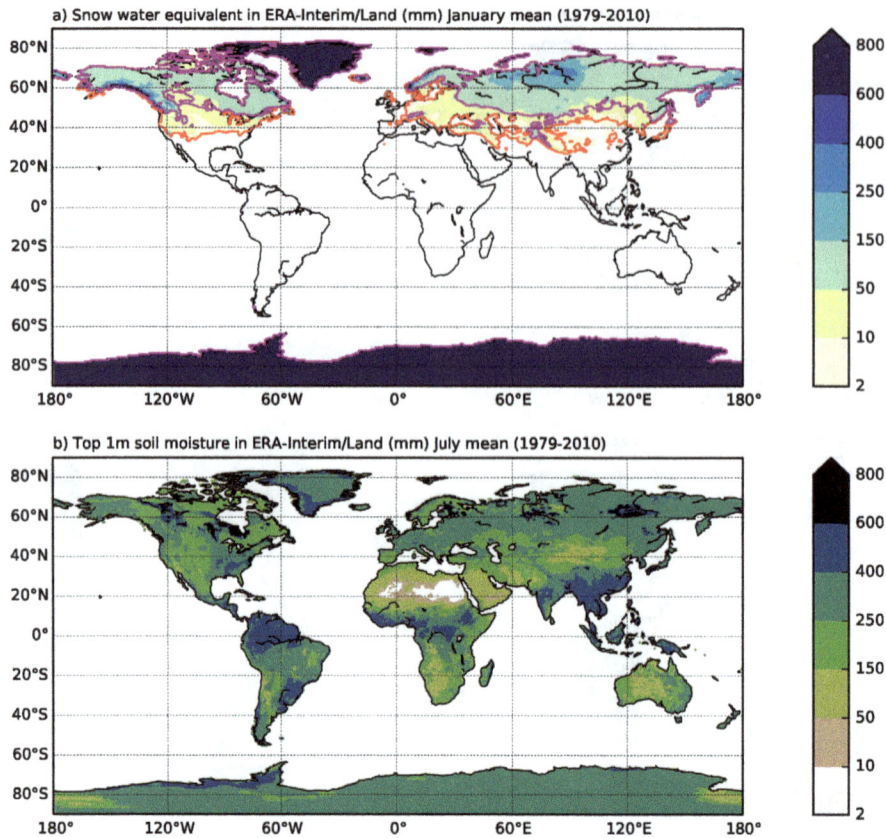

Figure 3. Same as Fig. 2, but for ERA-Interim/Land.

and 5a for snow water equivalent and in Figs. 4b and 5b for the top 1 m soil moisture. The snow changes in Fig. 5a are mainly a consequence of the new snow scheme and highlight both a snow mass increase at high latitudes and a slight reduction at midlatitudes. There is also a phase shift in the seasonal cycle at midlatitudes with less snow during accumulation and more snow in the melting season. The soil moisture presents large differences in Fig. 5b, which can be attributed to the soil hydrology revisions. Figure 5 is meant to illustrate that ERA-Interim and ERA-Interim/Land are significantly different with respect to land water resources. In these runs, observational constraints on the snow and soil water reservoirs, such as those applied by the screen-level data assimilation, are totally absent. However, the resulting water reservoirs of snow and soil moisture and the river discharges are shown to improve with respect to the original ERA-Interim data, without deteriorating the turbulent fluxes to the atmosphere. In the following sections, a selection of results is presented to demonstrate the added value of ERA-Interim/Land.

3.1 Land flux verification

In the following subsections, fluxes from the offline-driven land simulations are validated against two observation cat-

egories: the land-to-atmosphere turbulent heat and moisture fluxes and the river discharges.

3.1.1 Latent and sensible heat flux

Fluxes from 34 FLUXNET, CEOP and BERMS flux towers, as listed in Table 1, are used for verification in 2004. Correlation, mean bias and root mean squared errors are computed based on 10-day averages, so the verification is focusing on the seasonal and subseasonal timescales. Figure 6 shows the RMSEs of sensible and latent heat flux for the individual flux towers. The RMSEs of sensible heat flux are of the order of $20 \, \text{W} \, \text{m}^{-2}$, which is typical for point verification. The errors of latent heat flux are larger and vary from station to station. Positive and negative differences are seen in Fig. 6, and it is difficult to draw firm conclusions on the relative merit of ERA-Interim/Land compared to ERA-Interim. A major issue with point verification is that the station may not be representative of a large area. The vegetation cover around the station may also be different from the vegetation type as specified in the corresponding model grid box. The latter is probably the case for stations that show atypically large errors.

An overall quantitative estimate of the errors is reported in Table 3. Latent and sensible heat fluxes have RMSEs of $21.8 \, (\pm 0.9)$ and $21.3 \, (\pm 0.9) \, \text{W} \, \text{m}^{-2}$ with ERA-Interim/Land

Table 3. Summary of mean latent heat flux (LE) and sensible heat flux (H) statistics averaged over the 34 sites (units: $\mathrm{W\,m^{-2}}$). The CI of RMSE is based on the Chi-squared distribution. The R of model fluxes to observations include a 95 % CI calculated using a Fisher Z transform.

Model	LE RMSE	LE Bias	LE R	H RMSE	H Bias	H R
ERA-Interim/Land	21.8 ($\pm$0.9)	14.4	0.85 ($\pm$0.02)	21.3 ($\pm$0.9)	-2.6	0.83 ($\pm$0.02)
ERA-Interim	26.0 ($\pm$1.0)	18.2	0.83 ($\pm$0.02)	19.6 ($\pm$0.8)	-3.8	0.85 ($\pm$0.02)

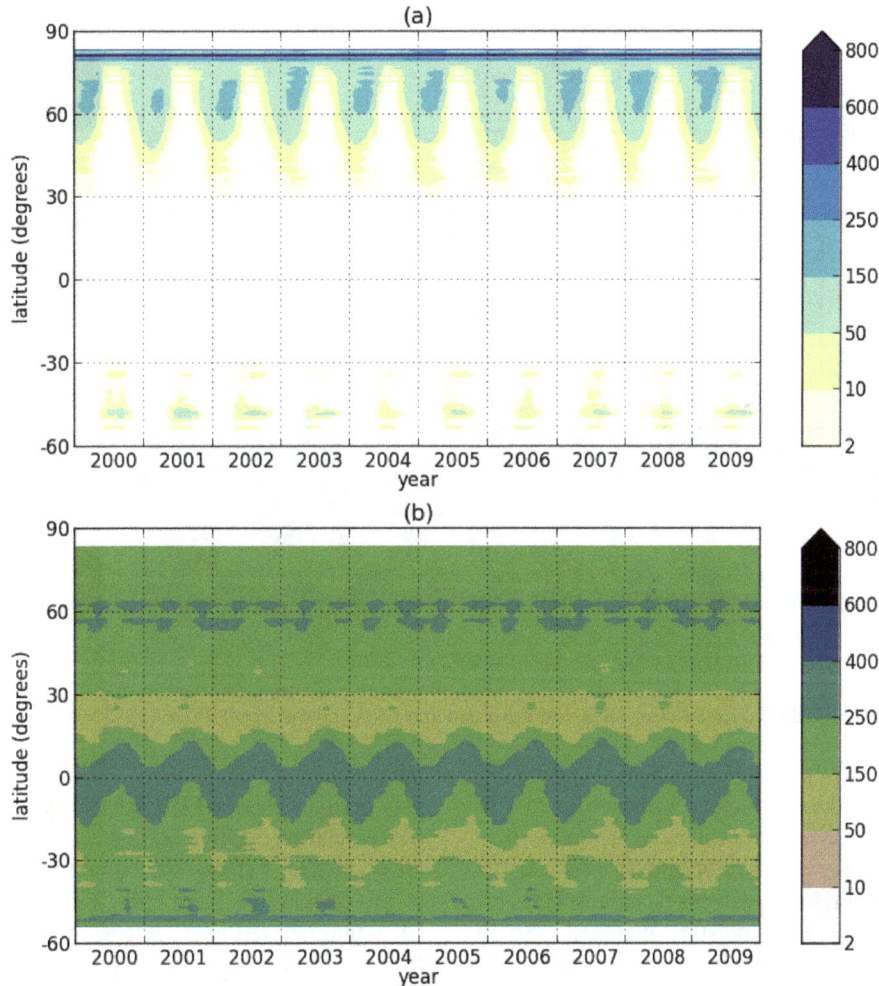

Figure 4. ERA-Interim Hovmöller diagram of the land water reservoirs (zonally averaged over land) for the 2001–2010 period: (**a**) SWE ($\mathrm{kg\,m^{-2}}$) and (**b**) TCSM ($\mathrm{kg\,m^{-2}}$).

and 26.0 ($\pm$1.1) and 19.6 ($\pm$0.8) $\mathrm{W\,m^{-2}}$ with ERA-Interim. Correlation is fairly high and typically 0.85. It can be concluded that, given the uncertainty estimates, the latent heat fluxes are better with ERA-Interim/Land, but the impact on sensible heat flux is not significant.

Prior to production, preliminary experimentation was performed with intermediate versions towards ERA-Interim/Land: (i) offline with the TESSEL model (which indicates the impact of land data assimilation in ERA-Interim),

and (ii) offline with HTESSEL but no GPCP corrections (which indicates the effect of the model changes). It turns out that the RMSEs of latent flux are 26.0 $\mathrm{W\,m^{-2}}$ with ERA-Interim, 30.4 $\mathrm{W\,m^{-2}}$ with version (i), 25.1 $\mathrm{W\,m^{-2}}$ with version (ii), and 21.8 $\mathrm{W\,m^{-2}}$ with ERA-Interim/Land. All these versions are significantly different on the basis of a typical uncertainty of 1 $\mathrm{W\,m^{-2}}$. Deleting the data assimilation increases the error from 26.0 to 30.4 $\mathrm{W\,m^{-2}}$, changing the model reduces the error from 30.4 to 25.1, and applying

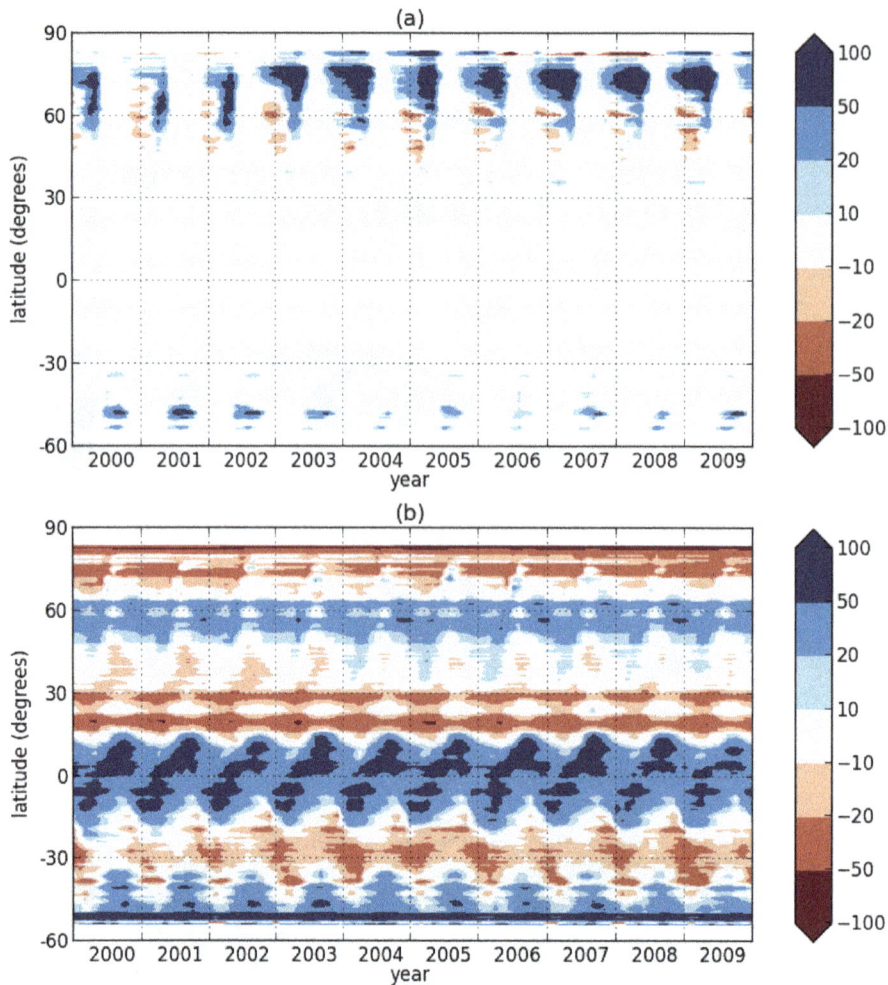

Figure 5. As Fig. 4, but for the difference between ERA-Interim/Land and ERA-Interim in (**a**) SWE (kg m^{-2}) and (**b**) TCSM (kg m^{-2}).

GPCP bias correction reduces the error further from 25.1 to 21.8 W m^{-2}. It is not surprising that soil moisture data assimilation with SYNOP observations is beneficial, because this type of indirect data assimilation reduces the atmospheric errors by construction through soil moisture increments. So, in ERA-Interim/Land relative to ERA-Interim, the lack of soil moisture data assimilation and the model improvement compensate each other in the flux tower verification. The GPCP bias correction contributes further to the improvement. Similar signals exist for sensible heat flux (not shown), except that for sensible heat flux the GPCP part is not significant.

3.1.2 River discharge

River discharge is used here to provide an integrated quantity of the continental water cycle for verifying improvements in the representation of land hydrology. For each discharge station, ERA-Interim and ERA-Interim/Land runoff are averaged over the corresponding catchment area and correlated with the observed monthly values covering the entire reanal-

ysis period. Then a PDF (probability density function) of the correlation coefficients is created by clustering over large areas. Figure 7 shows the cumulative distribution function of the correlations from ERA-Interim/Land (blue line) and ERA-Interim (red line). A general improvement is seen in ERA-Interim/Land, as the correlations are higher at all levels in nearly all cases (the blue line is nearly always to the right of the red line, indicating a higher frequency of high correlation).

The improvements in runoff are large for two reasons: (i) the revised hydrology, i.e. soil infiltration, soil properties and runoff formulation, and (ii) the GPCP bias correction in the tropics and the Southern Hemisphere, consistent with what is known of ERA-Interim precipitation errors (e.g. Betts et al., 2009; Agustì-Panareda et al., 2010). Both effects can be seen in Fig. 7. The improvements over Asia, North America and Europe are mainly the result of the model changes, whereas the impact over Africa, South and Central America and Australia are much larger as the result of the additional effect of GPCP bias correction.

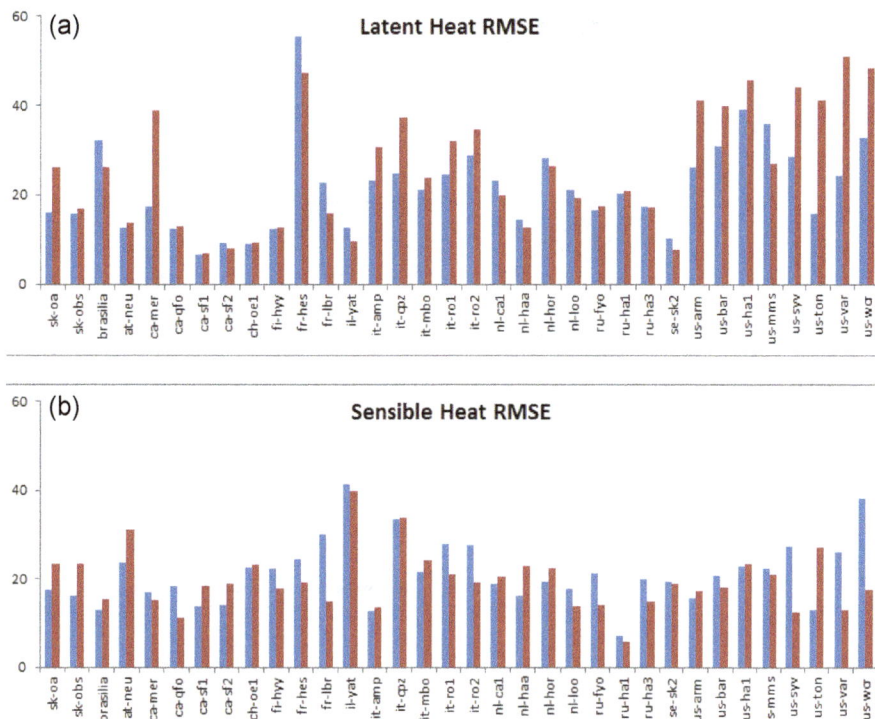

Figure 6. Root mean square error (W m^{-2}) based on hourly values in 2004 for **(a)** latent heat flux and **(b)** sensible heat flux with respect to observations at 34 sites (as in Table 1) for ERA-Interim/Land (blue) and ERA-Interim (red).

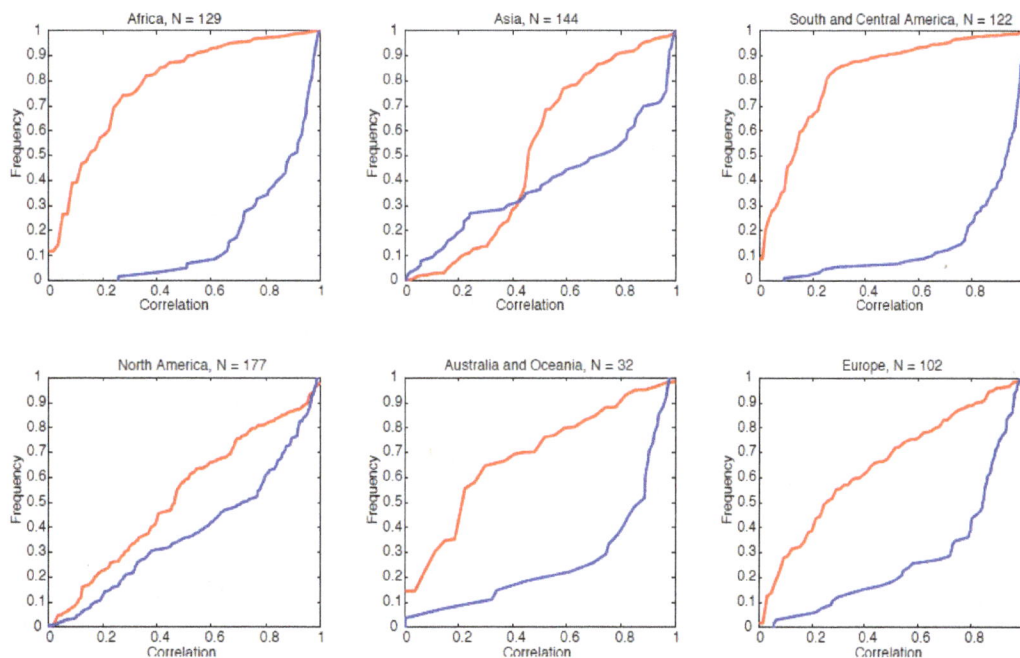

Figure 7. Cumulative distribution function of river discharge correlations of ERA-Interim (red) and ERA-Interim/Land (blue) with GRDC data clustered by continents.

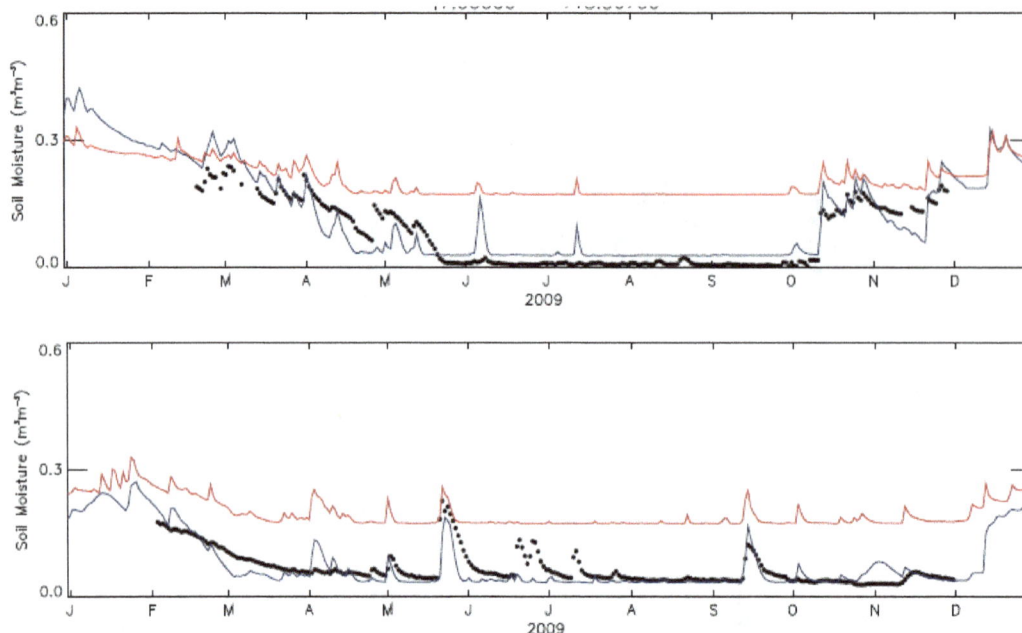

Figure 8. Evolution of volumetric soil moisture for the year 2009 at a site in Utah (latitude 47.000, longitude −118.567, top panel) and Washington (latitude 39.017, longitude −110.167, bottom panel). In situ observations are in black, ERA-Interim is in red, and ERA-Interim/Land estimates in blue.

Although there is still some way to go in effectively representing river discharge in large-scale land surface schemes, coupling such schemes to state-of-the-art river hydrology models can bring further improvement (Pappenberger et al., 2012). In the current evaluation it is particularly encouraging that the average improvement of river discharge correlations of ERA-Interim/Land over ERA-Interim occurs on all continents, which encompass different rivers and different water balance regimes.

3.2 Land water reservoir verification

The water reservoir verification aims at assessing the daily performance of ERA-Interim/Land in soil water content and the snow water equivalent, which are responding to the diurnal, synoptic and seasonal variations of fluxes. The deeper and slowly evolving soil moisture layers, such as the water table, are not considered in the present verification since they are not yet properly represented in the ECMWF model.

3.2.1 Soil moisture

The changes in land surface parameterization have largely preserved the mean annual soil moisture, which ranges around 0.23–0.24 m^3 m^{-3} as global land average over the ERA-Interim period. However, the spatial variability has greatly increased with the introduction of the revised soil hydrology (Balsamo et al., 2009). In order to verify the soil moisture produced by the offline simulations we make use of the ISMN ground-based observing networks. This has been

applied by Albergel et al. (2012b) to validate soil moisture from both ECMWF operational analysis and ERA-Interim.

Considering the field sites of the NRCS-SCAN network (covering the US) with a fraction of bare ground greater than 0.2 (according to the model), the RMSE of soil moisture decreases from 0.118 m^3 m^{-3} with ERA-Interim to 0.087 m^3 m^{-3} with ERA-Interim/Land, mainly due to the new formulation of bare soil evaporation. In the TESSEL formulation of ERA-Interim, minimum values of soil moisture are limited by the wilting point of the dominant vegetation type. However, ground data indicate much drier conditions, as is clearly observed at bare soil locations, e.g. at the Utah and Washington sites from May to September 2009 shown in Fig. 8. The new soil hydrology and bare ground evaporation allows the model to go below the wilting point, which is in much better agreement with the observations than in ERA-Interim.

The improved capability of ERA-Interim/Land to simulate soil moisture in bare soil areas is also clear in Fig. 9. It illustrates the gain in skill in reproducing the observed soil moisture in dry land as a function of vegetation cover. With the RMSE being positive, definite, and calculated against in situ soil moisture observations, the RMSE differences between ERA-Interim and ERA-Interim/Land indicate improvements realized by the latter. The RMSE difference is calculated for locations with varying vegetation fraction and the improvement is shown to be larger on points with sizeable bare soil. This is a demonstration that the enhanced match to the ob-

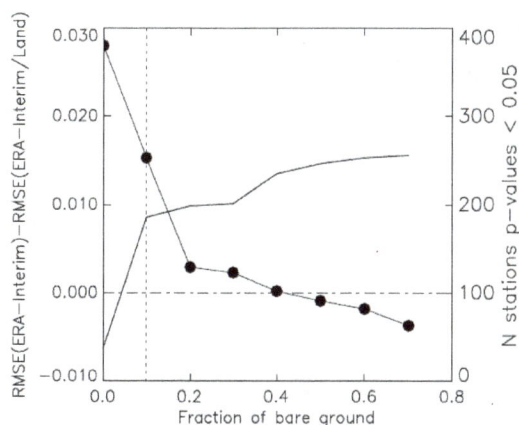

Figure 9. RMSE difference between ERA-Interim and ERA-Interim/Land (solid line, left y axis) as a function of the fraction of bare ground. The number of in situ stations (line with solid dots, right y axis) with significant correlations is also presented. Sensitivity to fraction of bare soil is only pronounced above the threshold indicated by the vertical dashed line.

served soil moisture is indeed the result of the bare soil evaporation revision as detailed in Albergel et al. (2012a).

The correlation of ERA-Interim/Land soil moisture with the various observed soil moisture networks varies depending on the network (Fig. 10, Table 2). In general, the correlations are similar to those with ERA-Interim and not significantly improved. However, the variability is increased as can be seen in the Taylor diagram of Fig. 11. The distance to the point marked "in situ" has been reduced with ERA-Interim/Land, because the standard deviation of observations is better reproduced.

The site verification of soil moisture presented in this section, has also been applied to an offline experiment where the only difference is that ERA-Interim forcing is not corrected with GPCP. It turns out that the results are indistinguishable. It can be concluded that monthly GPCP bias correction has no impact on soil moisture in the extratropics, in spite of the small beneficial impact on precipitation as was seen by Balsamo et al. (2010a) over the US.

Interestingly, Albergel et al. (2013) verified an ERA-Interim/Land variant (with no precipitation readjustment) and MERRA-Land for the full 1988–2010 period with all available in situ soil moisture observations. They find average correlations for superficial soil moisture (95 % confidence interval) of 0.66 (±0.038) for ERA-Interim/Land, and 0.69 (±0.038) for MERRA-Land. Root zone soil moisture correlations of 0.68 (±0.035) are found for ERA-Interim/Land and 0.73 (±0.032) for MERRA-Land. It is impossible to speculate on the origin of the differences between these two reanalyses because they are different on many aspects.

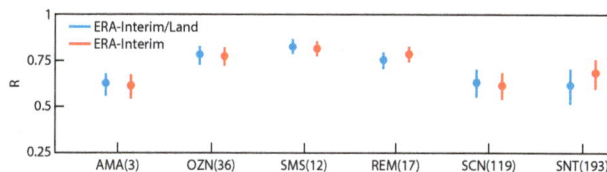

Figure 10. Correlation with observed ISMN soil moisture networks (as in Table 2) for ERA-Interim/Land (blue) and ERA-Interim (red). Only significant correlations with p values < 0.05 are considered and for each of the observing networks the bars indicate the 95 % confidence interval calculated using a Fisher Z transform.

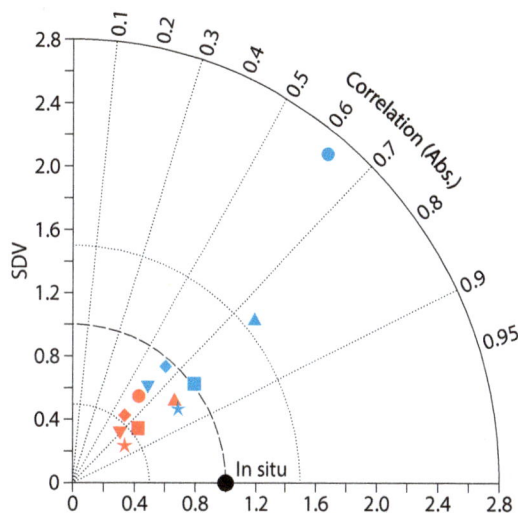

Figure 11. Taylor diagram illustrating the statistics from the comparison between ERA-Interim/Land (blue) and ERA-Interim (red), compared to situ observations for 2010. Each symbol indicates the correlation value (angle), the normalized SD (radial distance to the origin point), and the normalized centred root mean square error (distance to the point marked "in situ"). Circles are for the stations of the AMMA network (3 stations), square for the OZNET network (36 stations), stars for the SMOSMANIA network (12 stations), triangles for the REMEDHUS network (17 stations), diamonds for the SCAN network (119 stations) and inverted triangle for the SNOTEL network (193 stations). Only stations with significant correlation values are considered.

3.2.2 Snow

Dutra et al. (2010) attributed the largest improvement in the new snow scheme to the snow density representation. This could be confirmed with station data from the former USSR. At a large number of sites, snow density was measured in the snow season at typical northern latitudes from October to June from 1979 to 1993 (Brun et al., 2013). In ERA-Interim, as well as ERA-Interim/Land, snow density is not constrained by data assimilation due to a lack of observations that are exchanged routinely and therefore it relies solely on the capacity of the land surface model to represent the seasonal evolution, from about $100\,\mathrm{kg\,m^{-3}}$ at the beginning

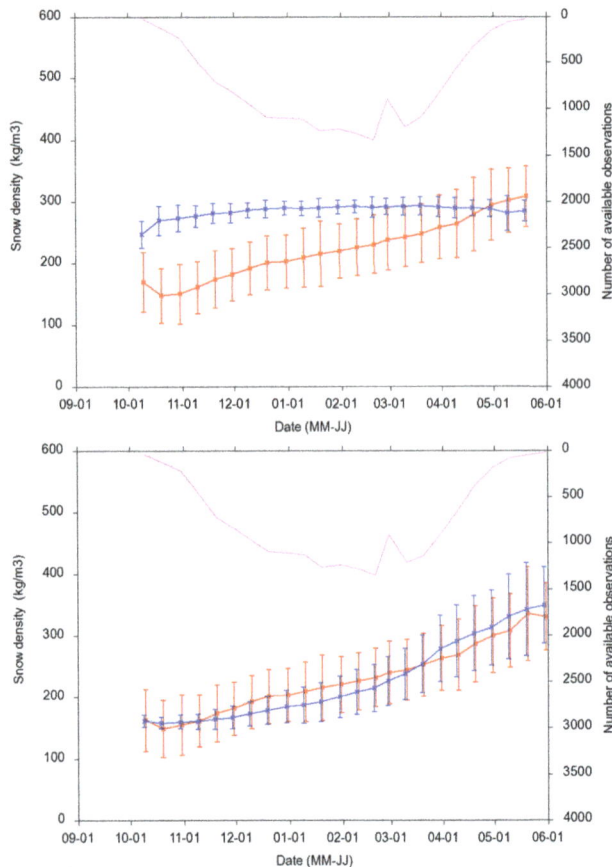

Figure 12. Snow density seasonal evolution as observed (red) and estimated (blue) by ERA-Interim (top panel) and by ERA-Interim/Land (bottom panel). Each point represents the station data from the former USSR, averaged over about 20 points along transects around each station, all stations and all years from 1997 to 1993. The vertical bar indicates ±1 standard deviation and the purple line indicates the number of observations with the right-hand scale from top to bottom. Observations are only included when both observations and model have snow. The snow season from October to June is considered only.

Figure 13. ERA-Interim snow depth RMSE/BIAS (solid/dashed red line) and ERA-Interim/Land snow depth RMSE/BIAS (solid/dashed blue line) with respect to the daily European SYNOP observations at 06:00 UTC. The number of stations with snow is indicated by squares (right y axis). Model snow depth combines the snow mass and density variables.

RMSEs than ERA-Interim, because the latter assimilates the same SYNOP observations and ERA-Interim/Land does not. The explanation is that the background field in ERA-Interim is so much worse than in ERA-Interim/Land that the analysis increments do not fully compensate for the poor background field. It is also remarkable that a good quality land snow mass analysis can be obtained without any constraint from direct snow mass observations. A good quality snowfall is obviously key to such a success.

Finally, the MODIS land surface albedo is used to verify ERA-Interim/Land, particularly in the snow representation in forest areas (Fig. 14) in northern Canada and Siberia, where conventional SYNOP observations are generally less informative. Figure 14c points to a substantially reduced albedo bias in ERA-Interim/Land attributed to the snow scheme revision described in Dutra et al. (2010) and in particular the snow–vegetation albedo retuning. The main improvement comes from the albedo optimization for vegetated areas. Particularly, forests tend to keep a low albedo with snow accumulating under the canopy rather than on it; however, in ERA-Interim forests with snow were specified to be too dark, not accounting for the openness of many forests, and ERA-Interim/Land has lighter snow-forest albedos. As albedo is an important component of the surface energy balance, it significantly affects the atmospheric heating and the timing of snowmelt in spring.

4   Discussion

Dedicated land surface reanalyses, such as ERA-Interim/Land described and evaluated here, are becoming established added-value products within the reanalysis efforts worldwide (Dee et al., 2014). They allow computa-

of the winter season to more than $300\,\mathrm{kg\,m^{-3}}$ towards the end of the snow season. Figure 12 clearly shows that the seasonal evolution of snow density of ERA-Interim/Land is much more realistic than in ERA-Interim, mainly because the density formulation in ERA-Interim relaxes too quickly to the $300\,\mathrm{kg\,m^{-3}}$ value. This is obviously also important for data assimilation of any snow depth observations, because snow depth has to be converted to snow mass making an assumption about snow density.

Verification of snow mass is difficult because, at best, snow depth is measured without information on density. Here routine SYNOP observations are used although the network is fairly sparse. Figure 13 shows the seasonal cycle of the RMSE of snow depth from ERA-Interim and ERA-Interim/Land over Europe (more than 600 observations daily). It is remarkable that ERA-Interim/Land has smaller

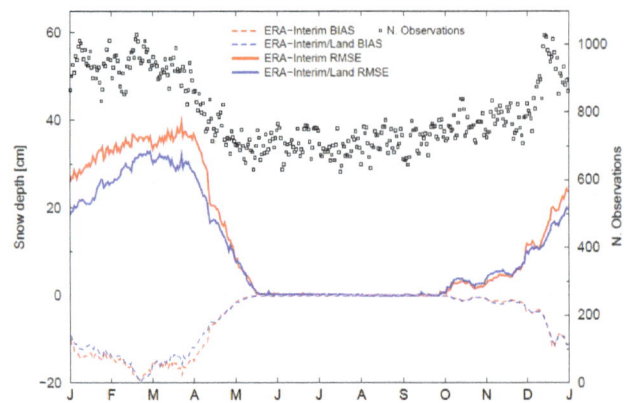

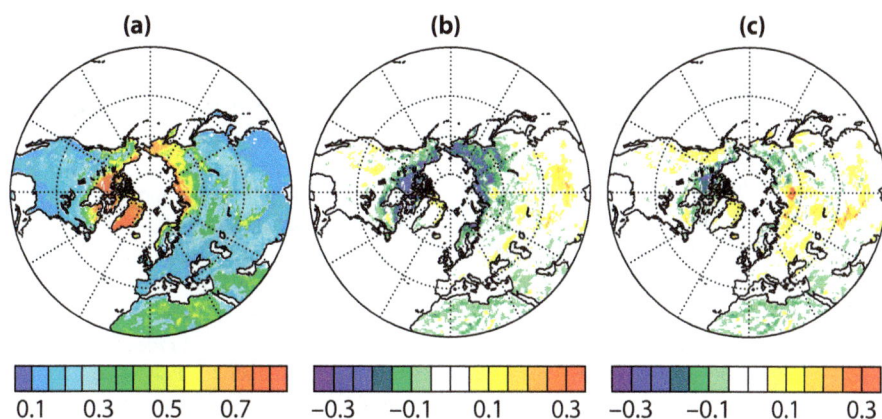

Figure 14. Mean observed Northern Hemisphere albedo during spring (MAM) derived from MODIS (**a**), difference between ERA-Interim and MODIS (**b**), and difference between ERA-Interim/Land and MODIS (**c**).

tionally efficient testing of new land surface developments, including improvements to the process representation and parameterization of the hydrological and biogeochemical cycles that contribute to fast-track land surface model developments as identified by van den Hurk et al. (2011). Future research into improved representation of the land surface is high priority, and work already underway in this area includes land carbon exchanges (Boussetta et al., 2013b), vegetation interannual variability, and hydrological applications such as global water-bodies reanalysis (e.g. Balsamo et al., 2012) and global flood risk assessment (e.g. Pappenberger et al., 2012). More sophisticated rescaling methods (e.g. Weedon et al., 2011, 2014) are envisaged to bias-correct the meteorological forcing and to permit a high-resolution downscaling of land reanalysis. In addition, consideration of land surface parameterization uncertainty could be used to further improve predictive skill (e.g. Cloke et al., 2011).

Important developments with advanced land data assimilation methods such as the extended or ensemble Kalman filters (Reichle et al., 2014; de Rosnay et al., 2013; Drusch et al., 2009) can be combined with offline surface simulations. The experimental equivalence of offline and atmospheric coupled land data assimilation (Balsamo et al., 2007; Mahfouf et al., 2008) offers also in this case a 2 orders of magnitude computational saving. This is expected to provide a fast land surface reanalysis as envisaged within the EU-funded ERA-CLIM2 project. Moreover, it can open up new possibilities of more advanced data assimilation schemes (e.g. Fowler and van Leeuwen, 2012), especially designed for non-linear systems.

5 Conclusions

This paper documents the configuration and the performance of the ERA-Interim/Land reanalysis in reconstructing the land surface state over the past 3 decades. ERA-Interim/Land is produced with an improved land surface scheme in offline simulations forced by ERA-Interim meteorological forcing. It has been demonstrated that the ERA-Interim/Land dedicated land surface reanalysis has added value over the standard land component of the ERA-Interim reanalysis product. The ERA-Interim/Land runs are an integral part of the ERA-Interim ongoing research efforts and respond to the wish to reactualize the land surface initial conditions of ERA-Interim, following several model parameterization improvements. The newly produced land surface estimates benefit from the latest land surface hydrology schemes used operationally at ECMWF for the medium-range, monthly, and seasonal forecasts. The ERA-Interim/Land added-value components encompass soil, snow and vegetation description upgrades, as well as a bias correction of the ERA-Interim monthly-accumulated precipitation based on GPCP v.2.1. In the Northern Hemisphere, the precipitation correction is shown to be effective in reducing the bias over the US and is rather neutral over Eurasia, while over tropical land clear benefits are seen in the river discharge.

The new land surface reanalysis has been verified against several data sets for the main water reservoirs (snow and soil moisture), together with the energy and water fluxes that have direct impact on the atmosphere. The verification makes use of both in situ observations and remote sensing products. A modest improvement has been seen in the latent heat fluxes, which turns out to be the result of a combination of deterioration due to the lack of soil moisture data assimilation, a substantial improvement due to model changes, and a small improvement due to GPCP precipitation bias correction. It is encouraging to see that the modelled runoff has been improved when compared to observed river discharge from the GRDC river network showing an enhanced correlation to the observations. The improvement compared to ERA-Interim is the combined effect of the GPCP precipitation correction and the land surface model improvements, and future work will extend the use of river discharge for supporting model development and disentangle the impact of different components (e.g. meteorological forcing and parameterization changes) in the framework of the EU-funded EartH2Observe project.

Variability of soil moisture is improved due to the hydrology improvements and the introduction of a soil texture map. Also, bare soil areas indicate a distinct improvement related to the handling of the low soil moisture regime. Both snow depth and snow albedo are shown to have a better seasonal cycle, mainly due to the new model formulations. The model improvement appears to overwhelm the lack of data assimilation.

While river discharge verification is not enough for a global water balance assessment, the results from the verification of evaporation fluxes (the other main outgoing land water flux) and of the two main water reservoirs (soil moisture and snow-pack), permit to qualify the ERA-Interim/Land enhanced accuracy as genuine. When water fluxes and water storage terms show consistent indication of improvements there are in fact good grounds to believe that the parameterization changes are physically meaningful and not the result of compensating errors.

Finally, it is worth noting that offline land reanalysis plays an important role in the model development cycle of the operational system at ECMWF. The forecasting system uses back integrations covering the last 30 years with ERA-Interim as initial condition to obtain a model climate as reference for anomalies. As soon as the land surface model is changed substantially, it becomes important to have a consistent initial condition, and the latter is obtained by offline reanalysis. It has been demonstrated that this procedure has a positive effect on the back integrations particularly for the longer lead times (Balsamo et al., 2012; Vitart et al., 2008; Molteni et al., 2011).

Ongoing work focuses on interannual variability of vegetation state (leaf area index), efforts to extend the current ERA-Interim/Land data set beyond 2010, and future ECMWF reanalyses (Dee et al., 2014).

Data set access

The ERA-Interim/Land data set is freely available and can be downloaded from: http://apps.ecmwf.int/datasets/.

Acknowledgements. The authors thank R. Riddaway from ECMWF for his valuable comments on the text, and C. O'Sullivan, A. Bowen, S. Witter for their help preparing the figures. This work used eddy covariance data acquired by the FLUXNET community that is gratefully acknowledged. TU Wien provided the ISMN data for soil moisture verification and we thank them for their important effort. We thank the GRDC for data provision of global river discharge. The ECMWF User-Support team is acknowledged for making the data easily accessible.

Edited by: B. Su

References

Adler, R. F., Huffman, G. J., Chang, A., Ferraro, R., Xie, P., Janowiak, J., Rudolf, B., Schneider, U., Curtis, S., Bolvin, D., Gruber, A., Susskind, J., Arkin, P., and Nelkin, E. J.: The Version 2.1 Global Precipitation Climatology Project (GPCP) Monthly Precipitation Analysis (1979–Present), J. Hydrometeorol., 4, 1147–1167, 2003.

Agustì-Panareda, A., Balsamo, G., and Beljaars, A.: Impact of improved soil moisture on the ECMWF precipitation forecast in West Africa, Geophys. Res. Lett., 37, L20808, doi:10.1029/2010GL044748, 2010.

Albergel, C., Rüdiger, C., Pellarin, T., Calvet, J.-C., Fritz, N., Froissard, F., Suquia, D., Petitpa, A., Piguet, B., and Martin, E.: From near-surface to root-zone soil moisture using an exponential filter: an assessment of the method based on in-situ observations and model simulations, Hydrol. Earth Syst. Sci., 12, 1323–1337, doi:10.5194/hess-12-1323-2008, 2008.

Albergel, C., Balsamo, G., de Rosnay, P., Muñoz-Sabater, J., and Boussetta, S.: A bare ground evaporation revision in the ECMWF land-surface scheme: evaluation of its impact using ground soil moisture and satellite microwave data, Hydrol. Earth Syst. Sci., 16, 3607–3620, doi:10.5194/hess-16-3607-2012, 2012a.

Albergel, C., de Rosnay, P., Balsamo, G., Isaksen, L., and Muñoz Sabater, J.: Soil moisture analyses at ECMWF: evaluation using global ground-based in-situ observations, J. Hydrometeorol., 13, 1442–1460, 2012b.

Albergel, C., de Rosnay, P., Gruhier, C., Muñoz-Sabater, J., Hasenauer, S., Isaksen, L., Kerr, Y., and Wagner, W.: Evaluation of remotely sensed and modelled soil moisture products using global ground-based in situ observations, Remote Sens. Environ., 118, 215–226, doi:10.1016/j.rse.2011.11.017, 2012c.

Albergel, C., Dorigo, W., Reichle, R. H., Balsamo, G., de Rosnay, P., Munoz-Sabater, J., Isaksen, L., de Jeu, R., and Wagner, W.: Skill and global trend analysis of soil moisture from reanalyses and microwave remote sensing, J. Hydrometeorol., 14, 1259–1277, doi:10.1175/JHM-D-12-0161.1, 2013.

Baldocchi, D., Falge, E., Gu, L., Olson, R., Hollinger, D., Running, S., Anthoni, P., Bernhofer, C., Davis, K., Evans, R., Fuentes, J., Goldstein, A., Katul, G., Law, B., Lee, X., Malhi, Y., Meyers, T., Munger, W., Oechel, W., Paw, K. T., Pilegaard, K., Schmid, H. P., Valentini, R., Verma, S., Vesala, T., Wilson, K., and Wofsy, S.: FLUXNET: A new tool to study the temporal and spatial variability of ecosystem-scale carbon dioxide, water vapor, and energy flux densities, B. Am. Meteorol. Soc., 82, 2415–2434, doi:10.1175/1520-0477(2001)082<2415:FANTTS>2.3.CO;2, 2001.

Balsamo, G., Mahfouf, J.-F., Bélair, S., and Deblonde, G.: A land data assimilation system for soil moisture and temperature: An information content study, J. Hydrometerorol., 8, 1225–1242, 2007.

Balsamo, G., Viterbo, P., Beljaars, A., van den Hurk, B., Hirschi, M., Betts, A. K., and Scipal, K.: A revised hydrology for the ECMWF model: Verification from field site to terrestrial water storage and impact in the Integrated Forecast System, J. Hydrometeorol., 10, 623–643, 2009.

Balsamo, G., Boussetta, S., Lopez, P., and Ferranti, L.: Evaluation of ERA-Interim and ERA-Interim-GPCP-rescaled precipitation over the U.S.A., ERA Report Series No. 5, ECMWF, Reading, UK, 10 pp., 2010a.

Balsamo, G., Pappenberger, F., Dutra, E., Viterbo, P., and van den Hurk, B.: A revised land hydrology in the ECMWF model: A step towards daily water flux prediction in a fully-closed water cycle, Hydrol. Process., 25, 1046–1054, doi:10.1002/hyp.7808, 2010b.

Balsamo, G., Boussetta, S., Dutra, E., Beljaars, A., Viterbo, P., and van den Hurk, B.: Evolution of land surface processes in the IFS, ECMWF Newslett., 127, 17–22, 2011.

Balsamo, G., Salgado, R., Dutra, E., Boussetta, S., Stockdale, T., and Potes, M.: On the contribution of lakes in predicting near-surface temperature in a global weather forecasting model, Tellus A, 64, 15829, doi:10.3402/tellusa.v64i0.15829, 2012.

Beljaars, A., Viterbo, P., Miller, M., and Betts, A.: The anomalous rainfall over the USA during July 1993: Sensitivity to land surface parametrization and soil moisture anomalies, Mon. Weather Rev., 124, 362–383, 1996.

Berrisford, P., Kallberg, P., Kobayashi, S., Dee, D., Uppala, S., Simmons, A. J., Poli, P., and Sato, H.: Atmospheric conservation properties in ERA-Interim, Q. J. Roy. Meteorol. Soc., 137, 1381–1399, doi:10.1002/qj.864, 2011.

Betts, A. K., Ball, J. H., Barr, A. G., Black, T. A., McCaughey, J. H., and Viterbo, P.: Assessing land surface-atmosphere coupling in the ERA-40 re-analysis with boreal forest data, Agr. Forest Meteorol., 140, 365–382, doi:10.1016/j.agrformet.2006.08.009, 2006.

Betts, A. K., Köhler, M., and Zhang, Y.-C.: Comparison of river basin hydrometeorology in ERA-Interim and ERA-40 re-analyses with observations, J. Geophys. Res., 114, D02101, doi:10.1029/2008JD010761, 2009.

Boussetta, S., Balsamo, G., Beljaars, A., and Jarlan, J.: Impact of a satellite-derived Leaf Area Index monthly climatology in a global Numerical Weather Prediction model, Int. J. Remote Sens., 34, 3520–3542, 2013a.

Boussetta, S., Balsamo, G., Beljaars, A., Agustì-Panareda, A., Calvet, J.-C., Jacobs, C., van den Hurk, B., Viterbo, P., Lafont, S., Dutra, E., Jarlan, L., Balzarolo, M., Papale, D., and van der Werf, G.: Natural land carbon exchanges in the ECMWF Integrated Forecasting system: Implementation and offline validation, J. Geophys. Res., 118, 1–24, 2013b.

Brun, E., Vionnet, V., Boone, A., Decharme, B., Peings, Y., Valette, R., Karbou, F., and Morin, S.: Simulation of northern Eurasian local snow depth, mass and density using a detailed snowpack model and meteorological reanalyses, J. Hydrometeorol., 14, 203–219, 2013.

Ceballos, A., Scipal, K., Wagner, W., and Martinez-Fernandez, J.: Validation of ERS scatterometer-derived soil moisture data in the central part of the Duero-Basin, Spain, Hydrol. Process., 19, 1549–1566, 2005.

Cloke, H. L., Weisheimer, A., and Pappenberger, F.: Representing uncertainty in land surface hydrology: fully coupled simulations with the ECMWF land surface scheme, Proceeding of ECMWF, WMO/WGNE, WMO/THORPEX and WCRP Workshop on "Representing model uncertainty and error in numerical weather and climate prediction models", ECMWF, Reading, UK, June 2011.

Cressman, G. P.: An operational objective analysis system, Mon. Weather Rev., 87, 367–374, 1959.

Decker, M., Brunke, M. A., Wang, Z., Sakaguchi, K., Zeng, X., and Bosilovich, M. G.: Evaluation of the reanalysis products from GSFC, NCEP, and ECMWF Using Flux Tower Observations, J. Climate, 25, 1916–1944, doi:10.1175/JCLI-D-11-00004.1, 2012.

Dee, D. P., Uppala, S. M., Simmons, A. J., Berrisford, P., Poli, P., Kobayashi, S., Andrae, U., Balmaseda, M. A., Balsamo, G., Bauer, P., Bechtold, P., Beljaars, A., van de Berg, L., Bidlot, J., Bormann, N., Delsol, C., Dragani, R., Fuentes, M., Geer, A. J., Haimberger, L., Healy, S. B., Hersbach, H., Hólm, E. V., Isaksen, L., Kallberg, P., Köhler, M., Matricardi, M., McNally, A. P., Monge-Sanz, B. M., Morcrette, J.-J., Park, B. K., Peubey, C., de Rosnay, P., Tavolato, C., Thépaut, J.-N., and Vitart, F.: The ERA-Interim reanalysis: configuration and performance of the data assimilation system, Q. J. Roy. Meteorol. Soc., 137, 553–597, doi:10.1002/qj.828, 2011.

Dee, D., Balmaseda, M., Balsamo, G., Engelen, R., Simmons, A., and Thépaut, J.-N.: Toward a consistent reanalysis of the climate system, B. Am. Meteorol. Soc., 95, 1235–1248, doi:10.1175/BAMS-D-13-00043.1, 2014.

de Rosnay, P., Drusch, M., Vasiljevic, D., Balsamo, G., Albergel, C., and Isaksen, L.: A simplified Extended Kalman Filter for the global operational soil moisture analysis at ECMWF, Q. J. Roy. Meteorol. Soc., 139, 1199–1213, doi:10.1002/qj.2023, 2013.

de Rosnay, P., Balsamo, G., Albergel, C., Muñoz-Sabater, J., and Isaksen, L.: Initialisation of land surface variables for Numerical Weather Prediction, Surv. Geophys., 35, 607–621, doi:10.1007/s10712-012-9207-x, 2014.

Dirmeyer, P. A.: A history and review of the Global Soil Wetness Project (GSWP), J. Hydrometeorol., 12, 729–749, doi:10.1175/JHM-D-10-05010.1, 2011.

Dirmeyer, P. A., Gao, X., and Oki, T.: The second Global Soil Wetness Project – Science and implementation plan, IGPO Int. GEWEX Project Office Publ. Series 37, Global Energy and Water Cycle Exp. (GEWEX) Proj. Off., Silver Spring, MD, 65 pp., 2002.

Dirmeyer, P. A., Gao, X., Zhao, M., Guo, Z., Oki, T., and Hanasaki, N.: GSWP-2: Multimodel analysis and implications for our perception of the land surface, B. Am. Meteorol. Soc., 87, 1381–1397, 2006.

Dorigo, W. A., Wagner, W., Hohensinn, R., Hahn, S., Paulik, C., Xaver, A., Gruber, A., Drusch, M., Mecklenburg, S., van Oevelen, P., Robock, A., and Jackson, T.: The International Soil Moisture Network: a data hosting facility for global in situ soil moisture measurements, Hydrol. Earth Syst. Sci., 15, 1675–1698, doi:10.5194/hess-15-1675-2011, 2011.

Dorigo, W. A., Xaver, A., Vreugdenhil, M., Gruber, A., Hegyiovaì, A., Sanchis-Dufau, A. D., Wagner, W., and Drusch, M.: Global automated quality control of in-situ soil moisture data from the International Soil Moisture Network, Vadose Zone J., 12, doi:10.2136/vzj2012.0097, 2013.

Douville, H., Royer, J. F., and Mahfouf, J.-F.: A new snow parameterization for the Meteo-France climate model, 1: Validation in stand-alone experiments, Clim. Dynam., 12, 21–35, 1995.

Drusch, M. and Viterbo, P.: Assimilation of screen-level variables in ECMWF's Integrated Forecast System: A study on the impact on the forecast quality and analyzed soil moisture, Mon. Weather Rev., 135, 300–314, 2007.

Drusch, M., Vasiljevic, D., and Viterbo, P.: ECMWF's global snow analysis: assessment and revision based on satellite observations, J. Appl. Meteorol., 43, 1282–1294, 2004.

Drusch, M., Scipal, K., de Rosnay, P., Balsamo, G., Andersson, E., Bougeault, P., and Viterbo, P.: Towards a Kalman Filter based soil moisture analysis system for the operational ECMWF Integrated Forecast System, Geophys. Res. Lett., 36, L10401, doi:10.1029/2009GL037716, 2009.

Dutra, E., Balsamo, G., Viterbo, P., Miranda, P., Beljaars, A., Schär, C., and Elder, K.: An improved snow scheme for the ECMWF land surface model: description and offline validation, J. Hydrometeorol., 11, 899–916, doi:10.1175/JHM-D-11-072.1, 2010.

Dutra, E., Viterbo, P., Miranda, P. M. A., and Balsamo, G.: Complexity of snow schemes in a climate model and its impact on surface energy and hydrology, J. Hydrometeorol., 13, 521–538, doi:10.1175/jhm-d-11-072, 2012.

Fowler, A. and van Leeuwen, P.: Measures of observation impact in non-Gaussian data assimilation, Tellus A, 64, 17192, doi:10.3402/tellusa.v64i0.17192, 2012.

Huffman, G. J., Adler, R. F., Bolvin, D. T., and Gu, G.: Improving the global precipitation record: GPCP Version 2.1, Geophys. Res. Lett., 36, L17808, doi:10.1029/2009GL040000, 2009.

Kållberg, P.: Forecast drift in ERA-Interim, ERA Report Series No. 10, ECMWF, Reading, UK, 9 pp., 2011.

Koster, R. D., Guo, Z., Dirmeyer, P. A., Bonan, G., Chan, E., Cox, P., Davies, H., Gordon, C. T., Kanae, S., Kowalczyk, E., Lawrence, D., Liu, P., Lu, C. H., Malyshev, S., McAvaney, B., Mitchell, K., Mocko, D., Oki, T., Oleson, K. W., Pitman, A., Sud, Y. C., Taylor, C. M., Verseghy, D., Vasic, R., Xue, Y., and Yamada, T.: GLACE: The Global Land-Atmosphere Coupling Experiment, Part I: Overview, J. Hydrometeorol., 7, 590–610, 2006.

Koster, R., Mahanama, S., Yamada, T., Balsamo, G., Boisserie, M., Dirmeyer, P., Doblas-Reyes, F., Gordon, T., Guo, Z., Jeong, J.-H., Lawrence, D., Li, Z., Luo, L., Malyshev, S., Merryfield, W., Seneviratne, S. I., Stanelle, T., van den Hurk, B., Vitart, F., and Wood, E. F.: The contribution of land surface initialization to sub-seasonal forecast skill: First results from the GLACE-2 Project, Geophys. Res. Lett., 37, L02402, doi:10.1029/2009GL041677, 2009.

Koster, R., Mahanama, S. P. P., Yamada, T. J., Balsamo, G., Berg, A. A., Boisserie, M., Dirmeyer, P. A., Doblas-Reyes, F. J., Drewitt, G., Gordon, C. T., Guo, Z., Jeong, J.-H., Lee, W.-S., Li, Z., Luo, L., Malyshev, S., Merryfield, W. J., Seneviratne, S. I., Stanelle, T., van den Hurk, B. J. J. M., Vitart, F., and Wood, E. F.: The second phase of the global land-atmosphere coupling experiment: soil moisture contributions to sub-seasonal forecast skill, J. Hydrometeorol., 12, 805–822, doi:10.1175/2011JHM1365.1, 2011.

Mahfouf, J.-F. and Noilhan, J.: Comparative study of various formulations of evaporation from bare soil using in situ data, J. Appl. Meteorol., 30, 351–362, 1991.

Mahfouf, J.-F., Bergaoui, K., Draper, C., Bouyssel, F., Taillefer, F., and Taseva, L.: A comparison of two off-line soil analysis schemes for assimilation of screen level observations, J. Geophys. Res., 114, D08105, doi:10.1029/2008JD011077, 2008.

Molteni, F., Stockdale, T., Balmaseda, M., Balsamo, G., Buizza, R., Ferranti, L., Magnusson, L., Mogensen, K., Palmer, T., and Vitart, F.: The new ECMWF seasonal forecast system (System 4), ECMWF Tech. Memo No. 656, ECMWF, Reading, UK, 2011.

Oki, T. and Kanae, S.: Global hydrological cycles and world water resources, Science, 313, 1068–1072, doi:10.1126/science.1128845, 2006.

Pappenberger, F., Cloke, H., Balsamo, G., Ngo-Duc, T., and Oki, T.: Global runoff routing with the hydrological component of the ECMWF NWP system, Int. J. Climatol., 30, 2155–2174, doi:10.1002/joc.2028, 2009.

Pappenberger, F., Dutra, E., Wetterhall, F., and Cloke, H. L.: Deriving global flood hazard maps of fluvial floods through a physical model cascade, Hydrol. Earth Syst. Sci., 16, 4143–4156, doi:10.5194/hess-16-4143-2012, 2012.

Pellarin, T., Laurent, J. P., Cappelaere, B., Decharme, B., Descroix, L., and Ramier, D.: Hydrological modelling and associated microwave emission of a semi-arid region in south-western Niger, J. Hydrol., 375, 262–272, 2009.

Polcher, J., McAvaney, B., Viterbo, P., Gaertner, M.-A., Hahamann, A., Mahfouf, J.-F., Noilhan, J., Phillips, T., Pitman, A., Schlosser, C., Schulz, J.-P., Timbal, B., Verseghy, D., and Xue, Y.: A proposal for a general interface between land surface schemes and general circulation models, Global Planet. Change, 19, 261–276, 1998.

Reichle, R. H., Koster, R. D., De Lannoy, G. J. M., Forman, B. A., Liu, Q., Mahanama, S. P. P., and Toure, A.: Assessment and enhancement of MERRA land surface hydrology estimates, J. Climate, 24, 6322–6338, doi:10.1175/JCLI-D-10-05033.1, 2011.

Reichle, R. H., De Lannoy, G. J. M., Forman, B. A., Draper, C. S., and Liu, Q.: Connecting satellite observations with water cycle variables through land data assimilation: Examples using the NASA GEOS-5 LDAS, Surv. Geophys., 35, 577–606, doi:10.1007/s10712-013-9220-8, 2014.

Rienecker, M. M., Suarez, M. J., Gelaro, R., Todling, R., Julio, B., Liu, E., Bosilovich, M. G., Schubert, S. D., Takacs, L., Kim, G.-K., Bloom, S., Chen, J., Collins, D., Conaty, A., da Silva, A., Gu, W., Joiner, J., Koster, R. D., Lucchesi, R., Molod, A., Owens, T., Pawson, S., Pegion, P., Redder, C. R., Reichle, R., Robertson, F. R., Ruddick, A. G., Sienkiewicz, M., and Woollen, J.: MERRA – NASA's modern-era retrospective analysis for research and applications, J. Climate, 24, 3624–3648, doi:10.1175/JCLI-D-11-00015.1, 2011.

Román, M. O., Schaaf, C. B., Woodcock, C. E., Strahler, A. H., Yang, X., Braswell, R. H., Curtis, P. S., Davis, K. J., Dragoni, D., Goulden, M. L., Gu, L., Hollinger, D. Y., Kolb, T. E., Meyers, T. P., Munger, J. W., Privette, J. L., Richardson, A. D., Wilson, T. B., and Wofsy, S. C.: The MODIS (Collection V005) BRDF/albedo product: Assessment of spatial representativeness over forested landscapes, Remote Sens. Environ., 113, 2476–2498, 2009.

Salomon, J. G., Schaaf, C. B., Strahler, A. H., Gao, F., and Jin, Y. F.: Validation of the MODIS bidirectional reflectance distribution function and albedo retrievals using combined observations from the Aqua and Terra platforms, IEEE T. Geosci. Remote, 44, 1555–1565, 2006.

Schaefer, G. L. and Paetzold, R. F.: SNOTEL (SNOwpack TELemetry) and SCAN (Soil Climate Analysis Network), in: Automated Weather Stations for Applications in Agriculture and Water Resources Management: Current Use and Future Perspectives, Hubbard, edited by: Kenneth G. and Sivakumar, M. V. K., Proceedings of an International Workshop, 6–10 March 2000, http://www.wamis.org/agm/pubs/agm3/WMO-TD1074.pdf, USA High Plains Climate Center Lincoln, Nebraska, and World Meteorological Organization, Geneva, Switzerland, 248 pp., 2001.

Simmons, A. J., Willett, K. M., Jones, P. D., Thorne, P. W., and Dee, D. P.: Low-frequency variations in surface atmospheric humidity, temperature and precipitation: Inferences from reanalyses and monthly gridded observational datasets, J. Geophys. Res., 115, 1–21, doi:10.1029/2009JD012442, 2010.

Smith, A. B., Walker, J. P., Western, A. W., Young, R. I., Ellett, K. M., Pipunic, R. C., Grayson, R. B., Siriwardena, L., Chiew, F. H. S., and Richter, H.: The Murrumbidgee soil moisture monitoring network data set, Water Resour. Res., 48, W07701, doi:10.1029/2012WR011976, 2012.

Szczypta, C., Calvet, J.-C., Albergel, C., Balsamo, G., Boussetta, S., Carrer, D., Lafont, S., and Meurey, C.: Verification of the new ECMWF ERA-Interim reanalysis over France, Hydrol. Earth Syst. Sci., 15, 647–666, doi:10.5194/hess-15-647-2011, 2011.

van den Hurk, B. and Viterbo, O.: The Torne-Kalix PILPS 2(e) experiment as a test bed for modifications to the ECMWF land surface scheme, Global Planet. Change, 38, 165–173, 2003.

van den Hurk, B., Viterbo, P., Beljaars, A., and Betts, A. K.: Offline validation of the ERA-40 surface scheme, ECMWF Tech. Memo. No. 295, ECMWF, Reading, UK, 2000.

van den Hurk, B., Best, M., Dirmeyer, P., Pitman, A., Polcher, J., and Santanello, J.: Acceleration of Land Surface Model Development over a Decade of Glass, B. Am. Meteorol. Soc., 92, 1593–1600, doi:10.1175/BAMS-D-11-00007.1, 2011.

Vitart, F., Buizza, R., Alonso Balmaseda, M., Balsamo, G., Bidlot, J.-R., Bonet, A., Fuentes, M., Hofstadler, A., Molteni, F., and Palmer, T.: The new VAREPS-monthly forecasting system: a first step towards seamless prediction, Q. J. Roy. Meteorol. Soc., 134, 1789-1799, 2008.

Viterbo, P. and Betts, A. K.: Impact of the ECMWF reanalysis soil water on forecasts of the July 1993 Mississippi flood, J. Geophys. Res.-Atmos., 104, 19361–19366, 1999.

Wang, Z. and Zeng, X. B.: Evaluation of snow albedo in land models for weather and climate studies, J Appl. Meteorol. Clim., 49, 363–380, doi:10.1175/2009jamc2134.1, 2010.

Weedon, G. P., Gomes, S., Viterbo, P., Shuttleworth, W. J., Blyth, E., Österle, H., Adam, J. C., Bellouin, N., Boucher, O., and Best, M.: Creation of the WATCH Forcing Data and its use to assess global and regional reference crop evaporation over land during the twentieth century, J. Hydrometeorol., 12, 823–848, 2011.

Weedon, G. P., Balsamo, G., Bellouin, N., Gomes, S., Best, M. J., and Vidale, P.-L.: The WFDEI meteorological forcing dataset: WATCH forcing data methodology applied to ERA-Interim reanalysis data, Water Resour. Res., 50, 7505–7514, doi:10.1002/2014WR015638, 2014.

Zhou, L., Dickinson, R. E., Tian, Y., Zeng, X., Dai, Y., Yang, Z.-L., Schaaf, C. B., Gao, F., Jin, Y., Strahler, A., Myneni, R. B., Yu, H., Wu, W., and Shaikh, M.: Comparison of seasonal and spatial variations of albedos from Moderate-Resolution Imaging Spectroradiometer (MODIS) and common land model, J. Geophys. Res., 108, 4488, doi:10.1029/2002jd003326, 2003.

4

Monitoring of riparian vegetation response to flood disturbances using terrestrial photography

K. Džubáková[1,2], **P. Molnar**[1], **K. Schindler**[3], **and M. Trizna**[2]

[1]Institute of Environmental Engineering, ETH Zurich, Switzerland
[2]Department of Physical Geography and Geoecology, Comenius University in Bratislava, Slovakia
[3]Institute of Geodesy and Photogrammetry, ETH Zurich, Switzerland

Correspondence to: P. Molnar (molnar@ifu.baug.ethz.ch)

Abstract. Flood disturbance is one of the major factors impacting riparian vegetation on river floodplains. In this study we use a high-resolution ground-based camera system with near-infrared sensitivity to quantify the immediate response of riparian vegetation in an Alpine, gravel bed, braided river to flood disturbance with the use of vegetation indices. Five large floods with return periods between 1.4 and 20.1 years in the period 2008–2011 in the Maggia River were analysed to evaluate patterns of vegetation response in three distinct floodplain units (main bar, secondary bar, transitional zone) and to compare the sensitivity of seven broadband vegetation indices. The results show both a negative (damage) and positive (enhancement) response of vegetation within 1 week following the floods, with a selective impact determined by pre-flood vegetation vigour, geomorphological setting and intensity of the flood forcing. The spatial distribution of vegetation damage provides a coherent picture of floodplain response in the three floodplain units. The vegetation indices tested in a riverine environment with highly variable surface wetness, high gravel reflectance, and extensive water–soil–vegetation contact zones differ in the direction of predicted change and its spatial distribution in the range 0.7–35.8 %. We conclude that vegetation response to flood disturbance may be effectively monitored by terrestrial photography with near-infrared sensitivity, with potential for long-term assessment in river management and restoration projects.

1 Introduction

Riparian vegetation is under natural conditions a dynamic component of the riverine environment, which together with floodplains and river marginal wetlands provides a range of important ecosystem services such as biodiversity, flood retention, nutrient sink, pollution control, groundwater recharge, timber production, and recreation (e.g. Tockner et al., 2008). The species composition and spatial distribution of riparian vegetation is largely determined by floodplain morphology and river flow regime (e.g. Bendix and Hupp, 2000; Merritt et al., 2010; Gurnell et al., 2012) as well as by plant tolerance and response to flood disturbance and water stress (e.g. Auble et al., 1994; Blanch et al., 1999; Glenz et al., 2006; Pasquale et al., 2012). The reciprocal interactions between hydromorphological processes and riparian vegetation lead on the long term to the formation of complex mosaics of landforms and their respective biological communities and habitat patches (e.g. Pringle et al., 1988; Gregory et al., 1991; Decamps, 1996; Latterell et al., 2006; Gurnell and Petts, 2006; Corenblit et. al., 2007; Gurnell and Petts, 2011).

The impact of floods on riparian vegetation is well documented in the literature. The most apparent is a direct negative impact when the vegetation is scoured (Bendix, 1999; Edmaier et al., 2011; Crouzy et al., 2013), covered by sediment and debris (Ballesteros et al., 2011), drowned (Friedman and Auble, 1999), or where it looses its connection to the water table due to channel displacement (Loheide and Booth, 2011). A less evident negative impact of floods is a general decrease in vegetation vigour associated with the post-stress reaction of plants. Plants under flood-induced stress have both short-term and long-term physiological and morphological responses (Kozlowski and Pallardy, 2002), such as root mortality or reduced photosynthetic activity, plant growth, dry matter production, and reproduction (e.g. Hatfield, 1997; Toda et al., 2005). However, floods can positively influence riparian vegetation by generation of new germination sites, by distribution of propagules and woody debris (Gurnell and Petts, 2006; Bertoldi et al., 2011a; Gurnell et al., 2012), and by enabling access to water and nutrients in usually disconnected parts of the floodplain (Amoros and Bornette, 2002). Some of these relationships have been replicated in flume experiments (e.g. Tal and Paola, 2010; Perona et al., 2012) and used in numerical modelling (e.g. Perona et al., 2009a, b).

The monitoring of riparian vegetation in floodplains can be achieved by a range of sensors and methods (see review in Carbonneau and Piégay, 2012). Changes in riparian vegetation cover at the large scale are commonly quantified with remotely sensed data, such as satellite imagery (Verrelst et al., 2008; Johansen et al., 2010; Bertoldi et al., 2011a, b; Caruso et al., 2013; Parsons and Thoms, 2013) and aerial photography (e.g. Bertoldi et al., 2011a, b; Mulla, 2012). These are usually suitable for applications to large rivers at irregular time sampling. More recently, unmanned aerial vehicles (UAVs) have been used for monitoring with high resolution and large coverage, at a sampling rate determined by user (Berni et al., 2009; Dunford et al., 2009; Zhang and Kovacs, 2012).

Another approach for detailed local analysis of riparian vegetation is terrestrial photography. Consumer grade cameras have recently been successfully used for monitoring plant conditions and phenology (e.g. Sakamoto et al., 2012; Sonnentag et al., 2012; Petach et al., 2014; Nijland et al., 2014). Similarly to UAV systems, terrestrial photography has the advantage of a high spatial resolution and a user-defined regular high sampling rate. In addition, terrestrial photography by fully automatic systems has minimal running costs after installation. Disadvantages are a restricted areal coverage and limits of oblique photography (e.g. Morgan et al., 2010; Crouzy et al., 2013). Since we study short-term floodplain response to floods we have opted for terrestrial monitoring by cameras, where we can obtain images shortly before and immediately after large flood events. Our photographic monitoring system records the imagery of a gravel bar in the visible and near-infrared range that is processed into broadband vegetation indices.

Broadband and narrowband vegetation indices (VIs) are standard methods used in remote sensing to identify vegetation and quantify properties such as leaf surface pigmentation, photosynthetic activity, and canopy structure. The detailed overview of VIs and their applications is well explained in the literature (e.g. Jones and Vaughan, 2010). The choice of a suitable vegetation index depends on target plant attributes (Sims and Gamon, 2003; Ortiz et al., 2011; Bargain et al., 2013), environment settings (Barati et al., 2011), and available spectral bands (Adam et al., 2010). In this study we have used a selection of the most common broadband indices (Table 1).

The main aims of this study are (1) to analyse the spatial distribution and intensity of the vegetation response to large floods, where we aim to capture not only severe vegetation damage and scouring, but also less apparent change of vegetation vigour; (2) to study the differences in vegetation response to floods in three distinct floodplain units (main bar, secondary bar, transitional zone), which are meaningful units with regard to the concept of the floodplain mosaic system; and (3) to study differences in the performance of several vegetation indices in identifying the direction and magnitude of floodplain change. The analysis was performed for five floods in a 4-year period (2008–2011) on a gravel bar of an Alpine braided river (Maggia River, Switzerland). The relatively numerous flood events within the 4 study years enabled us to assess the vegetation response of the same species composition to different flood stages and longer-term weather conditions.

The novelty of this work lies in (a) the use of near-infrared (NIR) sensitive camera monitoring of a complex alluvial system consisting of water, sediment and vegetated surfaces; (b) high spatial resolution of the images which allows identifying individual plants; and (c) continuous (daily) monitoring which allows for the spatial analysis of the short-term response before and after individual large floods in terms of both vegetation enhancement and damage.

2 Study area

Maggia is an Alpine river located in southeastern Switzerland, north of the city of Locarno. The river originates at an altitude of about 2500 m and flows south through the Maggia Valley into Lake Maggiore (193 m). The bedrock of the valley is formed by Penninic crystalline nappe predominantly covered by Holocene alluvial deposits. Within these settings Maggia evolved into a braided river system with a gravel cobble bed occasionally covered with fine sand sediment deposits on elevated alluvial bars. The average bed slope in the main valley is about 0.8 %.

The hydrological regime of the river is significantly influenced by hydropower infrastructure (dams, intakes, canals) constructed in the upper watershed in the 1950s. Since then, approximately 75 % of the natural river flow has been diverted to the power station Verbano at Lake Maggiore and only minimum flows are released into the main valley. At present, the bypassed section has an average daily streamflow of $4.1 \, \mathrm{m^3 \, s^{-1}}$, while it was close to $16 \, \mathrm{m^3 \, s^{-1}}$ prior to 1954 (Molnar et al., 2008). The 100-year flood peak is estimated at $768 \, \mathrm{m^3 \, s^{-1}}$ (Bignasco) at the upper end of our study reach. The hydropower system regulation practically removes the snowmelt spring–summer flow peak in the valley, but does not affect the largest floods appreciably, mainly due to the upstream location of reservoirs and their relatively low stor-

Table 1. Overview of the VIs used in this study. NIR, R, and G stand for the spectral reflectance in the near-infrared, visible red and visible green frequencies. L is a scaling constant (here $L = 0.5$).

	Vegetation index	Formula	Reference
RVI	Red VI	NIR/R	Birth and Mcvey (1968)
GRVI	Green ratio VI	NIR/G	Sripada et al. (2008)
NDVI	Normalized difference VI	$(NIR - R)/(NIR + R)$	Rouse et al. (1974)
GNDVI	Green normalized difference VI	$(NIR - G)/(NIR + G)$	Gitelson et al. (1996)
SAVI	Soil adjusted VI	$(1 + L)(NIR - R)/(NIR + R + L)$	Huete (1988)
GSAVI	Green soil adjusted VI	$(1 + L)(NIR - G)/(NIR + G + L)$	Sripada et al. (2008)
CVI	Chlorophyll VI	$NIR \cdot R/G^2$	Vincini et al. (2008)

age capacity. As a consequence, floods with a perceptible impact on riparian vegetation still occur on average more than once per year in the main valley (Perona et al., 2009a).

In this study we focused on the 500 m long and 300–400 m wide reach of the river in the main valley located between the villages Someo and Giumaglio. Three distinct floodplain units within the study reach were identified, namely main gravel bar (MB), secondary gravel bar (SB), and transitional zone (TZ) (Fig. 1). The units were delineated based on the floodplain morphology identification of river banks using a LiDAR (light detection and ranging) DEM (digital elevation model) of 2004 and image quality (the marginal zones of the floodplain were excluded due to the interference of surrounding forest). The main bar is the largest, most elevated unit. It is located in the centre of the floodplain in close proximity to the main channel. The secondary bar is at the edge of the floodplain. Both bars are separated by the transitional zone with very active channel dynamics. The secondary channel in the transitional zone is fully connected with the main channel only during flood events.

The vegetation composition within the study reach is heterogeneous (Fig. 2). The dominant willows (*Salix* species) are *Salix purpurea*, *Salix alba*, *Salix eleagnos*, often accompanied by poplars (*Populus nigra*) and alders (*Alnus incana*), occasionally by maples (*Acer pseudoplatanus*), lindens (*Tilia cordata*), knotweeds (*Fallopia sachalinensis*), and locusts (*Robinia pseudoacacia*). The tree height varies from 1 to 10 m. Sparse herbaceous cover grows sporadically on the inner part of the bars with sand accumulation. The variability in the vegetation composition within the three studied floodplain units is notable. *Salix* individuals are located at the upstream part of MB, and towards its inner part are often accompanied by *Populus*. Unlike on MB, *Salix* is predominantly mixed with *Fallopia* on SB. Although fewer in number, the largest diversity in species is found in TZ with *Alnus*, *Salix*, locally *Populus* and *Acer*.

3 Data and methods

3.1 Meteorological and hydrological data

Hourly records of solar radiation, air temperature, relative humidity, and rainfall used in this study were obtained from the weather station Locarno-Monti (MeteoSwiss – Federal Office of Meteorology and Climatology), located about 15 km downstream from the study reach. Hourly streamflow is gauged on the Maggia River at Bignasco, Ponte Nuovo station (FOEN – Federal Office for the Environment), approximately 7.8 km upstream of the study reach. There is an ungauged small tributary (Rovana) between the gauging station and our study reach; thus, the reported peak flows of the studied floods in our reach are a lower estimate.

We analysed the five largest summer floods occurring between 2008 and 2011 with return periods between 1.4 and 20.1 years (Table 2). The flood in 2008 submerged the upstream and middle part of the MB and the whole TZ; more voluminous floods in 2009 and 2010 progressed further and submerged the TZ and the majority of the bars. The most elevated areas of the MB and SB were submerged only in 2011. We defined the duration of the floods based on the discharge when the river inundates the predominantly unvegetated floodplain (180 m^3 s^{-1}). The flood peaks of the first four floods exceeded a discharge of 180 m^3 s^{-1} once for several hours; thus, they are considered to be single-peak floods. The flood in 2011 consisted of two flood peaks greater than 180 m^3 s^{-1} over a period of 5 days.

The meteorological conditions and streamflow before and after each flood are summarized in Fig. 3. The flood in May 2008 was the earliest in the season with the lowest air temperature (minimum 10 °C) and the highest relative humidity prior to the event. The rain gauge at Locarno-Monti did not capture the storm rainfall which occurred mostly in the headwaters of the catchment. With the flood peak of 192 m^3 s^{-1}, it was the smallest but at the same time the longest flood analysed. There were two floods with similar peaks in 2009. The summer of 2009 was very dry and hot, air temperatures prior to both floods reached or exceeded 30 °C, relative humidity was generally very low. The flood in June had intense rainfall (40 mm h^{-1}) measured in Locarno-Monti and the flood peak

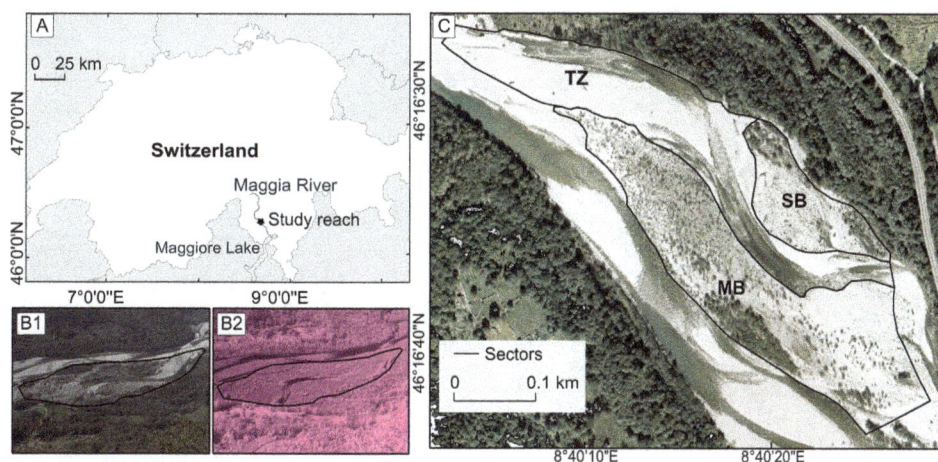

Figure 1. (a) Study reach location within Switzerland. **(b)** Maggia Valley view from the cameras (VIS left, and IR right). **(c)** Study reach subdivided into three units: main alluvial bar (MB), secondary alluvial bar (SB), transitional zone (TZ); flow is from top left to bottom right.

Figure 2. Typical vegetation composition of the Maggia floodplain. From left: gravel bar detail with **(a)** small herbaceous plants (inner zone of the main bar); **(b)** taller 1–3-year-old *Salix* saplings (upstream part of the main bar); **(c)** 2–3 m tall *Salix* trees which range up to 5–6 years in age (middle of the main bar); **(d)** tall *Salix*, *Populus* and *Alnus* trees which have been found to be up to 20 years in age (middle/downstream of the main bar close to the main channel, zone with the highest vegetation density). Flood debris is visible at the stems of larger trees.

Table 2. Analysed floods in this study in the period 2008–2011. The return period of the flood peaks is estimated from data for the period 1982–2011 at Bignasco (Source: MeteoSwiss and FOEN).

Flood date	No. of images before/after	Peak $m^3 s^{-1}$	Return period years
28 May 2008	2/3	192	1.4
6 June 2009	7/5	254	1.7
17 July 2009	6/5	272	1.9
12 June 2010	3/3	301	2.2
13 July 2011	4/6	598	20.1

reached $254 \, m^3 \, s^{-1}$. The subsequent flood in July was preceded by 3 days of moderately intense rainfall ($20 \, mm \, h^{-1}$) and reached a flood peak of $272 \, m^3 \, s^{-1}$. The flood in June 2010 occurred during a period with average air temperature around $20 \, °C$ and high relative humidity. With the flow reaching $301 \, m^3 \, s^{-1}$ it was the second largest analysed flood. The rain gauge in Locarno-Monti captured the storm event only partially, while the heaviest precipitation occurred in the upper catchment. The largest flood in June 2011 also occurred during a period with average air temperature slightly above $20 \, °C$. Intense rainfall covered the entire basin and was measured at Locarno-Monti with intensities of about $40 \, mm \, h^{-1}$. The flood peak reached $598 \, m^3 \, s^{-1}$.

3.2 Image collection and processing

The camera installation in the Maggia River consists of two digital cameras (Canon EOS 350D, 24 mm lens and 8 MP CCD sensor). The two cameras are placed next to each other in a weatherproof box. The box is placed on a steep rocky ridge above the river to give an unobstructed view of the floodplain at the highest angle we could safely get to. The depression angle to the centre of the image on the floodplain is $25°$, the horizontal distance to the study reach is between

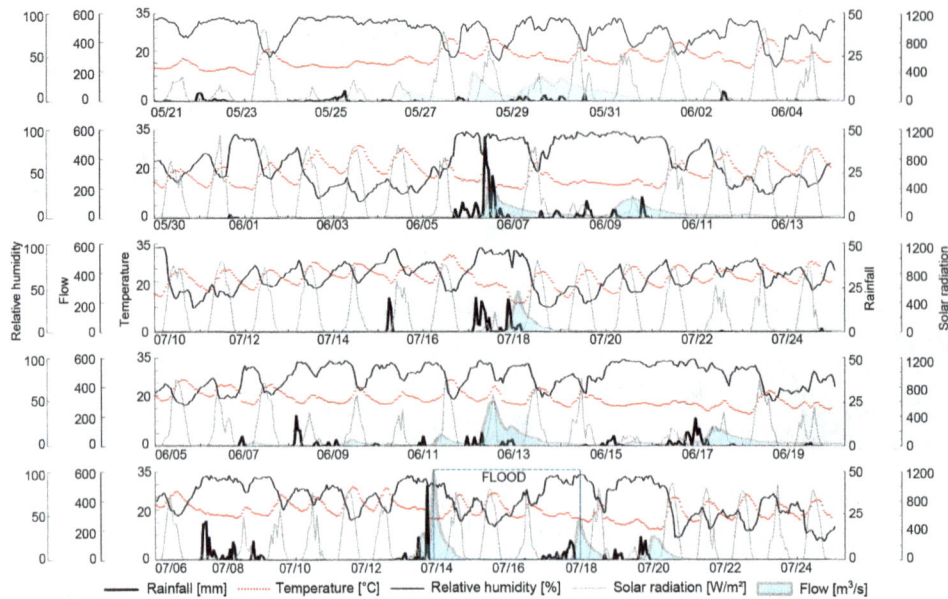

Figure 3. Meteorological and hydrological conditions 7 days before and after each flood. Floods are arranged according to Table 2, from top (2008) to bottom (2011). The first four floods were considered as single-peak events, and the flood in 2011 as a 5-day event with two peaks (delineated by box).

860 and 1460 m, the vertical distance is 537 m. The camera box is accessible only by foot, along a steep mountain path. Photographs have been triggered with a timer remote control every 24 h at 11:00 UTC (universal time coordinated) since summer 2008. The images are stored locally in the cameras on CF (CompactFlash) memory cards. The cameras are powered by Canon lithium-ion 700 mAh batteries. We visit the camera location three times a year to replace batteries, download the images, and perform basic maintenance.

The first camera is a regular camera recording the RGB visual bands. The second camera is adjusted to be sensitive in the near-infrared range by replacing the UV/IR blocking filter on the sensor with a clear filter and a 780 nm IR filter on the lens (Nijland et al., 2014). Sample images can be seen in Fig. 1b. Unlike studies which use cameras with automatic settings or webcams (e.g. Richardson et al., 2007, 2009; Mizunuma et al., 2011), we fixed all adjustable settings manually (except white balance) so that we could directly compare the digital numbers (DNs, brightness at sensor) in the RGB bands and the NIR band in all images without transformations. The white balance was adjusted to an uniform setting in post-processing for all images.

To fix the key camera settings (focus, aperture, exposure) to the best average lightning conditions in the valley we looked at the image DNs of the floodplain in the R–NIR space for a range of typical light conditions. We explored the aperture and time setting ranges to make sure that even for the brightest days we had only limited saturation of pixels in both bands (overexposure). This analysis led us to fix the

aperture on both cameras to $f = 11$ and the exposure time to 1/160 s for the RGB camera and 1/40 s for the NIR camera.

The images were converted to TIFF 48 bit format and registered using a cross-correlation algorithm which searches for the shift in horizontal and vertical directions. The images with significantly lower visibility due to rain, high relative humidity, or haze/mist were excluded from further analysis based on their colour histograms. Seven VIs (Table 1) were computed on the registered images and subsequently orthorectified. The orthorectification was performed in order to link the DEM and field observations. The grid resolution after orthorectification was 0.5 m; hence, individual shrubs and trees on the gravel bar are detectable. The herbaceous cover is captured in limited extent due to its sparse distribution.

Two orthorectification methods were tested. While planar orthorectification defined by five rectification points of distinct fluvial features resulted in an evenly distributed image distortion of 1–2 pixels (< 1 m), the orthorectification based on a LiDAR (2 m resolution) was better in areas with reliable LiDAR points but significantly distorted (∼ 2.5 m) in zones with decreased LiDAR accuracy. Since our study reach is a flat surface especially in areas with vegetation present, we decided to apply planar orthorectification. The image distortion is acceptable for studying individual riparian trees and patches which have footprints greater than 1 m.

3.3   Vegetation index analysis

We evaluated the flood impact on riparian vegetation by comparing VIs from a period before and after each flood event.

We were particularly interested in the direction of VI change indicating vegetation enhancement or damage. The normalized difference vegetation index (NDVI) is used as a reference index due to its common use for vegetation monitoring.

To obtain a statistically robust measure of vegetation change, we defined the before-flood $VI^{bf}(t)$ and post-flood $VI^{pf}(t)$ arrays as

$$VI^{bf}(t) = \text{median}(VI(t-k); \ k = 1, \ldots, 7), \tag{1}$$

$$VI^{pf}(t) = \text{median}(VI(t+k); \ k = 1, \ldots, 7), \tag{2}$$

where $VI(t)$ is the vegetation index on day t and the median is computed pixel-wise. We chose the pixel-wise computation of the median for a period k before and after each flood in order to reduce the potential impact of adverse light conditions and shadows on the images in individual days. We experimented with different k values and found that $k = 7$ days provided an acceptable smoothing without destroying the signal in the data. Although the images after a flood peak may be affected by the flood recession, most studied floods had recessions of less than 24 h, so this effect is not likely to persist for more than 1–2 days (images) and will not affect the pixel-based median of the estimated vegetation change.

Next we computed the difference between the two arrays to get the vegetation change array:

$$\Delta VI(t) = VI^{pf}(t) - VI^{bf}(t). \tag{3}$$

Negative values of ΔVI indicate a decrease in the vegetation index after the flood, e.g. by the erosion and damage of vegetation, while positive values indicate an increase in the vegetation index after the flood, e.g. rise in photosynthetic activity and growth. To analyse vegetation change only pixels representing vegetation cover prior to the flood (i.e. $VI^{bf} > 0.5$).

We compared the indices using the vegetation change array for each flood for a pair of indices, ΔVI_m and ΔVI_n (m, n are indexes from Table 1), and we estimated an index of disagreement as

$$ID(t)_{m,n} = \text{area}(\Delta VI_m(t) \cdot \Delta VI_n(t) < 0)/\text{total area}. \tag{4}$$

The index of disagreement between all floods gives us a relative assessment of the different information content contained in each VI. It should however be noted that with this analysis we do not intend to identify the single best vegetation index; rather, we want to compare the differences in the performance of selected vegetation indices, all of which have been used and validated in the literature.

3.4 Validation

We conducted a site-specific validation of our approach in two steps. First we conducted a ground validation of the NDVI index by comparing the NDVI computed from the images with an estimate derived from direct field measurement of reflectance of different surfaces on the main gravel bar with a spectroradiometer (ASD FieldSpec). Altogether, 18 sites (2 water, 5 gravel and sand covered floodplain surfaces, and 11 different vegetation types and fractions) were measured on a single day and average reflectance in RGB and NIR ranges was computed. The field measurement sites were localized on images taken at the same time (maximal deviation of 7 min), and a 3×3 pixel window was used to extract the RGB and NIR digital numbers in the images. A window was selected to avoid possible location errors on a pixel basis. Figure 4 shows a good fit of the spectral and image NDVIs with a linear correlation coefficient above 0.9. Note that the image colour scales lead to lower NDVIs than local-scale spectrometer measurements, but the relationship is linear (e.g. Petach et al., 2014). The footprint of the image is about 2–3 times the footprint of the spectral measurements, so the location error contributes to the noise in Fig. 4.

In a second step, we quantified the expected range of variability in ΔVI for vegetation change in periods with and without overbank floods. We selected 41 periods of 14 days during which flows did not exceed discharge with a 1 year return period (Q_1) and so are low flow periods with no overbank inundation. For each period we computed VI^{bf}, VI^{pf}, and their corresponding ΔVI (index applied: NDVI), and we compared the spatial standard deviation of ΔVI on vegetated surfaces (NDVI > 0.5 in 2008) for these low flows with those of our selected five largest floods. The results in Fig. 5 show that indeed the largest floods do exhibit a higher variability in ΔVI than ordinary flows in the individual years. The most significant response is visible for the 2011 flood. Furthermore, the standard deviation in ΔVI for small floods is on the order of a 0.01–0.04, which gives us a reference beyond which we can expect VI change to be capturing significant vegetation change induced by floods.

4 Results

4.1 Comparison of vegetation indices

To quantify the vegetation response (i.e. vegetation damage or enhancement) to floods we report the overall comparison of the VIs by the index of disagreement in Table 3. The results show that the selected indices overall agree well in the prediction of vegetation change; the pair-wise differences between the indices are between 0.7 and 35.8 %. The ratio and normalized indices based on the same visible band(s) tend to have more similar results. For example, the RVI and GRVI differ by only less than 1 % from their normalized derivatives NDVI and GNDVI, but by 28.0–35.8 % from the soil-adjusted derivatives SAVI and GSAVI. Because of its widespread use, further detailed evaluation of vegetation response to floods was conducted using NDVI as a reference.

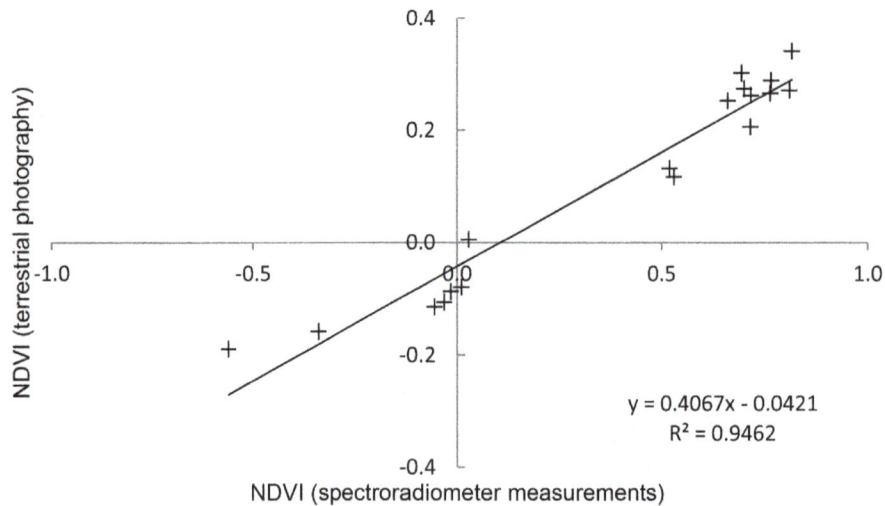

Figure 4. Comparison of NDVI computed from spectroradiometer field measurements and from terrestrial photographs for 18 control points in September 2014.

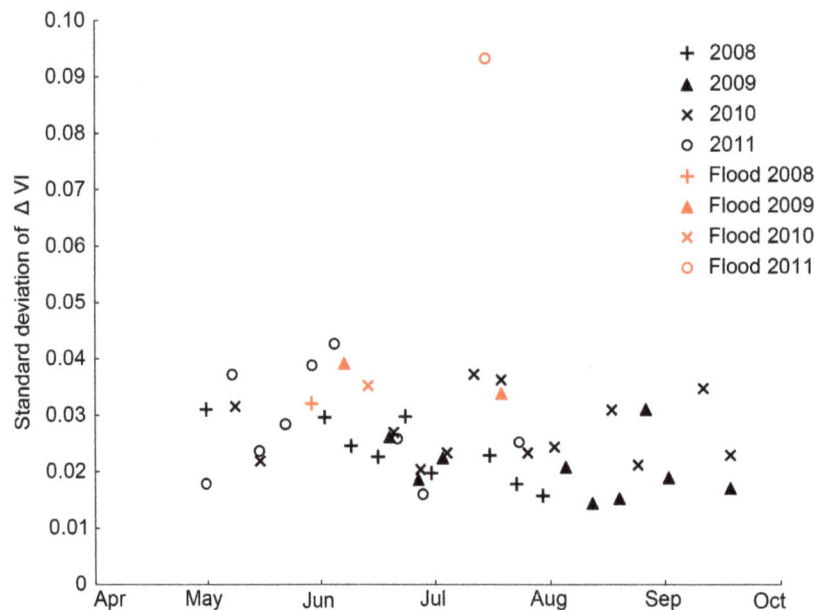

Figure 5. Comparison of the spatial standard deviation of ΔVI change in response to floods and to normal hydrological conditions without occurrence of overbank flow (displayed VI values correspond to NDVI). ΔVI was computed for periods of 14 days with discharge less than Q_1 and for the five major floods studied. Only pixels representing vegetation were considered (NDVI > 0.5).

4.2 Vegetation response in time

The complex nature of flood-induced vegetation change conditioned on the pre-flood vegetation vigour and river morphology is shown in Fig. 6. Here we plot the histograms (boxplots) of vegetation change ΔVI as a function of the pre-flood vegetation vigour VI^{bf} for the five studied floods and three floodplain units. In all of these analyses we considered only the pixels representing vegetation, which we selected based on the pre-flood vegetation vigour ($VI^{bf} > 0.5$). The threshold value was set visually to maximize the number of pixels representing vegetation and minimize the number of pixels representing soil.

The vegetation composition for the floodplain units, expressed as histograms of VI^{bf} in the insets in Fig. 6, shows that most of the vegetation is growing on the main bar, and considerably less on the secondary bar and in the transitional zone. All three floodplain units exhibit modes at $VI^{bf} = 0.6 - 0.75$, which correspond to healthy and large individual plants. Vegetation with $VI^{bf} = 0.75 - 0.85$, i.e. the

Table 3. Index of disagreement ID in percentage (%) of the total number of pixels where two VIs disagree on the direction of vegetation change, i.e. vegetation damage or enhancement.

	NDVI	GNDVI	RVI	GRVI	SAVI	GSAVI	CVI
NDVI		14.8	0.7	14.8	29.2	35.4	26.8
GNDVI	14.8		15.0	0.5	27.8	30.1	15.0
RVI	0.7	15.0		14.8	29.6	35.8	27.0
GRVI	14.8	0.5	14.8		28.0	30.4	15.1
SAVI	29.2	27.8	29.6	28.0		10.5	30.4
GSAVI	35.4	30.1	35.8	30.4	10.5		28.6
CVI	26.8	15.0	27.0	15.1	30.4	28.6	

highest computed VI, is present in all three floodplain units, especially in the transitional zone. The area covered by vegetation is relatively stable in all floodplain units from 2008 to 2011, and there is limited evidence for widespread scouring or abundant vegetation growth. Scouring of a small extent is visible between the flood in 2008 and 2009.

Based on the results in Fig. 6 we can conclude that the intensity of vegetation response to the first four floods is considerably smaller than to the largest flood in July 2011. Changes of up to $\Delta VI = -0.8$ were found after the flood in 2011, which indicates complete removal of vegetation locally. The transitional zone experienced the greatest average vegetation erosion, followed by the secondary bar. Interestingly, the erosive effects of the 2011 flood are more concentrated on vegetation with lower pre-flood VI^{bf}. This suggests a greater sensitivity of younger or less vigorous plants to flood erosion in comparison to well-established individuals. The changes in ΔVI in the 2011 flood are well beyond the ranges of normal variability shown in Fig. 5.

The smaller floods also exhibit vegetation change locally, both damage and enhancement; however, on average the change in ΔVI is small. Overall, the transitional zone is the most dynamic, with damage prevalent on plants with lower pre-flood VI^{bf}. Larger and stronger vegetation with higher VI^{bf} is generally less affected by flood disturbance. However, despite the fact that damage by flood erosion is the prevalent process, it is important that vegetation enhancement also appears locally after floods.

4.3 Spatial distribution of vegetation response

The spatial distribution of vegetation response ΔVI for each flood is shown together with the coherence among indices in identifying vegetation damage in Fig. 7. The intensity of the vegetation response differs between the floodplain units. The main bar has a moderate response with ΔVI mostly between -0.2 and 0.2 (outliers excluded) for the first four floods and between -0.8 and 0.2 for the flood in 2011. Vegetation enhancement is characteristic of the central parts of the main bar. The secondary bar has a slightly smaller vegetation response than the vegetation on the main bar. The exception is the response after the flood in 2011, where significant dam-

age is evident for low VI^{bf}. Unlike the vegetation response on the bars, the ΔVI range in the transitional zone fluctuates considerably more, from -0.4 to 0.2 for the first four floods, and from -0.8 to 0.2 for the flood in 2011. The transitional zone is an area of flow divergence and channel shifting during large floods.

The flood in May 2008 with its long duration early in the vegetation season caused a similar intensity but a slightly different spatial distribution of vegetation response compared to the following floods. The different vegetation response might have been impacted by the presence of plants in close proximity to the main channel and on the top of the transitional zone that were scoured in autumn 2008. Particularly interesting is the impact of the shortest flood analysed (in July 2009), which occurred only 1 month after the flood in June 2009. It was the only flood with widespread vegetation enhancement, most likely associated with an increased water supply (precipitation, groundwater rise). The largest flood in 2011 is the only analysed flood which caused severe vegetation damage, local scour, mostly on the upper part of the main alluvial bar and in the transitional zone. A detail of the scour and deposition of sediment is evident in Fig. 8. Despite the predominantly destructive impact of this flood by scour, the innermost elevated parts of the main bar also show significant vegetation enhancement, most likely caused by wetting of the inundated surfaces.

5 Discussion

Terrestrial photography is a viable approach for the continuous monitoring of riparian vegetation as attested by emerging recent studies (e.g. Richardson et al., 2009; Bertoldi et al., 2011b; Mizunuma et al., 2011; Welber et al., 2012; Crouzy et al., 2013; Sakamoto et al., 2012; Sonnentag et al., 2012; Petach et al., 2014; Nijland et al., 2014; Pasquale et al., 2014). We consider such monitoring to be a valuable low-cost alternative for the continuous repeated measurement and analysis of change in riverine environments which are considered worldwide to be among the most threatened ecosystems (Nilsson and Berggren, 2000; Tockner and Stanford, 2002). The application of vegetation indices to analyse

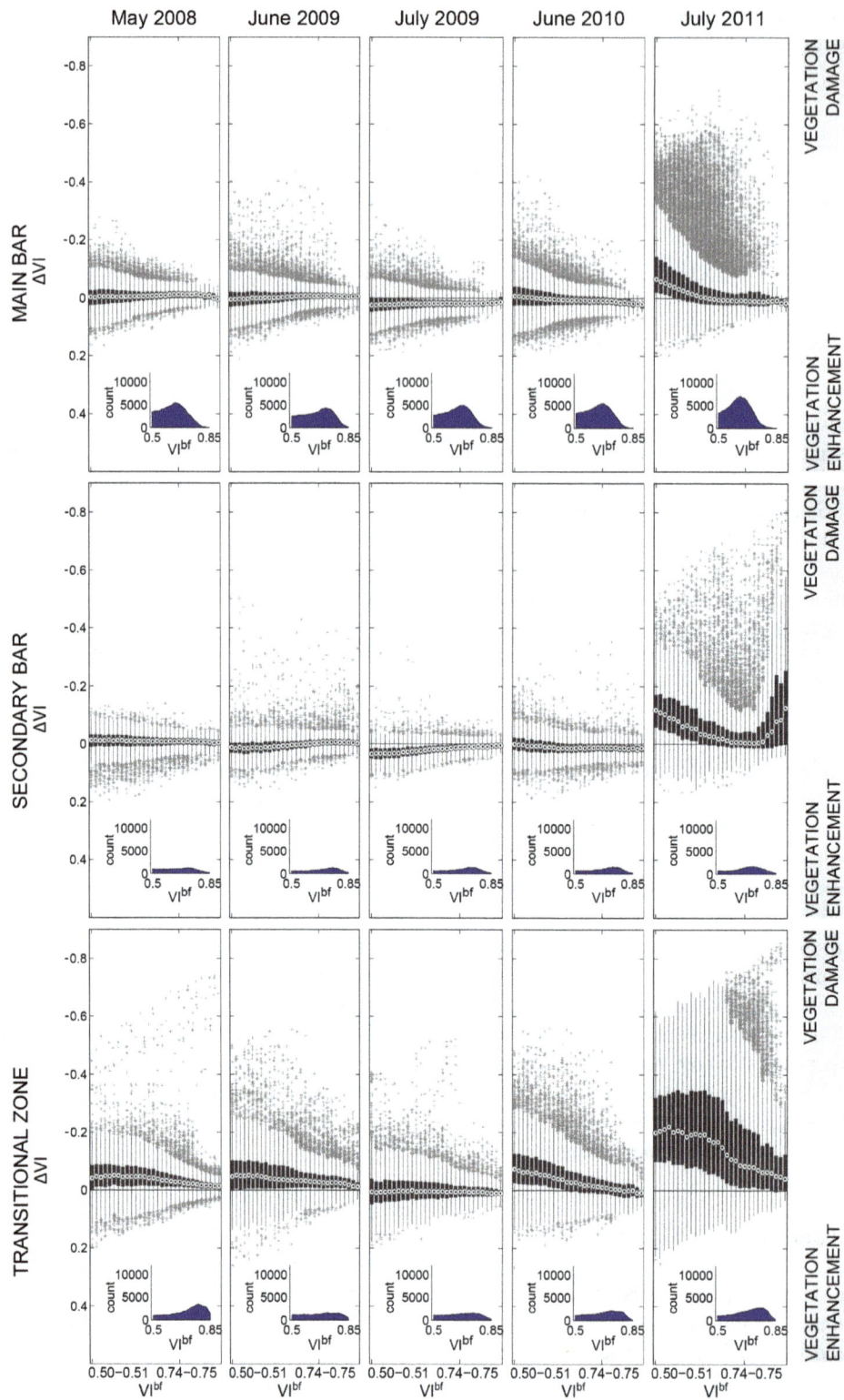

Figure 6. Boxplots for vegetation response ΔVI conditioned on the VI before the flood VI^{bf} (displayed VI values correspond to NDVI). ΔVI < 0 indicates vegetation damage and ΔVI > 0 vegetation enhancement. The boxplots are displayed for VI^{bf} values of 0.5–0.85. Points are drawn as outliers if they are larger than $q_3 + 1.5(q_3 - q_1)$ or smaller than $q_1 - 1.5(q_3 - q_1)$, where q_1 and q_3 are the 25th and 75th percentiles, respectively. Subplots display the distribution of VI before flood VI^{bf} for the corresponding flood and floodplain unit. Insets show the histograms of VI^{bf}.

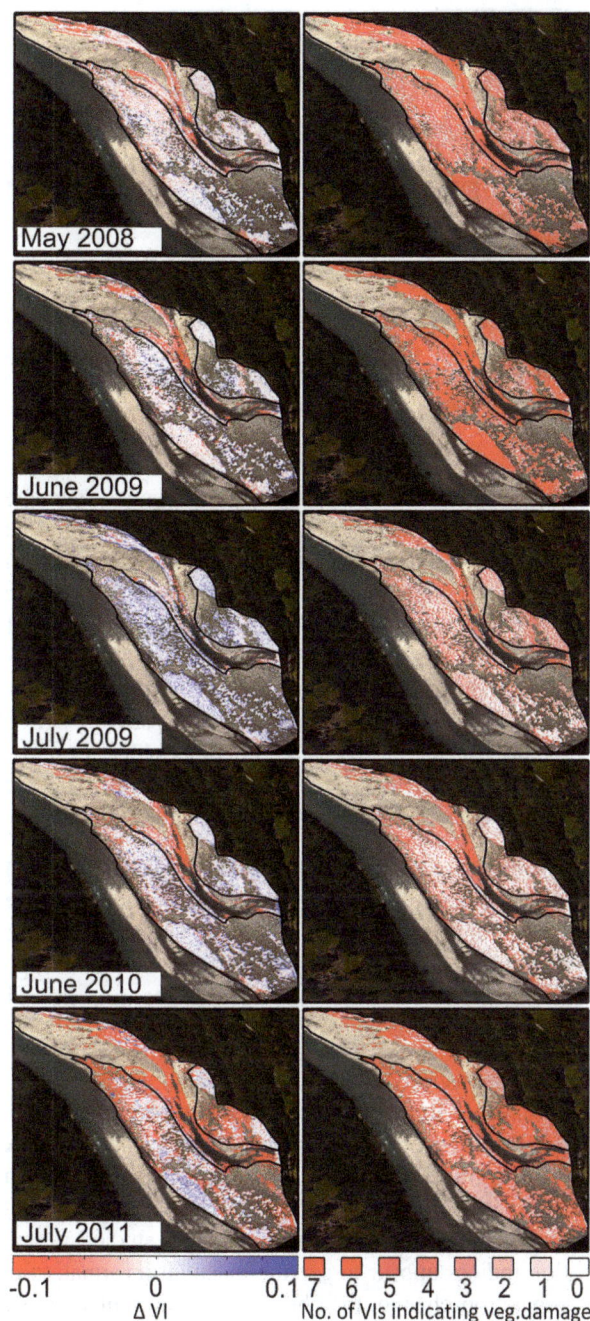

Figure 7. Left-column panels: spatial distribution of vegetation response ΔVI to each flood (displayed VI values correspond to the NDVI). ΔVI < 0 (red colour) indicates vegetation damage and ΔVI > 0 (blue colour) vegetation enhancement. Right-column panels: spatial distribution of number of studied vegetation indices predicting vegetation damage after each flood. Threshold for vegetation delineation: NDVI > 0.5. Floods are listed according to the time of their occurrence. Base image: camera image from 1 June 2009; the reference pre-flood and post-flood images are added in the Supplement. The black lines delineate floodplain units from left: main bar, transitional zone, secondary bar.

change of vegetation vigour after floods in our study raised some questions connected to the particularities of the riverine environment.

The vegetation indices were estimated for a heterogeneous and highly dynamic riverine environment characterized by a variable surface wetness, high gravel reflectance, extensive water–soil–vegetation contact zones, and riparian vegetation with different density and reflectance properties. This is a very challenging environment compared to the usual settings in the literature on camera monitoring of vegetation, where a particular species or canopies are being studied in isolation (e.g. Ahrends et al., 2008; Richardson et al., 2009; Mizunuma et al., 2011; Nagai et al., 2012) or in homogeneous soil substrate with relatively low reflectance (e.g. Viña et al., 2011; Mulla, 2012). The complexity of the environment is reflected in the variability of the estimated vegetation response by the different indices (disagreement between 0.7 and 35.8 %). However, the spatial prediction of change shows substantial coherence (see Fig. 7), including the largest flood in 2011, which is a promising result for applications in riparian environments.

Considering the general trend of vegetation response, the prevailing damage of vegetation with low VI^{bf} and some enhancement of vegetation with high VI^{bf} by floods indicate connections between vegetation stability, growth, and vigour. Smaller plants, predominantly *Salix* individuals, on surfaces exposed to more frequent and damaging stress during floods find it more difficult to recover between floods (Perona et al., 2012), while more protected locations on the gravel bar and floodplain provide a better environment for plants to germinate and grow (zones generally populated by *Salix*, *Populus*, occasionally by *Alnus*, *Tilia*, or *Acer*). This work supports the understanding of spatial distribution of riparian vegetation within the floodplain (e.g. Gurnell et al., 2012).

The floodplain units displayed different vegetation composition and response to floods. The main bar, populated by *Salix* and *Populus* individuals, was the most vegetated area with the most variable spatial pattern of vegetation response to flood disturbances. The *Salix* and *Fallopia* individuals on the secondary bar had generally lower index values than the vegetation on the main bar despite the fact that it is flooded less often than the vegetation on the main bar. The transitional zone was found to be the zone with the most diverse composition (*Salix, Alnus, Populus*, and others), but at the same time the most sensitive vegetation to floods, especially due to lateral erosion of the secondary channel (observed during the field campaign). The results are in accord with the understanding of the floodplain as a mosaic system, where each floodplain unit is determined by its specific morphological, hydrological, and biotic site conditions (Bendix and Hupp, 2000; Jacobson, 2013). More importantly, our study suggests that the mosaic system perspective on vegetation response is perhaps not only valid in a long-term perspective as shown in previous literature, but also on short flood-response timescales.

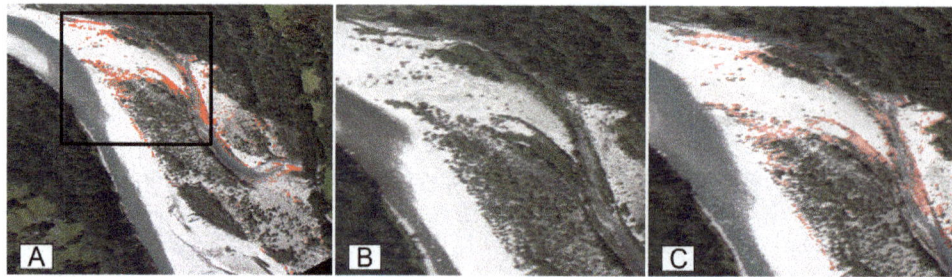

Figure 8. Spatial detail of the upstream section of the study reach with predicted vegetation scour (non-transparent/transparent red colour) after the flood in July 2011 shown together with the actual distribution of vegetation on the surface: **(a)** full view of the study reach with estimated vegetation scour, **(b)** detail of pre-flood distribution of vegetation, image from 11 July 2011, **(c)** detail of post-flood distribution of vegetation, image from 22 July 2011.

Next to the flood's mechanical forcing, there are additional factors impacting vegetation vigour. Since the floodplain of the Maggia River is built by coarse alluvial deposits, floods are accompanied by increased groundwater levels and lateral river–aquifer flows. We expect this additional subsurface water supply to have a considerable influence on the vegetation activity following floods, also in non-submerged areas of the floodplain. We think that an indication of river–aquifer flows is a more diverse vegetation composition in the transitional zone, especially in close proximity to the secondary channel. Another probable reason for differences in vegetation activity is different plant traits (e.g. plant structure, size, ability to adapt) determining vegetation capacity to withstand flood forcing. Additional complexity is added to this picture by the sediment. The presence of fine material in the substrate and a coarse gravel layer on the surface inhibiting evaporation have been shown to be critical for maintaining a high floodplain wetness after inundation (Meier and Hauer, 2010) and will likely impact the degree of vegetation enhancement following floods. In addition to increased substrate moisture, favourable pre- and post-flood weather conditions, i.e. a sunny dry period, may distinctly support vegetation enhancement (e.g. flood in July 2009). The relation between these two effects is highly complex and will be evaluated in the future by including local soil moisture and groundwater level observations.

6   Conclusions

This study demonstrated the use of a high-resolution ground-based infrared-sensitive camera monitoring of riparian vegetation in an Alpine, gravel bed, braided river. The focus was on quantifying the response of riparian vegetation to flood disturbance by standard broadband vegetation indices.

The results offer new insight into the complexity of riparian vegetation dynamics within a floodplain. The main results from the study of the five largest floods with return periods between 1.4 and 20.1 years in the period 2008–2011 in a reach of the Maggia River in Switzerland are the following.

1. Riparian vegetation displays both a negative (damage) and positive (enhancement) response within a short period after floods. There is evidence for a selective impact based on the morphological setting and flood forcing, with destructive effects on smaller or weaker plants and enhancement for stronger individuals higher up on the floodplain. In general, the most impacted plants are young *Salix* individuals on the upstream part of the floodplain, as well as considerably older vegetation (*Salix*, *Populus*, and *Alnus*) in close proximity to the secondary channel where lateral erosion takes place.

2. The intensity and spatial distribution of vegetation damage provides a coherent picture of the floodplain response in three distinct units (main bar, secondary bar, transitional zone), each with a different inundation potential and flood stress. A significant scouring effect is apparent only for the largest flood in 2011.

3. We demonstrated that standard vegetation indices provide means to quantify the vegetation response even in this heterogeneous environment characterized by a mixture of gravel and water surfaces and riparian vegetation with different density and reflectance properties. Overall, we conclude that although all studied vegetation indices appear to capture essential information on vegetation change, the choice of a representative vegetation index is a decision dependent on the composition of the riparian surface, vegetation types, and ultimately the purpose of the monitoring. Future work should be directed at the validation of such an index performance in the riverine environment where local effects of wet/dry sediment reflectance, vegetation type and composition, height and sparseness, light conditions, and others should be better understood (e.g. Parsons and Thoms, 2013; Nijland et al., 2014).

One of the main aims of this paper was to present an analysis of a ground-based infrared-sensitive camera monitoring set-up which provides high spatial and temporal resolution of riparian vegetation change at gravel-bar and river-reach scales. The resolution provides a considerable advantage over remote sensing by satellites, with the downside connected to the broadband nature of the photographic data. A practical advantage of such a system is a comparatively low purchasing and maintenance cost. We are convinced that such systems are suitable for long-term monitoring of riparian areas and have high potential for river management, particularly for regulated rivers or rivers with restoration projects.

Acknowledgements. This research was funded by the Scientific Exchange Program Sciex-NMS grant 12.111, the Slovak Research and Development Agency grant APVV-0625-11, and the Slovak Scientific Grant Agency VEGA project 1/0937/11. Meteorological data were provided by MeteoSwiss (Federal Office of Meteorology and Climatology). Hydrological data were provided by FOEN (Federal Office for the Environment). The ASD FieldSpec spectrometer was provided by the University of Zurich Remote Sensing Lab. We thank Ondrej Budac (EPFL, Switzerland) for his technical support. We are grateful for comments and questions raised by four anonymous reviewers which significantly improved our work.

Edited by: N. Romano

References

Adam, E., Mutanga, O., and Rugege, D.: Multispectral and hyperspectral remote sensing for identification and mapping of wetland vegetation: a review, Wetl. Ecol. Manag., 18, 281–296, 2010.

Ahrends, H. E., Bruegger, R., Stoeckli, R., Schenk, J., Michna, P., Jeanneret, F., Wanner, H., and Eugster, W.: Quantitative phenological observations of a mixed beech forest in northern Switzerland with digital photography, J. Geophys. Res., 113, G04004, doi:10.1029/2007JG000650, 2008.

Amoros, C. and Bornette, G.: Connectivity and biocomplexity in waterbodies of riverine floodplains, Freshwater Biol., 47, 761–776, 2002.

Auble, G. T., Friedman, J. M., and Scott, M. L.: Relating riparian vegetation to present and future streamflows, Ecol. Appl., 4, 544–554, 1994.

Ballesteros, J. A., Bodoque, J. M., Diez-Herrero, A., Sanchez-Silva, M., and Stoffel, M.: Calibration of floodplain roughness and estimation of flood discharge based on tree-ring evidence and hydraulic modelling, J. Hydrol., 403, 103–115, 2011.

Barati, S., Rayegani, B., Saati, M., Sharifi, A., and Nasri, M.: Comparison the accuracies of different spectral indices for estimation of vegetation cover fraction in sparse vegetated areas, The Egyptian Journal of Remote Sensing and Space Science, 14, 49–56, 2011.

Bargain, A., Robin, M., Méléder, V., Rosa, P., Le Menn, E., Harin, N., and Barillé, L.: Seasonal spectral variation of Zostera noltii and its influence on pigment-based Vegetation Indices, J. Exp. Mar. Biol. Ecol., 446, 86–94, 2013.

Bendix, J.: Stream power influence on southern Californian riparian vegetation, J. Veg. Sci., 10, 243–252, 1999.

Bendix, J. and Hupp, C. R.: Hydrological and geomorphological impacts on riparian plant communities, Hydrol. Process., 14, 2977–2990, 2000.

Berni J., Zarco-Tejada P. J., Suarez, L., and Fereres, E. : Thermal and Narrowband Multispectral Remote Sensing for Vegetation Monitoring From an Unmanned Aerial Vehicle, IEEE T. Geosci. Remote Sens., 47, 722–738, doi:10.1109/TGRS.2008.2010457, 2009.

Bertoldi, W., Drake, N. A., and Gurnell, A. M.: Interactions between river flows and colonizing vegetation on a braided river: exploring spatial and temporal dynamics in riparian vegetation cover using satellite data, Earth Surf. Process. Land., 36, 1474–1486, 2011a.

Bertoldi, W., Gurnell, A. M., and Drake, N. A.: The topographic signature of vegetation development along a braided river: Results of a combined analysis of airborne lidar, color air photographs and ground measurements, Water Resour. Res., 47, W06525, doi:10.1029/2010WR010319, 2011b.

Birth, G. S. and Mcvey, G. R.: Measuring the colour of growing turf with a reflectance spectrometer, Agron. J., 60, 640–643, 1968.

Blanch, S. J., Ganf, G. G., and Walker, K. F.: Tolerance of riverine plants to flooding and exposure indicated by water regime, Regul. River., 15, 43–62, 1999.

Carbonneau, P. E. and Piégay, H.: Fluvial Remote Sensing for Science and Management, Wiley-Blackwell, Chichester, 440 pp., 2012.

Caruso, B. S., Pithie, C., and Edmondson, L.: Invasive riparian vegetation response to flow regimes and flood pulses in a braided river floodplain, J. Environ. Manage., 125, 156–168, 2013.

Corenblit, D., Tabacchi, E., Steiger, J., and Gurnell, A. M.: Reciprocal interactions and adjustments between fluvial landforms and vegetation dynamics in river corridors: A review of complementary approaches, Earth Sci. Rev., 84, 56–86, 2007.

Crouzy, B., Edmaier, K., Pasquale, N., and Perona, P.: Impact of floods on the statistical distribution of riverbed vegetation, Geomorphology, 202, 51–58, 2013.

Decamps, H.: The renewal of floodplain forests along rivers: a landscape perspective, Verh. Int. Verein. Limnol., 26, 35–59, 1996.

Dunford, R., Michel, K., Gagnage, M., Piégay, H., and Trémelo, M.-L.: Potential and constraints of Unmanned Aerial Vehicle technology for the characterization of Mediterranean riparian forest, Int. J. Remote Sens., 30, 4915–4935, doi:10.1080/01431160903023025, 2009.

Edmaier, K., Burlando, P., and Perona, P.: Mechanisms of vegetation uprooting by flow in alluvial non-cohesive sediment, Hydrol. Earth Syst. Sci., 15, 1615–1627, doi:10.5194/hess-15-1615-2011, 2011.

Friedman, J. M. and Auble, G. T.: Mortality of riparian box elder from sediment mobilization and extended inundation, Regul. River., 15, 463–476, 1999.

Gitelson, A. A., Kaufman, Y. J., and Merzlyak, M. N.: Use of a green channel in remote sensing of global vegetation from EOS-MODIS, Remote Sens. Environ., 58, 289–298, 1996.

Glenz, C., Schlaepfer, R., Iorgulescu, I., and Kienast, F.: Flooding tolerance of Central European tree and shrub species, Forest Ecol. Manage., 235, 1–13, doi:10.1016/j.foreco.2006.05.065, 2006.

Gregory, S. V., Swanson, F. J., McKee, W. A., and Cummins, K. W.: An ecosystem perspective of riparian zones, BioScience, 41, 540–551, 1991.

Gurnell, A. M. and Petts, G.: Trees as riparian engineers: The Tagliamento River, Italy, Earth Surf. Process. Land., 31, 1558–1574, 2006.

Gurnell, A. M. and Petts, G.: Hydrology and ecology of river systems, in: Treatise on Water Science, edited by: Wilderer, P., Elsevier, Oxford, 237–269, ISBN: 9780444531995, 2011.

Gurnell, A. M., Bertoldi, W., and Corenblit, D.: Changing river channels: The roles of hydrological processes, plants and pioneer fluvial landforms in humid temperate, mixed load, gravel bed rivers, Earth Sci. Rev., 11, 129–141, 2012.

Hatfield, J. L.: Plant-water interactions, in: Plants for Environmental Studies, edited by: Wuncheng, W., Gorsuch, J. W., and Hughes, J., CRC Press, 81–100, 1997.

Huete, A. R.: A soil adjusted vegetation index (SAVI), Remote Sens. Environ., 25, 295–309, 1988.

Jacobson, R. B.: Riverine Habitat Dynamics, in: Treatise on Geomorphology, edited by: Shroder, J. F., Academic Press, San Diego, 6–19, ISBN: 9780080885223, 2013.

Johansen, K., Phinn, S., and Witte, C.: Mapping of riparian zone attributes using discrete return LiDAR, QuickBird and SPOT-5 imagery: Assessing accuracy and costs, Remote Sens. Environ., 114, 2679–2691, 2010.

Jones, H. G. and Vaughan, R. A.: Remote sensing of vegetation: principles, techniques, and applications. Oxford university press, ISBN: 9780199207794, 2010.

Kozlowski, T. T. and Pallardy, S. G.: Acclimation and adaptive responses of woody plants to environmental stresses, Bot. Rev., 68, 270–334, 2002.

Latterell, J. J., Scott Bechtold, J., O'Keefe, T. C., Pelt, R., and Naiman, R. J.: Dynamic patch mosaics and channel movement in an unconfined river valley of the Olympic Mountains, Freshwater Biol., 51, 523–544, 2006.

Loheide, S. P. and Booth, E. G.: Effects of changing channel morphology on vegetation, groundwater, and soil moisture regimes in groundwater-dependent ecosystems, Geomorphology, 126, 364–376, 2011.

Meier, C. I. and Hauer, F. R.: Strong effect of coarse surface layer on moisture within gravel bars: Results from an outdoor experiment, Water Resour. Res., 46, W05507, doi:10.1029/2008WR007250, 2010.

Merritt, D. M., Scott, M. L., Poff, L. N., Auble, G. T., and Lytle, D. A.: Theory, methods and tools for determining environmental flows for riparian vegetation: riparian vegetation-flow response guilds, Freshwater Biol., 55, 206–225, 2010.

Mizunuma, T., Koyanagi, T., Mencuccini, M., Nasahara, K. N., Wingate, L., and Grace, J.: The comparison of several colour indices for the photographic recording of canopy phenology of Fagus crenata Blume in eastern Japan, Plant Ecology & Diversity, 4, 67–77, doi:10.1080/17550874.2011.563759, 2011.

Molnar, P., Favre, V., Perona, P., Burlando, P., Randin, C, and Ruf, W.: Floodplain forest dynamics in a hydrologically altered mountain river, Peckiana, 5, 17–24, 2008.

Morgan, J. L., Gergel, S. E., and Coops, N. C.: Aerial Photography: A Rapidly Evolving Tool for Ecological Management. BioScience, 60, 47–59, doi:10.1525/bio.2010.60.1.9, 2010.

Mulla, D. J.: Twenty five years of remote sensing in precision agriculture: Key advances and remaining knowledge gaps, Biosyst. Eng., 114, 358–371, 2012.

Nagai, S., Saitoh, T. M., Koayashi, H., Ishihara, M., Suzuki, R., Motohka, T., Nasahara, K. N., and Muraoka, H.: In situ examination of the relationship between various vegetation indices and canopy phenology in an evergreen coniferous forest, Japan, Int. J. Remote Sens., 33, 6202–6214, 2012.

Nijland, W., de Jong, R., de Jong, S. M., Wulder, M. A., Bater, C. W., and Coops, N. C.: Monitoring plant condition and phenology using infrared sensitive consumer grade digital cameras, Agr. Forest Meteorol., 184, 98–106, 2014.

Nilsson, C. and Berggren, K.: Alterations of Riparian Ecosystems Caused by River Regulation, BioScience, 50, 783–792, 2000.

Ortiz, B. V., Thomson, S. J., Huang, Y., Reddy, K. N., and Ding, W.: Determination of differences in crop injury from aerial application of glyphosate using vegetation indices, Comput. Electron. Agr., 77, 204–213, 2011.

Parsons, M. and Thoms, M. C.: Patterns of vegetation greenness during flood, rain and dry resource states in a large, unconfined floodplain landscape, J. Arid Environ., 88, 24–38, 2013.

Pasquale, N., Perona, P., Francis, R., and Burlando, P.: Effects of streamflow variability on the vertical root density distribution of willow cutting experiments, Ecol. Eng., 40, 167–172, 2012.

Pasquale, N., Perona, P., Wombacher, A., and Burlando, P.: Hydrodynamic model calibration from pattern recognition of non-orthorectified terrestrial photographs, Computers & Geosciences, 62, 160–167, 2014.

Perona, P., Molnar, P., Savina, M., and Burlando, P.: An observation-based stochastic model for sediment and vegetation dynamics in the floodplain of an Alpine braided river, Water Resour. Res., 45, W09418, doi:10.1029/2008WR007550, 2009a.

Perona, P., Camporeale, C., Perucca, E., Savina, M., Molnar, P., Burlando, P., and Ridolfi, L.: Modelling river and riparian vegetation interactions and related importance for sustainable ecosystem management, Aquat. Sci., 71, 266–278, 2009b.

Perona, P., Molnar, P., Crouzy, B., Perucca, E., Jiang, Z., McLelland, S., and Gurnell, A. M.: Biomass selection by floods and related timescales: Part 1. Experimental observations, Adv. Water Resour., 39, 85–96, 2012.

Petach, A. R., Toomey, M., Aubrecht, D. M., and Richardson, A. D.: Monitoring vegetation phenology using an infrared-enabled security camera, Agr. Forest Meteorol., 195-196, 143-151, 2014.

Pringle, C. M., Naiman, R. J., and Bretschko, G.: Patch dynamics in lotic systems: The stream as a mosaic, J. North American Benthological Soc., 7, 503–524, 1988.

Richardson, A. D., Jenkins, J. P., Braswell, B. H., Hollinger, D. Y., Ollinger, C. V., and Smith, M.-L.: Use of digital webcam images to track spring green-up in a deciduous broadleaf forest, Oecologia, 152, 323–334, 2007.

Richardson, A. D., Braswell, B. H., Hollinger, D. Y., Jenkins, J. P., and Ollinger, S. V.: Near-surface remote sensing of spatial and temporal variation in canopy phenology, Ecol. Appl., 19, 1417–1428. doi:10.1890/08-2022.1, 2009.

Rouse Jr., J. W., Haas, R. H., Deering, D. W., Schell, J. A., and Harlan, J. C.: Monitoring the vernal advancement and retrogradation (green wave effect) of natural vegetation, Greenbelt, MD: NASA/GSFC Type III Final Report, 1974.

Sakamoto, T., Gitelson, A. A., Nguy-Robertson, A. L., Arkebauer, T. J., Wardlow, B. D., Suyker, A. E., Verma, S. B. and Shibayama, M.: An alternative method using digital cameras for continuous monitoring of crop status, Agr. Forest Meteorol., 154–155, 113–126, 2012.

Sims, D. A. and Gamon, J. A.: Estimation of vegetation water content and photosynthetic tissue area from spectral reflectance : a comparison of indices based on liquid water and chlorophyll absorption features, Remote Sens. Environ., 84, 526–537, 2003.

Sonnentag, O., Hufkens, K., Teshera-Sterne, C., Young, M., Friedl, M., Braswelle, B. H., Milliman, T., O'Keefe, J., and Richardson, A. D.: Digital repeat photography for phenological research in forest ecosystems, Agr. Forest Meteorol., 152, 159–177, 2012.

Sripada, R. P., Schmidt, J. P., Dellinger, A. E., and Beegle, D. B.: Evaluating Multiple Indices from a Canopy Reflectance Sensor to Estimate Corn N Requirements, Agron. J., 100, 1553–1561, 2008.

Tal, M. and Paola, C.: Effects of vegetation on channel morphodynamics: results and insights from laboratory experiments, Earth Surf. Process. Land., 35, 1014–1028, 2010.

Tockner, K. and Stanford, J. A.: Riverine flood plains: present state and future trends, Env. Conservation, 29, 308–330, doi:10.1017/S037689290200022X, 2002.

Tockner, K., Bunn, S. E., Gordon, C., Naiman, R. J., Quinn, G. P., Standord, J. A., and Polunin, N. V. C.: Flood plains: critically threatened ecosystems, in: Aquatic ecosystems: trends and global prospects, edited by: Polunin, N. V., Edinburgh, Cambridge University Press, 45–61, 2008.

Toda, Y., Ikeda, S., Kumagai, K., and Asano, T.: Effects of flood flow on flood plain soil and riparian vegetation in a gravel river, J. Hydraulic. Eng.-ASCE, 131, 950–960, 2005.

Verrelst, J., Schaepman, M. E., Koetz, B., and Kneubühler, M.: Angular sensitivity analysis of vegetation indices derived from CHRIS/PROBA data, Remote Sens. Environ., 112, 2341–2353, 2008.

Viña, A., Gitelson, A. A., Nguy-Robertson, A. L., and Peng, Y.: Comparison of different vegetation indices for the remote assessment of green leaf area index of crops, Remote Sens. Environ., 115, 3468–3478, 2011.

Vincini, M., Frazzi, E., and D'Alessio, P.: A broad-band leaf chlorophyll vegetation index at the canopy scale, Precis. Agric., 9, 303–319, 2008.

Welber, M., Bertoldi, W., and Tubino, M.: The response of braided planform configuration to flow variations, bed reworking and vegetation: the case of the Tagliamento River, Italy, Earth Surf. Process. Land., 37, 572–582, 2012.

Zhang, C. and Kovacs, J. M.: The application of small unmanned aerial systems for precision agriculture: a review, Precis. Agric., 13, 693–712, doi:10.1007/s11119-012-9274-5, 2012.

5

Protecting environmental flows through enhanced water licensing and water markets

T. Erfani[1,2,*], **O. Binions**[2,*], **and J. J. Harou**[1,2]

[1] School of Mechanical, Aerospace and Civil Engineering, University of Manchester, Manchester, M13 9PL, UK
[2] Department of Civil, Environmental and Geomatic Engineering, University College London, Chadwick Building, Gower Street, London, WC1E 6BT, UK
[*] These authors contributed equally to this work.

Correspondence to: J. J. Harou (julien.harou@manchester.ac.uk)

Abstract. To enable economically efficient future adaptation to water scarcity some countries are revising water management institutions such as water rights or licensing systems to more effectively protect ecosystems and their services. However, allocating more flow to the environment can mean less abstraction for economic production, or the inability to accommodate new entrants (diverters). Modern licensing arrangements should simultaneously enhance environmental flows and protect water abstractors who depend on water. Making new licensing regimes compatible with tradable water rights is an important component of water allocation reform. Regulated water markets can help decrease the societal cost of water scarcity whilst enforcing environmental and/or social protections. In this article we simulate water markets under a regime of fixed volumetric water abstraction licenses with fixed minimum flows or under a scalable water license regime (using water "shares") with dynamic environmental minimum flows. Shares allow adapting allocations to available water and dynamic environmental minimum flows vary as a function of ecological requirements. We investigate how a short-term spot market manifests within each licensing regime. We use a river-basin-scale hydroeconomic agent model that represents individual abstractors and can simulate a spot market under both licensing regimes. We apply this model to the Great Ouse River basin in eastern England with public water supply, agricultural, energy and industrial water-using agents. Results show the proposed shares with dynamic environmental flow licensing system protects river flows more effectively than the current static minimum flow requirements during a dry historical year, but that the total opportunity cost to water abstractors of the environmental gains is a 10–15 % loss in economic benefits.

1 Introduction

Recent projections show that the amount of available water runoff currently appropriated for human needs globally is around 50 %, and likely to rise to 70 % by 2025 (Postel et al., 1996; Postel, 1998). Current water diversion practices lead to degradation of river environments in some areas, resulting in regional water scarcity and conflicts (Smakhtin et al., 2004).

Water trading developed in some countries as a response to water scarcity with the aim of allocating water efficiently (Bjornlund, 2003; Howe et al., 1986; Thobanl, 1997). In the US, Chile, South Africa and Australia trading is permitted or encouraged in some regions. In the US and Australia, government-allocated funds are used to buy back water allocations to leave water in the environment (Brewer et al., 2008; Wheeler et al., 2013; Wilkinson, 2008). These methods are short-term solutions to immediate water scarcity problems and such uses of public funds can be a contentious issue. Reforms of water allocation systems are under way in countries such as the United States, South Africa, Australia, Russia and England and Wales to ensure environmental protection in the longer term (Gleick et al., 2011; Stern, 2013; Young, 2012). In England there are significant institutional barriers to water trading (Environment Agency and Ofwat, 2008; Hodgson, 2006).

The ability of water markets to help users adapt to water scarcity challenges is heavily dependent on the water resource management institutions (Grafton et al., 2011). The issues of fairness in water allocation between environmental and human uses, and between varying human uses have become controversial as economic considerations and market reallocation may not result in a socially just outcome (Syme et al., 1999). Without appropriate regulatory ability to preserve shared ecosystem services there is a risk that overabstraction will continue or worsen under market systems.

The objectives of water resource allocation systems are to regulate access to water resources, ensuring flexibility, security of access, predictability, and fairness, and to reflect public values and opportunity costs (Howe et al., 1986). More recently, environmental protection has been added to those goals. One of the methods used to preserve adequate river flows is to set a minimum flow below which water abstraction must reduce or cease (Acreman, 2005). These static threshold or minimum flow methods of maintaining river flows often do not achieve ecologically or economically efficient results (Arthington et al., 2006; Katz, 2011). The aquatic environment relies on a natural hydrological cycle, but human water abstractions alter the natural flow variability which is important to sustaining riverine species, and minimum flow regimes do not support natural flow regimes (Poff et al., 1997). Hence, fixed volumetric allowances have evolved into allowances with reference to river flow conditions such as "per cent of flow" regime, with abstractions limited to a sustainable share of the natural river flows (Richter et al., 2012). Environmental flow methods are used to determine the sustainable levels of abstractions. Over 200 environmental flow approaches have been developed to provide the policy-makers with tools to redesign water allocation systems ensuring that river ecology is protected, whilst taking into account human water needs (Acreman and Dunbar, 2004). Environmental flow methodologies have been developed and applied in 44 countries, spanning six world regions, with the US as the most active proponent of the approach (Tharme, 2003).

Allocation of water across individual water abstractors, similarly, should be linked to water availability. Examples of these new systems can be found in Australia (Libecap et al., 2010; Young, 2012), Chile and Mexico (Hodgson, 2006). Water allocations in this system are according to available water and river flow conditions. The shares are translated into volumetric licenses for each abstractor.

In redesigning a water allocation system policy makers need to assess how well the new system meets the objectives outlined above, and whether it promotes economically efficient allocation whilst preventing negative externalities of water diversions on the environment or other users. River basin modelling and integrated assessment (Loucks et al., 1981; Jakeman and Letcher, 2003) can provide insights into potential environmental and water allocation outcomes of the proposed changes. Hydroeconomic models that incorporate

hydrology, institutions and economics are particularly relevant (Harou et al., 2009). Traditional hydroeconomic models can simulate aggregate regional results of water trading (Draper et al., 2003; Ward et al., 2006). To determine market outcomes at the scale of individual water diverters, however, it is important to simulate the transactions between individual water users. Cheng et al. (2009) developed a flow-path model formulation that allows tracking transactions between users. Erfani et al. (2013) presented an efficient variant used by Erfani et al. (2014) to model a surface water spot market.

This paper extends the generic water market simulation model proposed by Erfani et al. (2014) to assess possible outcomes of water trading under a share-based licensing system where allocations (water rights) are updated according to current flow conditions and dynamically updated environmental flows (Environment Agency, 2013; Young and McColl, 2005). The new model is applied to a case-study basin in eastern England. The performance of the proposed licensing system is compared to the currently used licensing system which uses static minimum environmental flows and volumetric licenses. The current system allocates fixed water volumes whilst the proposed system scales weekly licensed volumes proportionally to each abstractor's shares depending on flow conditions. The contribution of this paper is to represent a novel modern water management licensing system within a hydroeconomic water market simulator to assess the hydrological and economic impacts of the new policies on a real-world complex multi-sector water resource system.

The next section describes the generalised river basin model formulation used to model both licensing regimes. Section 3 outlines the case study and additional constraints to represent the Ouse Basin and its regulatory environment. Section 4 presents results followed by a discussion in Sect. 5 and conclusions in Sect. 6.

2 Methods

The model presented in this paper is an extension of the hydroeconomic model of Erfani et al. (2014) which uses economic optimisation to simulate and track pair-wise water market transactions between individual water users. This paper introduces dynamic environmental flows and scalable "share" licenses into the pair-wise transaction tracking hydroeconomic water market simulator to evaluate how they perform in a water trading context. The short-term spot water market considered here is a system where each user can observe the bid and ask prices of others, as could exist with an online transaction system. Model constraints are used to represent the physical, regulatory and water-user-specific realities to try and incorporate plausible trading behaviours. The model formulation described in Sect. 2.1 and Appendix B summarise previous work by Erfani et al. (2014). In this paper model extensions to model dynamic environmental flows and scalable water licenses are presented in Sects. 2.2 and 2.3.

2.1 Pair-wise trading model

The Erfani et al. (2014) model (see Appendix B for equations) uses economic optimisation subject to constraints to simulate a short-term (spot) market for water. The river network is modelled as a network of nodes (e.g. demands, storage reservoirs, junctions where flow links converge or diverge) and conveyance links (e.g. river "reaches", i.e. segments) and water balance is ensured at each junction or storage node (see e.g. Loucks et al., 1981; Loucks et al., 2005). Economic benefit functions that quantify the economic gains from water diverted must be provided for each demand node at each time step. The maximised objective function is the sum of economic benefits from water use across all users in each individual time step, the net of transaction costs. This objective function identifies trades that make economic sense whilst meeting constraints that ensure regulations are followed and plausible agent behaviours are considered. For example, it includes a penalty function for deviating from the target level of reservoir storage. Water-user nodes consume some water using their own license or by buying from other license holders, and can sell the rest to others. Since most abstractors' water use is not fully consumptive, some water is returned to the river as return flow. The sum of volumes of water abstracted and sold by the users cannot exceed both their annual and weekly licensed allocation.

2.2 Dynamic environmental flows

The total amount of water across all users allowed for abstraction is the difference between the natural flow (excluding human water diversions and additions) and the minimum flow (MinFlow) at the downstream gauges. MinFlow is used in the following equation:

$$\sum_{\substack{i \\ CO_{il}=1}} x_{il}^k + \text{inFl}_l \geq \text{MinFlow}_j \forall l \in \text{Junction}, \quad (1)$$

$$j \in \text{Gauge}, k \in \text{river},$$

for both the fixed and dynamic environmental flow water management systems. inFl_l is the external inflow at junction node l. The junction node l is connected to the gauge j to record how much water passes by the gauge j. With fixed volumetric water abstraction licenses, water available for abstraction is set using a fixed value of minimum flows (MinFlow_j) regardless of the available flow recorded at the gauges. In the case of dynamic environmental flows, MinFlow_j is a function of naturalised river flows (flow without human water abstractions). Naturalised river flows are estimated from the river flow through the gauging stations, and the MinFlow_j is the sustainable minimum level of river flows.

2.3 License scaling

Under license scaling, the river basin is divided into sub-catchments separated by river flow gauging stations. The water available for abstraction at each gauge j is divided between the upstream license holders in that subcatchment proportionally to their shares.

$$\text{WaterAbstracted}_i^k \leq \begin{cases} \theta_j \times WkLi_i, & \text{nFlGA}_j - \text{MinFlow}_j \leq \sum_{\substack{l \in \text{User} \\ \text{WlGA}_{lj}=1}} WkLi_l \\ WkLi_i & \text{otherwise} \end{cases}$$

$$\forall i \in \text{User} k \in \text{river},$$

where

$$\theta_j = \frac{\text{nFlGA}_j - \text{MinFlow}_j}{\sum_{\substack{l \in \text{User} \\ \text{WlGA}_{lj}=1}} WkLi_l}. \quad (2)$$

In the above equation, $\text{nFlGA}_j - \text{MinFlow}_j$ is the water available for abstraction for license holders upstream of gauge j.

3 Case study

3.1 Water management in England and Wales

In England water diversions ("abstractions") are regulated by the Environment Agency (EA). The abstraction licenses incur yearly charges based on the volumetric size of the license, and not on the actual abstraction volumes. The licenses state maximum daily and yearly abstraction volumes. Environmental protection is enforced through license-specific hands-off flow (HoF) restrictions which refer to minimum flow required through the relevant gauging station, below which the license is temporarily suspended. There are emergency provisions set out in Sect. 57 of the Water Resources Act 1991 which reduce spray irrigation in times of drought. Water trading is allowed, but rarely carried out. There is no water license spot market; each transaction has to be assessed and approved by the EA over several months.

The current system was set up in the 1960s and is not designed to manage competing water uses effectively. HoFs were introduced in an attempt to protect the environment from overabstractions and were applied to newly issued licenses, with no change in allocations for legacy licenses. There is a lack of appropriate incentives or price signals for efficient water use and there are institutional barriers to water trading (Defra, 2011). The current licensing system in many areas results in overabstraction and environmental damage: 18 % of river catchments are classed as overlicensed, and a further 15 % as overabstracted (Environment Agency, 2008). In around a quarter of the water bodies in England and 7% of the water bodies in Wales, new consumptive abstractions cannot be provided with reliable water supply (Environment

Agency, 2011). Nationally, over a third of licenses are not utilised and kept as a reserve in case of a drought, making 20 % of the licensed volume unused, but which could have otherwise been licensed to new uses requiring water (Environment Agency and Ofwat, 2012). Water trading could provide flexibility in regional water resource management and is being considered in individual water resource management zones (Acreman and Ferguson, 2010).

In response to the shortcomings of the current English abstraction licensing system, it is currently being reformed. The aim is to allow water abstractors to more easily manage changes in water availability and regulators to better guarantee environmental flows (Environment Agency and Ofwat, 2012). The new regime is due to be implemented by the mid- to late-2020s. In the meantime, the EA has been assessing sensitivity of rivers to abstractions through the Restoring Sustainable Abstraction program, and making changes to licenses on a case-by-case basis to help prevent further damage.

Water licensing changes in England and Wales are designed to comply with the European Water Framework Directive (WFD). The aim of the WFD is to bring the quality of rivers to "good ecological status". Methods to define environmental flow requirements have been developed to enable policy makers to move away from the "minimum flow" approach to a river management approach that takes into consideration human water needs (Acreman and Dunbar, 2004). These informed the environmental flow indicator (EFI) approach to dynamic environmental flows developed by the Environment Agency. The EFI approach uses flow duration curves to fix the percentage of flow that can be abstracted at different flow levels. Each river in England and Wales has been assigned with an "abstraction sensitivity band" according to its sensitivity to changes in flow. With reference to the abstraction sensitivity, the percentage of flow allowed for abstraction is assigned to each river (Environment Agency, 2013).

3.2 Modelling the Great Ouse River basin

To investigate the outcomes of potential license reform options, we apply the proposed model to the 3000 km^2 Great Ouse River basin in eastern England (Fig. 1). The largest towns are Milton Keynes and Bedford. The basin is characterised by gently rolling land in the upper part and flood plains and meadows in the lower part. Average annual rainfall varies from 540 mm in the east to 670 mm in the west (Environment Agency, 2005).

There are 94 active surface water licenses belonging to users from four sectors: energy, agriculture, public and private water supply and industry. Approximate locations of users are shown in Fig. 1. Around 95 % of yearly surface water abstractions are appropriated by the public water supply (PWS) company and either stored in the reservoir (marked PWS Reservoir in Fig. 1) or input into the treatment and dis-

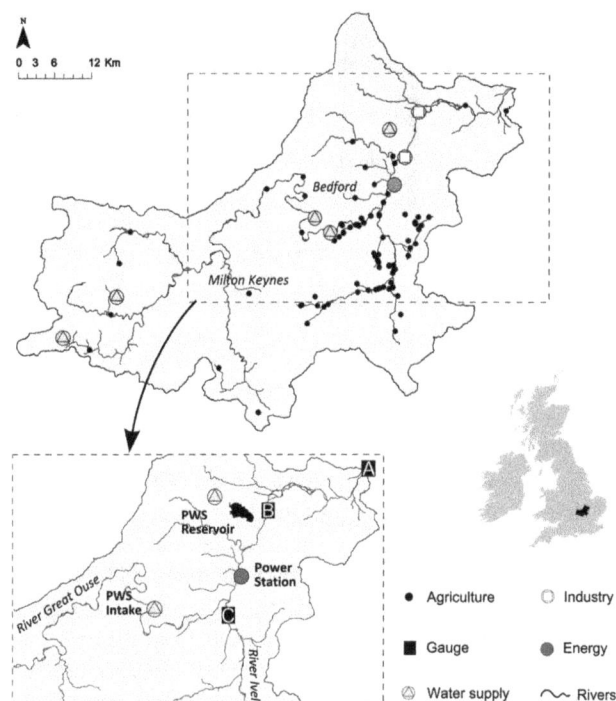

Figure 1. Map of the Great Ouse River basin showing approximate locations of water users and main river flow gauging stations: A – last flow gauge in the basin (sink), B – Offord gauge defining license scaling for PWS reservoir and power station, C – gauge defining license scaling for agricultural users located in the River Ivel tributary.

tribution network (abstraction point labelled PWS Intake in Fig. 1). The second-largest water abstractor is the power station, which uses 4 % of the total volume abstracted for cooling.

3.3 Case-study-specific constraints

In addition to the mass balance constraints described in Sect. 2, the constraints summarised below are used to represent regulatory rules and water-use behaviours in the basin. Incorporating rules is possible since the optimisation model is solved separately for each weekly time step; abstractors have limited hydrological foresight.

3.3.1 Current license restrictions

Water abstraction restrictions under the current system outlined in Sect. 3.1 are implemented to model the fixed volumetric water management system only. This is represented in the model by constraints on license usage. When the river flows are below the threshold limit defined by HoF, the license is temporarily suspended, prohibiting abstractions or trading of this license. The rule specified in Sect. 57 of the 1991 Water Resources Act reduces spray irrigation water diversions when river flow reaches low levels. In our model a

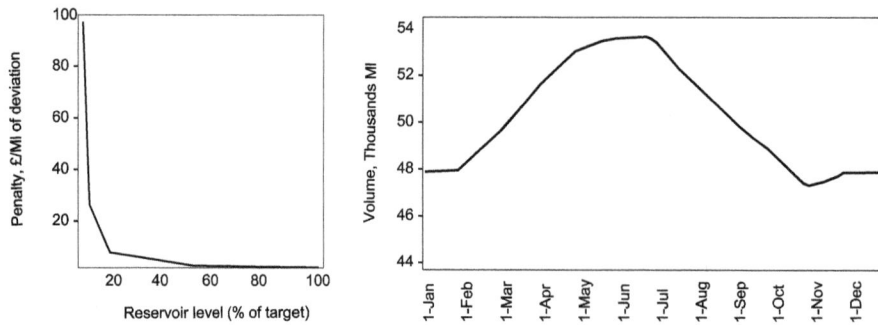

Figure 2. Reservoir storage deviation penalty (left panel), PWS reservoir storage targets (right panel).

50 % rationing is imposed on farmers when river flows are below the flow historically exceeded 95 % of the time at the downstream gauges (see Appendix B for equations).

To model PWS reservoir operation rules, the following set of instructions is employed for both the fixed and dynamic water management licensing regimes. If reservoir storage is below the minimum volume, withdrawals from the reservoir stop. Storage target seeking behaviour by utilities is modelled by penalising storage target deviations in the objective function. As the reservoir levels get progressively lower, the more water-saving initiatives are implemented, and the lower proportion of the demand is satisfied, resulting in lower benefits from water use for the water company and the consumers. This loss of benefits is reflected in the reservoir deviation penalty factor α (y axis in Fig. 2, left panel):

$$\text{Penalty}_j = \alpha \left| \text{tRes}_j - \sum_{k \in \text{Owner}} \text{Res}_j^k \right|, \qquad (3)$$

where tRes_j is the seasonal storage target level (Fig. 2, right panel).

Water companies can implement demand reduction measures during droughts and temporarily restrict non-critical water uses to ensure that key water demands are satisfied. To reflect this in the model, the volume of water abstracted from the reservoir is reduced when storage levels are low using a hedging constraint (Appendix B).

When the PWS reservoir storage volume is low and demand reductions are implemented, the PWS intake license manager is not expected to sell any water. This leads to the following trading rule: when the reservoir level is 50 % below target (half of target volume or less), water sales by PWS the following week are prohibited until the level recovers. Sensitivity analysis was carried out to test the impact on trading of more conservative policies where trading is stopped earlier (Sect. 4.4).

3.3.2 Water trading

Agricultural users require water for the irrigation season and will in many instances be unwilling to sell their license be-

fore it. To represent varying degrees of water market participation, a limit on volumes sold by agricultural users was set in both the fixed and dynamic water management system modelling. For this a constraint (Appendix B) implicitly sets aside a portion of the yearly license for own use and ensures the user does not sell prematurely, exposing themselves to requiring water purchases later in the year. A "trade reluctance coefficient" is used to represent the degree to which farmers keep licensed water for their own use, and can be customised for each user enabling the analyst to consider diverse market participation. If the coefficient is set to 0, the user always prefers to trade whenever it is economically beneficial, regardless of likely own future water needs. Conversely, users with a coefficient of 1 are conservative and will not sell water until they fully satisfy their yearly demand (at the end of the irrigation season).

3.4 Parameterising dynamic environmental flows

The model was applied to the Great Ouse River basin using hydrological data from one of the driest years on record, characterised by low river flows for the first 8 months of the year, followed by wet autumn and winter ("naturalised" flow in Fig. 3). Using the EFI method discussed in Sect. 2.3, and taking into consideration the abstraction sensitivity band of the river basin, the allowable water abstractions are calculated as proportions of naturalised flow (Table 1). In Table 1, percentile of naturalised flow is the percentage of time that flow historically exceeded a given flow value provided by England's Environment Agency (Environment Agency, 2013). Please see Klaar et al. (2014) for further information on the EFI approach.

4 Model results

In the following we review model results focusing on how the two licensing systems diverge in protecting environmental flows, water allocated to each sector, and plausible trades under a short-term spot water market.

Table 1. Allowable river diversions under the EFI system, defined as a percentage of river flow (source: Environment Agency, 2013).

Percentile of natural flow at downstream gauge of the subcatchment	Q_{30}	Q_{50}	Q_{70}	Q_{95}
Percentage of naturalised flow allowed for abstraction (%)	26	24	20	15

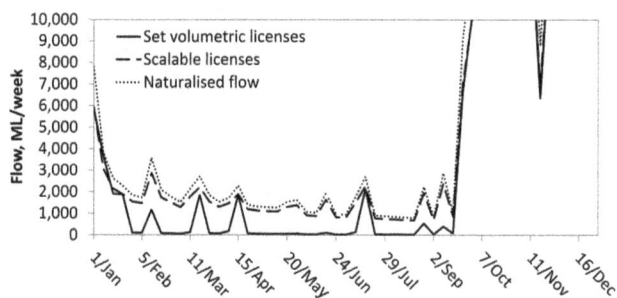

Figure 3. River flow at the last gauge in the basin (marked A in Fig. 1).

Table 2. Numbers of trades, buyers and sellers, and volumetric annual totals under the two licensing regimes for a simulated historical dry year (ML, millions of litres).

	Volumetric licenses	Sharing system
Number of trades	299	678
Sellers	48	90
Buyers	19	32
Total volumes traded, ML	2750	4860

4.1 Protection of the environment

Figure 3 compares modelled river flow exiting the river basin under the two licensing systems. The current system leads to large variability in river flows through the year, decreasing to low levels incompatible with recent regulations such as the European Water Framework Directive (Acreman and Ferguson, 2010). This is the result of the asymmetric impact of environmental HoF conditions on individual licenses. HoFs were assigned to new licenses in the past to prevent overabstraction of rivers but were not applied retrospectively to early licenses granted in the 1960s (see Sect. 3.1). As a result, some (large) licenses are not affected by HoFs and the system is not effective at ensuring environmentally acceptable abstractions during the drought.

The drought river flows are improved with the proposed licensing system (Fig. 3) and its higher environmental allocation. Whereas the current licensing system brings the flow to nearly zero for over 40 % of the dry year, the proposed licensing system never falls below 680 mL per week.

4.2 Water diversions

With the more stringent environmental protection enforced by the proposed scalable licensing system all users face a lower amount of water available for diversion. The total annual volume of water diverted decreases by over 40 % (from 88 000 to 50 000 mL). All water users except industry decrease their diversions: PWS reservoir by 44 %, PWS intake − 8 %, power station − 38 %, agriculture sector − 26 %, private water supply − 14 %. Industrial use increases its abstraction by less than 1 %.

The large decrease in the PWS reservoir abstraction is the main enabler of the higher river flows under the proposed system (Fig. 3). Under the current system there are no hands-off flow conditions imposed on the PWS licenses and the reservoir diverts heavily during the drought to stay within 50 % of its storage targets (Fig. 4, top panel). Under license scaling, the reservoir's weekly water license is scaled down to less than a quarter of the reservoir's historical weekly diversion for most weeks of the year, causing a rapid decrease in reservoir storage volumes that almost empties the reservoir (Fig. 4, lower panel).

4.3 License trading results

Figures 5 and 6 show which sectors are buying and selling water under the current and proposed licensing systems respectively. Figure 5 shows that because of PWS's lack of HoF conditions, they are able to sell water to the energy sector throughout the year. Under the proposed system, as modelled (Fig. 6), where sectors are on equal footing, these rents are not available and PWS stops selling water at the end of April, at which point the energy sector begins buying from farmers (with higher transaction costs due to the larger number of transactions involved).

Lower diversion allowances under the proposed system lead to a more active water market, with the number of trades more than doubling (127 % increase) and the volume traded increasing by 77 % (see Table 2). Trading between users from different sectors also increases. Figure 7 shows the proportion of total yearly volumes transferred between sectors. Under the current licensing system the largest transactions by volume are from the PWS to the power station (94 %). Under the proposed shares-based system the power station is the largest buyer until autumn (Fig. 6), purchasing from both the PWS intake and agricultural businesses, and followed by transfers from the power station to the PWS reservoir in autumn and winter. Agricultural users also sell to the PWS reservoir towards the end of the year, after the growing season. The purchases by the PWS reservoir are made to refill

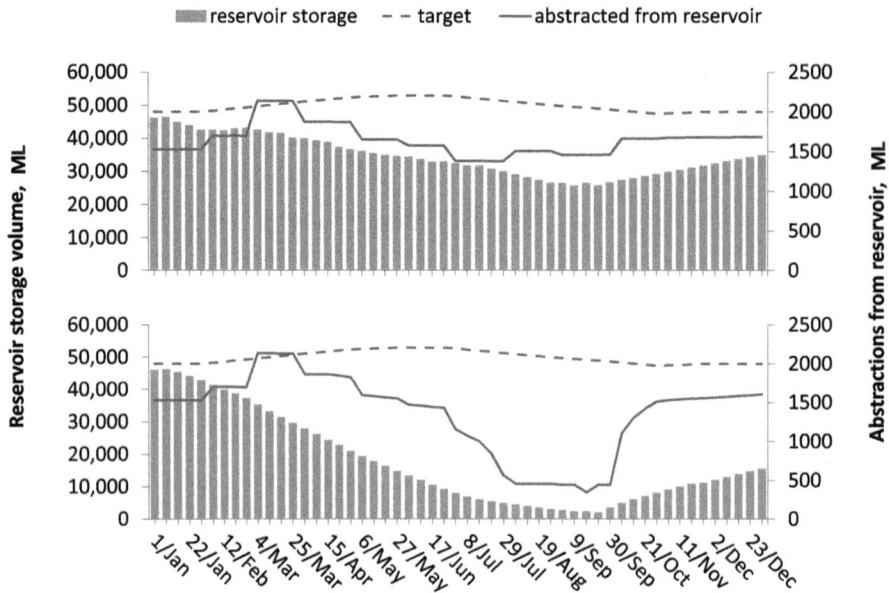

Figure 4. PWS reservoir storage and abstraction profiles for current (top panel) and proposed (bottom panel) licensing systems.

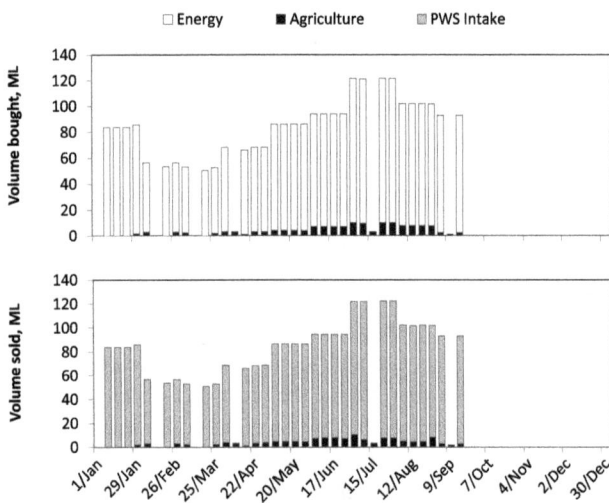

Figure 5. Water volumes bought (top panel) and sold (bottom panel) in millions of litres (ML) per week by sector under the current licensing (volumetric) system.

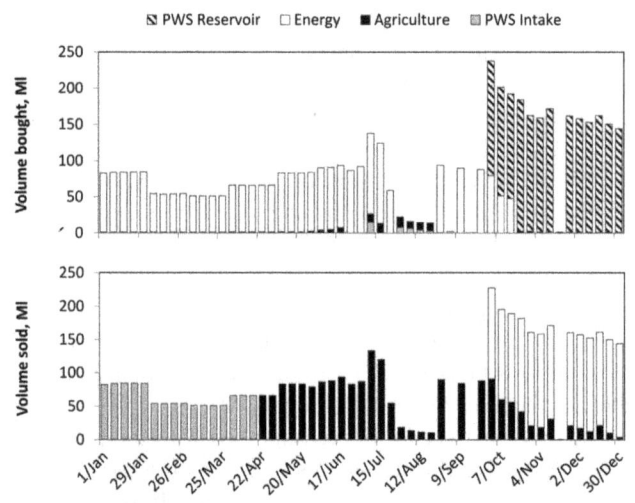

Figure 6. Water volumes bought (top panel) and sold (bottom panel) in millions of litres per week by sector under the proposed (scalable) licensing system.

the reservoir which was depleted through the year under the proposed licensing system.

In the current volumetric licensing simulation, license holders generally either sell or buy water throughout the year, and rarely switch from one status to the other. In the proposed shares-based system, however, some users who buy at the beginning of the year become sellers at the end of the year, and vice versa. Under the current system some license holders are affected by the drought more than others because of the stricter HoF conditions on their licenses, and are therefore systematically disadvantaged during droughts. With the pro-

posed shares system, as simulated, all users are affected by reductions in the available resource, and short-term leasing enables them to manage their water needs effectively: selling in weeks when they have no or low demand for water and purchasing from other users when they have relatively high economic water demands unmet by their allocation.

Under proposed licensing, when the PWS reservoir storage volume reaches half of the target level by mid-April, PWS intake ceases selling water due to the trading constraint outlined in Sect. 3.3.2, PWS intake becomes a buyer in July–August, purchasing small volumes from private water sup-

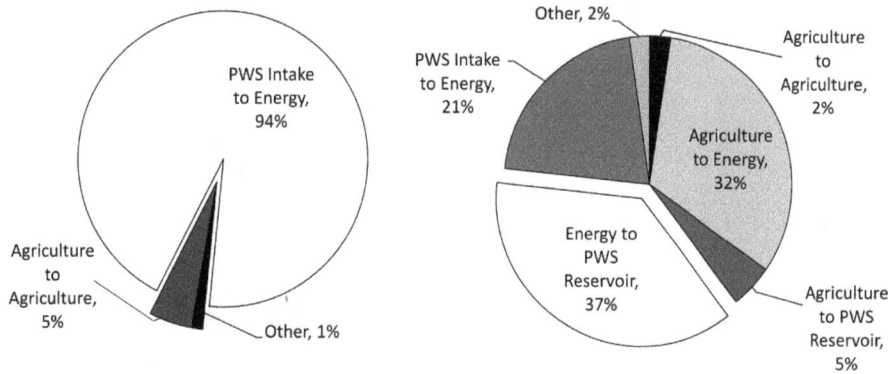

Figure 7. Proportion of the total annual volume of trade transactions between sectors under the current (left panel) and proposed (right panel) licensing systems as simulated in a dry year.

ply license holders and farmers taking advantage of the first opportunity to start filling its reservoir. Under the current licensing system water trading stops as river flow recovers in mid-September, whereas in the shares-based system, trading continues until the end of the year. The reason for this is the large impact of license scaling on the PWS reservoir, as discussed in Sect. 4.2. Low storage volumes activate demand reduction measures which impose an opportunity cost on the water company and its customers and the marginal value of stored water increases (as represented by the penalty function defined in Sect. 3.3.2). The reservoir is refilled late in the year using its own license and purchases from other users.

4.4 PWS trading rule sensitivity analysis

PWS is the largest abstractor in the catchment, the largest single seller of water licences in the current system simulation, and the second-largest in the proposed system of scaled licences. A sensitivity analysis has been carried out to test the effect of the PWS trading rule outlined in Sect. 3.3.1 on model results. A more conservative attitude to trading is considered where the PWS intake stops selling water if storage is 30 or 10 % below the storage target. The impact of these two scenarios on trades and sector benefits were assessed and compared to the original case where PWS stops trading if storage goes below 50 % of the storage target.

Figure 8 shows changes in volumes sold, by sector, under the two licensing systems with stricter PWS trading rules. Under the current system, as the PWS intake reduces its yearly volumes sold, agriculture increases its sales. The volumes sold by agriculture with the 30 % rule are 5 times the volume sold under the 50 % rule. The number of sellers is increased from 48 to 69 with 30 % below the storage target trading rule, and to 75 with 10 % below the storage target trading rule. The overall total volumes sold per year reduce by around 50 % as the trading rule is changed from 50 to 10 % below the storage target.

Under the proposed system, the PWS intake is not the largest seller, and the reduction in its volumes sold does not

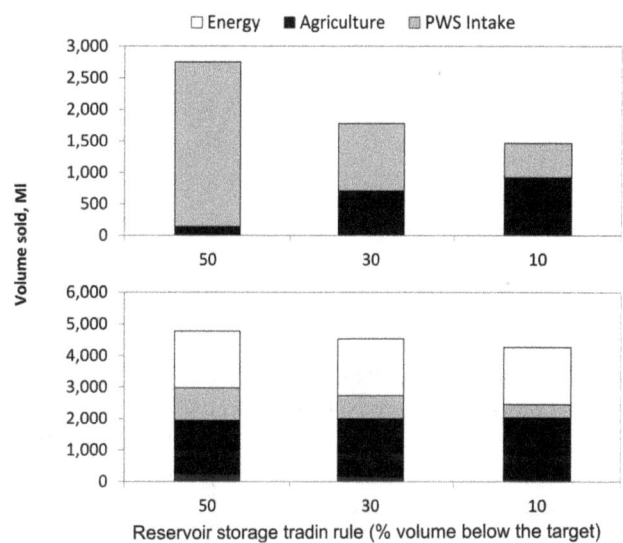

Figure 8. Effect of more conservative PWS trading rule on volumes sold under the current licensing system (top panel) and the proposed system (bottom panel). The water market under the proposed system is more active and less affected by the change in PWS trading policy.

produce as large of an effect on overall trading results. The number of sellers remains at 90 because all water users with water to sell are already participating in the market. The volumes sold by agriculture increase slightly (around 5 % increase as the rule changes from 50 to 10 % below the storage target). The overall total volumes sold reduce by around 10 %.

The agriculture sector's and the power station's benefits from water use reduce as the PWS intake adopts a more conservative selling rule. Due to higher selling volumes by agriculture, its benefits from water use reduce by 25 % (current system) and 23 % (proposed system). The overall reduction in the volumes sold means a reduction in the power station's ability to supplement its allocated water volumes by buying

- - - Current licensing system   —— License scaling

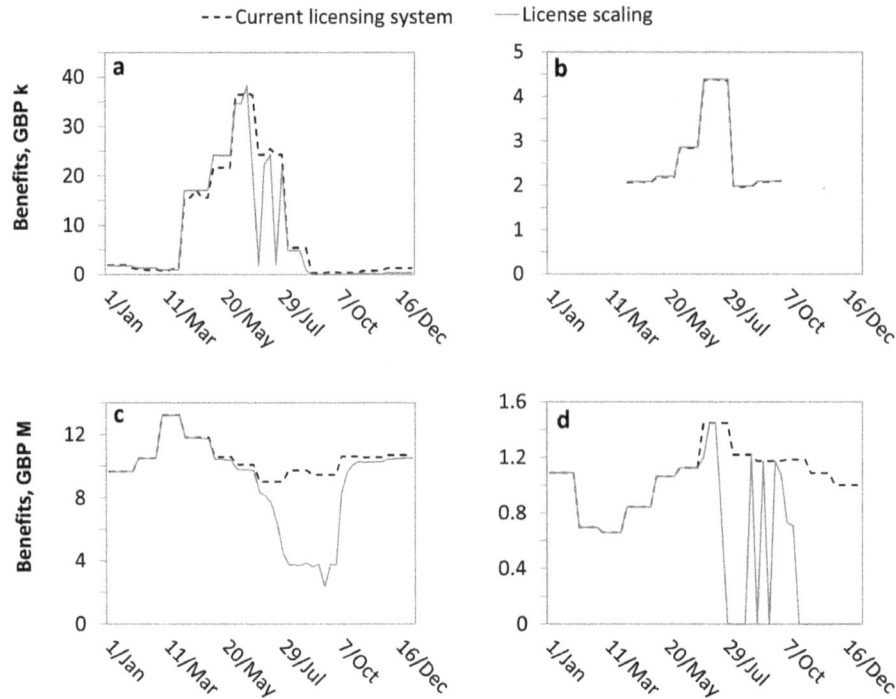

Figure 9. Comparison of gross economic benefits by sector from water use under the current and proposed licensing systems acting in conjunction with a short-term water market. The results are aggregated by sector: (**a**) agriculture; (**b**) industry; (**c**) water supply; (**d**) energy. The top panels show benefits in thousands of pounds and the bottom one in millions.

from other sectors, and its benefit reduces by 30 % (current system) and 21 % (proposed system).

These results suggest that the water market under the proposed licensing system can be less responsive to a single large user's attitudes to trading.

5   Discussion

The model tracks individual transactions allowing the analyst to assess how water markets could operate under different water licensing regimes and how individual abstractors could be affected. The aim of the model is to inform the policy makers of the potential outcomes of water management regulations and assess the effectiveness of a proposed licensing system in increasing environmental protection whilst reducing economic costs of water scarcity.

Gross economic benefits from water use are estimated for each abstraction license holder using their economic water demand functions (see Erfani et al. (2014) for details). For each week, the benefits generated from water use by each abstractor were aggregated into sectoral benefits. Figure 9 compares the economic benefits by sector generated from water use for the two licensing systems in conjunction with the modelled water market. The energy sector sees the largest decrease in benefits due to the increased environmental protection of the proposed licensing system mostly due to the sector's inability to buy water from public water under drought.

The water company also incurs significant summer losses as it introduces restrictions. English policymakers are currently discussing the possible "grandfathering" of the current system priorities in the new licensing system; this would result in a multitiered scaling system where certain sectors have priority over others. As the details of such a system were not yet established, these have not been modelled, but if PWS were given priority within the scaled system, its loss of benefits would likely decrease.

Compared to the current licensing arrangement plus a market, the loss in benefits through the dry year across all sectors in our study case is estimated at GPB 94 million (a 15 % reduction, from GBP 611 million for current licensing to GBP 517 million with the proposed system, both with the modelled surface water market). Erfani et al. (2014) estimated the total annual economic benefits for the same system and year with current licensing but without trading at GBP 575 million. In this case, the estimated opportunity cost of improved environmental flows is GBP 58 million, a 10 % loss in economic benefits.

These opportunity costs for improving environmental flows may appear large. Our analysis uses catchment inflows from one of the driest years on record so this cost can be considered an upper bound on potential costs imposed on water users for enhanced environmental performance. Also, if the power station had an alternative supply to its surface water licence, its opportunity costs would decrease. To put the value in perspective, a survey by NERA (2007) estimated the present value of improvements in the water environment of all water bodies in the UK to be between GBP 18 and 29 billion (benefits incurred for an indefinite period), or between GBP 618 and 1020 million per year. Garrod and Willis (1996) estimated the annual value of alleviating low flows for the River Darent (river catchment area is 14 % of the Great Ouse) at around GBP 37 million (GBP 2011).

Our model uses a single-objective ("aggregate") optimisation formulation that maximises the total social welfare of all water users to simulate the water market. Single-objective optimisation emulates centralised water allocation but is appropriate to model regulated water markets, "as long as interactions between agents and competition for resources can be interpreted in a competitive market paradigm" (Britz et al., 2013).

Several model limitations and simplifications should be mentioned. Groundwater resources were excluded from the model because we focus on the effects of changing the surface water licensing system. Abstraction license holders sometimes possess more than one license, sometimes for both surface water and groundwater abstractions. In this case, they will likely draw strategically from across their asset base (e.g. a water company will cost minimise when choosing sources), and such strategic abstraction would increase in a market – this is not considered in our current model where each abstraction point is modelled independently. Some abstractors, particularly agricultural ones, have small "winter storage" reservoirs to enable inter-temporal water management. Such users would likely switch between different water sources during droughts and involve reservoirs in sophisticated and diverse ways. At the time of the analysis, we did not have data on the locations and capacities of small reservoirs and so this detail is left to later work. Most strategic behaviours across different assets and over time (long-term decisions) are not reflected and are beyond the scope of this paper.

Economic water demands were estimated using data in the literature and are indicative of the water values across the different uses in our catchment. In our model, water diversions and trading are driven primarily by the spatially and temporally varying values of water as encoded in weekly demand curves for each abstractor. In reality, economic considerations are not the only drivers for human behaviour. Actual water markets would depend on the pre-existing social networks within the basin, preferences and attitudes towards trading, as well as perceptions of fairness and justice (Syme et al., 1999). Such motivations were not represented in our hydroeconomic model because they are not known. We take steps to represent some attitudes to trading by introducing a trade reluctance coefficient for agricultural users and embedding water company operating rules regarding their assets by a rule on trading. Furthermore, in our model the propensity of different agents to trade with each other can be calibrated on a pair-wise basis using transaction costs. In our applica-

tion we set transaction costs by abstractor sector but a more detailed study of transaction costs could be performed.

6 Conclusions

This paper uses a hydroeconomic model to assess the performance of two water licensing regimes in conjunction with surface water markets. The first regime is the minimum-flow-based system with fixed volumetric licenses currently used in England and Wales. The second one is a proposed licensing system based on scalable licenses where shares are translated into actual permissible allocation volumes depending on minimum environmental flows that are set dynamically to adapt to naturalised flow conditions. The model was applied to the Great Ouse River basin in eastern England over a historically dry year.

Results suggest the proposed dynamic environmental flow with scalable licensing system is better able to prevent very low flows during droughts than the current abstraction regime based on volumetric licensing. Flows under the proposed system do not reduce below 680 mL per week, whereas under current licensing flows reduce down to nearly 0 for over 4 months of the year. With more water left to the environment, less water is available to satisfy human water demands, leading to a more active water market. The number of trades under the scalable licenses system is more than double the number under the current system and the volume traded is 77 % greater. Also, the water market under the proposed system was less sensitive to a reduction in trading by one large agent (we tested this for PWS). Still the more active water market is not able to fully compensate for the loss of abstraction (increases in environmental flows); the opportunity cost of the increased environmental quality in the dry year is a loss of about 15 % compared to the current licensing system with a water market, or 10 % when compared to current system without a market (the current situation).

As pressure on water resources increases, water licensing systems will be expected to balance human and environmental water uses in increasingly effective and sophisticated ways. The English water allocation regime is currently being redesigned to protect environmental flows whilst minimising the societal economic cost of water scarcity. Water markets are viewed as part of the solution as they allow short-term economically efficient reallocation of water during scarcity events. In designing new water allocation institutions regulators will want to assess how new water allocation systems could work in conjunction with water trading to manage droughts. Customised hydroeconomic models, such as the one applied in this paper, help simulate coupled human-environmental systems, predict plausible behaviours and impacts, and assess proposed policies.

Appendix A: Nomenclature

Table A1. Nomenclature.

Junction	No-demand and non-storage nodes which join tow or more links in the network
User	The set of all licensed river abstractors including agriculture, industry, water supply and energy
Owner	The set of all water right holders, reservoirs and the river
x_{ij}^k	Decision variable, the water flowing from node i to j with license holder k
inFl_i	External hydrological inflow at junction node i
CO_{ij}	Connectivity matrix which contains 1 if node i is connected to node j, 0 if no connection
pRes_j^k	Reservoir j storage carried over from previous time step with water license k
Res_j^k	Reservoir j storage with water license k
tRes_j	Reservoir j target
$\text{WaterAbstracted}_i^k$	Water consumed by user i which is either bought from owner k or abstracted from river using user i's license
$\text{Trade}_i^{k \in \text{river}}$	Water license leased for one time step by user i
ReturnFlow_{ij}	Water returned back to the river at downstream junction node j of user i based on the consumption factor of user i
DW_{ij}	Junction node j downstream of user i
consFactor_i	Fraction of water evaporated relative to diverted for user i
Discharge_j	Discharge sink j at the mouth of the river
WkLi_i	Weekly license allowance for user i to abstract water from river
YrLi_i	Yearly license allowance for user i to abstract water from river
Deviation_j	Deviation of reservoir j from its target storage volume
flGA_j	Flow at gauge j
AlGA_j	Allowable flow at gauge j
RuGA_{ij}	Information with regards to the hands off flow condition which equals one if user i abstraction is controlled with the level of flow at gauge j
$Q_{95}\text{GA}_j$	Q_{95} flow level at gauge j
UpGA_{ij}	Agriculture user i upstream of gauge j
WaterUse_i^t	Water used by user i at time t including the abstraction and trading
sL_i^t	Selling limit for user i at time t
EWN_i	Historical expectation of water needs for user i

Appendix B: Formulation details

In this appendix we reproduce the formulation from Erfani et al. (2014) for reader convenience. Section headers specify which section of the current paper the equation relates to.

B1 Formulations for Sect. 2.1

The pair-wise trading model follows the multi-commodity modelling framework with an extra index k on the flow variable to represent water ownership (Erfani et al., 2013). The objective function of the model is

$$\text{NetBenefit} = \sum_{i \in \text{User}} \text{totalBenefit}_i$$
$$- \sum_{i \in \text{User}} \text{totalCost}_i - \sum_{j \in \text{Reservoir}} \text{Penalty}_j, \quad \text{(B1)}$$

subject to the following mass balance constraints:

$$\sum_{\substack{j \\ \text{CO}_{ji}=1}} x_{ji}^k + \text{inFl}_i + \sum_{l \in \text{User}} \text{ReturnFlow}_{li} = \sum_{\substack{j \\ \text{CO}_{ij}=1}} x_{ij}^k \quad \text{(B2)}$$

$$\forall i \in \text{Junction} k \in \text{Owner}, \text{DW}_{li} = 1, \quad \text{(B3)}$$

$$\sum_{\substack{i \\ \text{CO}_{ij}=1}} x_{ij}^k + \text{pRes}_j^k = \text{Res}_j^k + \sum_{\substack{i \\ \text{CO}_{ji}=1}} x_{ji}^k \forall j \in \quad \text{(B4)}$$

$$\text{Reservoir} k \in \text{Owner}, \quad \text{(B5)}$$

$$\sum_{\substack{j \\ \text{CO}_{ji}=1}} x_{ji}^k = \sum_{\substack{j \\ \text{CO}_{ij}=1}} x_{ij}^k + \text{WaterAbstracted}_i^k + \text{Trade}_i^{k \in \text{river}} \quad \text{(B6)}$$

$$\forall i \in \text{User} k \in \text{Owner}, \quad \text{(B7)}$$

$$\text{Trade}_i^{k \in \text{river}} = \sum_{\substack{j \\ \text{CO}_{ij}=1}} x_{ij}^i \quad \text{(B8)}$$

$$\forall i \in \text{User}, \quad \text{(B9)}$$

$$\text{ReturnFlow}_{ij} = \sum_{k \in \text{Owner}} (1 - \text{consFactor}_i) \, \text{WaterAbstracted}_i^k$$

$$\forall i \in \text{User}, \text{DW}_{ij} = 1, \quad \text{(B10)}$$

$$\sum_{\substack{k \in \text{river}}} \sum_{\substack{i \\ \text{CO}_{ij}=1}} x_{ij}^k = \text{Discharge}_j \quad \text{(B11)}$$

$$\forall j \in \text{Discharge}. \quad \text{(B12)}$$

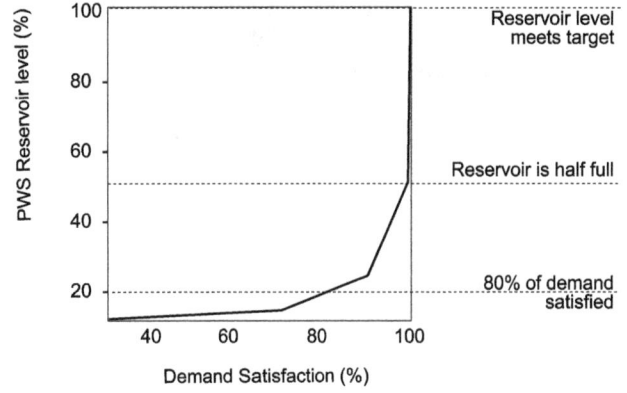

Figure B1. Public water supply company hedging rule.

B2 Formulations for Sect. 3.3.1

At the beginning of each week the river flow is checked and if the value is below the HoF limit, the license is suspended for the upcoming week (Erfani et al., 2014). This is imposed using the following constraint:

$$(\text{flGA}_j \leq \text{AlGA}_j) \longrightarrow \left(\text{WaterAbstracted}_i^{k \in \text{river}} + \text{Trade}_i^{k \in \text{river}} = 0\right)$$

$$\forall i \in \text{user}, j \in \text{Gauge}, \text{RuGA}_{ij} = 1. \quad \text{(B13)}$$

In addition, the 50 % rationing is imposed on farmers using the following set of check constraints:

$$(\text{flGA}_j \leq \text{Q95GA}_j) \longrightarrow \left(\text{WaterAbstracted}_i^{k \in \text{river}} + \text{Trade}_i^{k \in \text{river}} \leq 0.5 \times \text{WkLi}_i\right)$$

$$\forall i \in \text{Agriculture}, j \in \text{Gauge}. \quad \text{(B14)}$$

For the PWS reservoir, the volumetric capacity constraint is as follows:

$$2627 \leq \sum_{k \in \text{Owner}} \text{Res}_j^k \leq 55\,450, \quad \text{(B15)}$$

$$\forall j \in \text{Reservoir}. \quad \text{(B16)}$$

The hedging constraint for water company demand reduction is represented by

$$\text{WaterAbstracted}_i^j = F \left(\sum_{k \in \text{Owner}} \text{Res}_j^k \right), \quad \text{(B17)}$$

$$\forall i \in \text{user} j \in \text{Reservoir} \text{CO}_{li} = 1, \quad \text{(B18)}$$

where $F(.)$ is the function shown in Fig. B1 which represents the relationship between the reservoir level, as a percentage of the target, and the proportion of demand that is satisfied.

B3 Formulations for Sect. 3.3.2

At each weekly time period t of the modelling, agricultural willingness to sell their license is represented using

$$\text{Trade}_i^{k \in \text{river}} \leq \max 0, \text{sL}_i^t$$

$$\forall i \in \text{user}, \quad \text{(B19)}$$

This limit (sL) applies until the farmer abstracts a proportion c_i of their expected water needs (EWN_i) which is based on their historical yearly water use. For each user i,

$$sL_i^t = \begin{cases} sL_i^{t-1} - \text{WaterUse}_i^t, & \text{Sum of abstraction up to time } t \leq c_i \times \text{EWN}_i, \\ \text{WkLi}_i, & \text{otherwise} \end{cases} \qquad (B20)$$

where WaterUse is the sum of water diverted and sold, c_i is a value ranging from 0 to 1, and

$$sL_i^0 = \text{YrLi}_i - c_i \times \text{EWN}_i. \qquad (B21)$$

Acknowledgements. This work was undertaken within the "Transforming Water Scarcity though Trade" (TWSTT) project funded by the UK Engineering and Physical Science Research Council (EPSRC grant EP/J005274/1). The research was partially funded by Anglian Water and DEFRA through the Cambridge Programme for Sustainability Leadership (CPSL) and by the "Valuing Water" RG07 project funded by UK Water Industry Research (UKWIR). Hydrological data and software assistance was provided by John F. Raffensberger, Ralph Ledbetter and George A. F. Woolhouse. We thank them and other colleagues who contributed to this work including Jean Spencer, Martin Silcock, Henry Leveson-Gower, Keith Weatherhead, Mike Young, Darren Lumbroso, Steven Wade, and others. An anonymous reviewer provided comments that improved the paper. All errors or omissions are the authors' alone.

Edited by: A. Guadagnini

References

Acreman, M.: Linking science and decision-making: features and experience from environmental river flow setting, Environ. Modell. Softw., 20, 99–109, doi:10.1016/j.envsoft.2003.08.019, 2005.

Acreman, M. C. and Dunbar, M. J.: Defining environmental river flow requirements – a review, Hydrol. Earth Syst. Sci., 8, 861–876, doi:10.5194/hess-8-861-2004, 2004.

Acreman, M. C. and Ferguson, A. J. D.: Environmental flows and the European Water Framework Directive, Freshwater Biol., 55, 32–48, doi:10.1111/j.1365-2427.2009.02181.x, 2010.

Arthington, A. H., Bunn, S. E., Poff, N. L., and Naiman, R. J.: The challenge of providing environmental flow rules to sustain river ecosystems, Ecol. Appl., 16, 1311–1318, 2006.

Bjornlund, H.: Efficient water market mechanism to cope with water scarcity, Int. J. Water Resour. Develop., 19, 553–567, 2003.

Brewer, J., Glennon, R., Ker, A., and Libecap, G.: 2006 Presidential Address Water Markets in the West: Prices, Trading, and Contractual Forms, Econ. Inquiry, 46, 91–112, 2008.

Britz, W., Ferris, M., and Kuhn, A.: Modeling water allocating institutions based on Multiple Optimization Problems with Equilibrium Constraints, Environ. Modell. Softw., 46, 196–207, doi:10.1016/j.envsoft.2013.03.010, 2013.

Cheng, W.-C., Hsu, N.-S., Cheng, W.-M., and Yeh, W. W. G.: A flow path model for regional water distribution optimization, Water Resour. Res., 45, W09411, doi:10.1029/2009WR007826, 2009.

Defra: Water for life, TSO, Norwich, UK, 2011.

Draper, A., Jenkins, M., Kirby, K., Lund, J., and Howitt, R.: Economic-Engineering Optimization for California Water Management, J. Water Resour. Pl. Manage., 129, 155–164, doi:10.1061/(ASCE)0733-9496(2003)129:3(155), 2003.

Environment Agency: The Upper Ouse and Bedford Ouse Catchment Abstraction Management Strategy, Environment Agency, Bristol, UK, March 2005.

Environment Agency: Water resources in England and Wales – current state and future pressures, Environment Agency, Bristol, UK, December 2008.

Environment Agency: The case for change – current and future water availability, Bristol, UK, 2011.

Environment Agency: Environmental Flow Indicator, edited by: Environment Agency, Bristol, UK, 2013.

Environment Agency and Ofwat: Review of barriers to water rights trading, London, UK, 2008.

Environment Agency and Ofwat: The case for change: Reforming water abstraction management in England, Environment Agency, Bristol, UK, 2012.

Erfani, T., Huskova, I., and Harou, J. J. Tracking trade transactions in water resource systems: A node-arc optimization formulation, Water Resour. Res., 49, 3038–3043, doi:10.1002/wrcr.20211, 2013.

Erfani, T., Binions, O., and Harou, J. J., Simulating water markets with transaction costs, Water Resour. Res., 50, 4726–4745, doi:10.1002/2013WR014493, 2014.

Garrod, G. D. and Willis, K. G.: Estimating the Benefits of Environmental Enhancement: A Case Study of the River Darent, J. Environ. Pl. Manage., 39, 189–203, 1996.

Gleick, P., Allen, L., Christian-Smith, J., Cohen, M. J., Cooley, H. R., Heberger, M., Morrison, J., Palaniappan, M., and Schulte, P.: The World's Water, Volume 7, The Biennial Report on Freshwater Resources, Island Press, Washington, D.C., 2011.

Grafton, R. Q., Libecap, G., McGlennon, S., Landry, C., and O'Brien, B.: An integrated assessment of water markets: a cross-country comparison, Rev. Environ. Econ. Policy, 5, 219–239, 2011.

Harou, J. J., Pulido-Velazquez, M., Rosenberg, D. E., Medellín-Azuara, J., Lund, J. R., and Howitt, R. E.: Hydro-economic models: Concepts, design, applications, and future prospects, J. Hydrol., 375, 627–643, doi:10.1016/j.jhydrol.2009.06.037, 2009.

Hodgson, S.: Modern water rights: Theory and practice, Food and Agriculture Organization of the United Nations, Rome, 2006.

Howe, C. W., Schurmeier, D. R., and Shaw, W. D.: Innovative Approaches to Water Allocation: The Potential for Water Markets, Water Resour. Res., 22, 439–445, doi:10.1029/WR022i004p00439, 1986.

Jakeman, A. J. and Letcher, R. A.: Integrated assessment and modelling: features, principles and examples for catchment management, Environ. Modell. Softw., 18, 491–501, doi:10.1016/S1364-8152(03)00024-0, 2003.

Katz, D.: Water markets and environmental flows in theory and in practice, Ecol. Appl., 12, 1247–1260, 2011.

Klaar, M. J., Dunbar, M. J., Warren, M., and Soley, R.: Developing hydroecological models to inform environmental flow standards: a case study from England, Wiley Interdisciplinary Reviews: Water, 1, 207–217, doi:10.1002/wat2.1012, 2014.

Libecap, G., Grafton, R. Q., Landry, C., O'Brien, R. J., and Edwards, E. C.: Water Scarcity and Water Markets: A Comparison of Institutions and Practices in the Murray-Darling Basin of Australia and the Western US, ICER Working Paper No. 28/2010, 2010.

Loucks, D. P., Stedinger, J. R., and Haith, D. A.: Water resources systems planning and analysis, Prentice-Hal, Englewood Cliffs, NJ, 1981.

Loucks, D. P., Van Beek, E., Stedinger, J. R., Dijkman, J. P., and Villars, M. T.: Water resources systems planning and management: an introduction to methods, models and applications, UNESCO, Paris, 2005.

NERA: The Benefits of Water Framework Directive Programmes of Measures in England and Wales, A Final Report to DEFRA re CRP Project 4b/c, London, UK, 2007.

Poff, N. L., Allan, J. D., Bain, M. B., Karr, J. R., Prestegaard, K. L., Richter, B. D., Sparks, R. E., and Stromberg, J. C.: The Natural Flow Regime, BioScience, 47, 769–784, doi:10.2307/1313099, 1997.

Postel, S. L.: Water for Food Production: Will There Be Enough in 2025?, BioScience, 48, 629–637, doi:10.2307/1313422, 1998.

Postel, S. L., Daily, G. C., and Ehrlich, P. R.: Human appropriation of renewable fresh water, Science-AAAS-Weekly Paper Edition, 271, 785–787, 1996.

Richter, B. D., Davis, M. M., Apse, C., and Konrad, C.: A Presumptive standard for environmental flow protection, River Res. Appl., 28, 1312–1321, doi:10.1002/rra.1511, 2012.

Smakhtin, V. U., Revenga, C., and Döll, P.: Taking into account environmental water requirements in global-scale water resources assessments, Comprehensive Assessment Secretariat, Colombo, Sri Lanka, 2004.

Stern, J.: Water Rights and Water Trading in England and Wales, The Foundation for Law, Justice and Society, Oxford, UK, 2013.

Syme, G. J., Nancarrow, B. E., and McCreddin, J. A.: Defining the components of fairness in the allocation of water to environmental and human uses, J. Environ. Manage., 57, 51–70, doi:10.1006/jema.1999.0282, 1999.

Tharme, R. E.: A global perspective on environmental flow assessment: emerging trends in the development and application of environmental flow methodologies for rivers, River Res. Appl., 19, 397–441, doi:10.1002/rra.736, 2003.

Thobanl, M.: Formal Water Markets: Why, When, and How to Introduce Tradable Water Rights, World Bank Res. Observ., 12, 161–179, doi:10.1093/wbro/12.2.161, 1997.

Ward, F., Booker, J., and Michelsen, A.: Integrated Economic, Hydrologic, and Institutional Analysis of Policy Responses to Mitigate Drought Impacts in Rio Grande Basin, J. Water Resour. Pl. Manage., 132, 488–502, doi:10.1061/(ASCE)0733-9496(2006)132:6(488), 2006.

Wheeler, S., Garrick, D., Loch, A., and Bjornlund, H.: Evaluating water market products to acquire water for the environment in Australia, Land Use Policy, 30, 427–436, doi:10.1016/j.landusepol.2012.04.004, 2013.

Wilkinson, M.: Farmers won't go with flow, in: Sydney Morning Herald, Sydney, 2008.

Young, M.: Towards a Generic Framework for the Abstraction and Utilisation of Water in England and Wales, UCL Environment Institute Report, UCL Environment Institute, London, UK, 2012.

Young, M. and McColl, J.: Defining Tradable Water Entitlements and Allocations: A Robust System, Can. Water Resour. J., 30, 65–72, doi:10.4296/cwrj300165, 2005.

6

Correction of systematic model forcing bias of CLM using assimilation of cosmic-ray Neutrons and land surface temperature: a study in the Heihe Catchment, China

X. Han[1,2,3], **H.-J. H. Franssen**[2,3], **R. Rosolem**[4], **R. Jin**[1,5], **X. Li**[1,5], **and H. Vereecken**[2,3]

[1]Key Laboratory of Remote Sensing of Gansu Province, Cold and Arid Regions Environmental and Engineering Research Institute, Chinese Academy of Sciences, Lanzhou 730000, PR China
[2]Forschungszentrum Jülich, Agrosphere (IBG 3), Leo-Brandt-Strasse, 52425 Jülich, Germany
[3]Centre for High-Performance Scientific Computing in Terrestrial Systems: HPSC TerrSys, Geoverbund ABC/J, Leo-Brandt-Strasse, 52425 Jülich, Germany
[4]Department of Civil Engineering, University of Bristol, Bristol BS8 1TR, UK
[5]CAS Center for Excellence in Tibetan Plateau Earth Sciences, Chinese Academy of Sciences, Beijing 100101, PR China

Correspondence to: X. Han (hanxj@lzb.ac.cn)

Abstract. The recent development of the non-invasive cosmic-ray soil moisture sensing technique fills the gap between point-scale soil moisture measurements and regional-scale soil moisture measurements by remote sensing. A cosmic-ray probe measures soil moisture for a footprint with a diameter of ~ 600 m (at sea level) and with an effective measurement depth between 12 and 76 cm, depending on the soil humidity. In this study, it was tested whether neutron counts also allow correcting for a systematic error in the model forcings. A lack of water management data often causes systematic input errors to land surface models. Here, the assimilation procedure was tested for an irrigated corn field (Heihe Watershed Allied Telemetry Experimental Research – HiWATER, 2012) where no irrigation data were available as model input although for the area a significant amount of water was irrigated. In the study, the measured cosmic-ray neutron counts and Moderate-Resolution Imaging Spectroradiometer (MODIS) land surface temperature (LST) products were jointly assimilated into the Community Land Model (CLM) with the local ensemble transform Kalman filter. Different data assimilation scenarios were evaluated, with assimilation of LST and/or cosmic-ray neutron counts, and possibly parameter estimation of leaf area index (LAI). The results show that the direct assimilation of cosmic-ray neutron counts can improve the soil moisture and evapotranspiration (ET) estimation significantly, correcting for lack of information on irrigation amounts. The joint assimilation of neutron counts and LST could improve further the ET estimation, but the information content of neutron counts exceeded the one of LST. Additional improvement was achieved by calibrating LAI, which after calibration was also closer to independent field measurements. It was concluded that assimilation of neutron counts was useful for ET and soil moisture estimation even if the model has a systematic bias like neglecting irrigation. However, also the assimilation of LST helped to correct the systematic model bias introduced by neglecting irrigation and LST could be used to update soil moisture with state augmentation.

1 Introduction

Soil moisture plays a key role for crop and plant growth, water resources management and land surface–atmosphere interaction. Therefore accurate soil moisture retrieval is important. Point-scale measurements can be obtained by methods like time domain reflectometry (TDR) (Robinson et al., 2003) and larger-scale, coarse soil moisture information from remote sensing sensors (Entekhabi et al., 2010; Kerr et al., 2010). Wireless sensor networks (WSNs) allow characterization of soil moisture at the catchment scale with many local connected sensors at separated locations (Bogena et al., 2010). TDR only measures the point-scale soil moisture, and the maintenance of WSN is expensive. Recently, neutron count intensity measured by aboveground cosmic-ray probes was proposed as an alternative information source on soil moisture. Neutron count intensity is measured non-invasively at an intermediate scale between the point-scale and the coarse remote sensing scale (Zreda et al., 2008). A network of cosmic-ray sensors (CRSs) has been set up over North America (Zreda et al., 2012).

Cosmic rays are composed of primary protons mainly. The fast neutrons generated by high-energy neutrons colliding with nuclei lead to "evaporation" of fast neutrons, and the generated and moderated neutrons in the ground can diffuse back into the air, where their intensity can be measured by the cosmic-ray soil moisture probe. Soil moisture affects the rate of moderation of fast neutrons and controls the neutron concentration and the emission of neutrons into the air. Dry soils have low moderating power and are highly emissive; wet soils have high moderating power and are less emissive. The neutrons are mainly moderated by the hydrogen atoms contained in the soil water and emitted to the atmosphere, where the neutrons mix instantaneously at a scale of hundreds of meters. The measurement area of a cosmic-ray soil moisture probe represents a circle with a diameter of ~ 600 m at sea level (Desilets and Zreda, 2013), and the measurement depth decreases nonlinearly from ~ 76 (dry soils) to ~ 12 cm (saturated soils) (Zreda et al., 2008). The measured cosmic-ray neutron counts show an inverse correlation with soil moisture content. The cosmic-ray neutron intensity could be reduced to 60 % of surface cosmic-ray neutron intensity by increasing the soil moisture from 0 to 40 % (Zreda et al., 2008). The soil moisture estimation on the basis of cosmic-ray-probe-based neutron counts over a horizontal footprint of hectometers has received considerable attention in the scientific literature in recent years (Desilets et al., 2010; Zreda et al., 2008, 2012).

Hydrogen atoms are present as water in the soil, lattice soil water, belowground biomass, atmospheric water vapor, snow water, aboveground biomass, intercepted water by vegetation and water on the ground. These additional hydrogen sources contribute to the measured neutron intensity. The role of these additional hydrogen sources should be included in the analysis of the cosmic-ray measurements in order to isolate the main contribution from soil moisture. Formulations for handling water vapor (Rosolem et al., 2013), for lattice water and organic carbon (Franz et al., 2013) and for a litter layer present on the soil surface (Bogena et al., 2013) have been developed.

The positive impact of soil moisture data assimilation has been shown in several studies. Importantly, surface soil moisture could be used to obtain better characterization of the root zone soil moisture (Barrett and Renzullo, 2009; Crow et al., 2008; Das et al., 2008; Draper et al., 2011; Li et al., 2010). It has also shown that the assimilation of soil moisture observations can be used to correct rainfall errors (Crow et al., 2011; Yang et al., 2009). Often a systematic bias between measured and modeled soil moisture content can be found; soil moisture estimation can be significantly improved using joint state and bias estimation (De Lannoy et al., 2007; Kumar et al., 2012; Reichle, 2008). Also studies on data assimilation of remotely sensed land surface temperature products show a positive impact on the estimation of soil moisture, latent heat flux and sensible heat flux (Ghent et al., 2010; Xu et al., 2011). Also in these studies it was found that bias, in these cases soil temperature bias, of land surface models can be removed with land surface temperature assimilation (Bosilovich et al., 2007; Reichle et al., 2010). Other studies have updated both land surface model states and parameters with soil moisture and land surface temperature data (Bateni and Entekhabi, 2012; Han et al., 2014a; Montzka et al., 2013; Pauwels et al., 2009). The assimilation of measured cosmic-ray neutron counts in a land surface model was successfully tested, but these studies focused on state updating alone (Rosolem et al., 2014; Shuttleworth et al., 2013). In this paper we focus on the assimilation of measured cosmic-ray neutron counts for improving soil moisture content characterization at the field scale. This paper focuses on the case of model input being biased. Land surface models still are affected by limited knowledge on water resources management, and for regions in China (and elsewhere) typically no information on irrigation amounts is available as irrigation is mainly by the flooding system. We analyze whether measured neutron counts are able to correct for such biases. This case is not only relevant for neglecting irrigation in China, but also for other water resources management issues (e.g., groundwater pumping) which are neglected in the simulations. Neglecting irrigation in land surface models results in a large bias in the simulated soil moisture content because of a lack of water input. The bias in soil moisture content also results in a too-small latent heat flux and too-high sensible heat flux. We hypothesize that data assimilation also can play an important role for removing such biases in data-deficient areas. One possible strategy in data assimilation studies for handling this type of bias, which is not followed in this paper, is to calibrate the simulation model (e.g., land surface model) prior to data assimilation to remove biases (Kumar et al., 2012) and use the corrected simulation model in the context of sequential data assimilation. A different strategy was followed in this paper, and no a priori bias correction was carried out because this type of problem (neglecting water resources management) does not allow for such an a priori bias correction. The bias can be attributed to the model structure, model parameters, atmospheric forcing or observation data, and the bias-aware assimilation requires the assumption that the bias comes from a particular source. If the source of bias is not attributed to the right source, model predictions cannot be improved (Dee, 2005). Therefore bias-blind assimilation was used for safety, and the bias estimation was not handled explicitly. Instead, we investigated whether neutron counts measured by cosmic-ray probe were able to correct for the bias. The aim is to improve the soil moisture profile estimation in a crop land with seed corn as the main crop type.

In CLM, land surface fluxes are calculated based on the Monin–Obukhov similarity theory. The sensible heat flux is formulated as a function of temperature and leaf area index

(LAI), and the latent heat flux is formulated as a function of the temperature and leaf stomatal resistances. The leaf stomatal resistance is calculated from the Ball–Berry conductance model (Collatz et al., 1991). The updates of soil temperature and vegetation temperature are derived based on the solar radiation absorbed by top soil (or vegetation), longwave radiation absorbed by soil (or vegetation), sensible heat flux from soil (or vegetation) and latent heat flux from soil (or vegetation). Measured land surface temperature is composed of the ground temperature and vegetation temperature. Therefore a difference between measured and calculated land surface temperature can be adjusted by changing land surface fluxes. As land surface fluxes are sensitive to soil moisture content, land surface temperature is sensitive to soil moisture content.

Therefore, the land surface temperature (LST) products measured by the Moderate-Resolution Imaging Spectroradiometer (MODIS) Terra (MOD11A1) and Aqua (MYD11A1) are also assimilated jointly to improve the soil temperature profile estimation because the evapotranspiration (ET) is sensitive to the soil temperature. Two Terra LST products can be obtained per day at 10:30/22:30 and two Aqua LST products can be obtained per day at 01:30/13:30. Soil moisture, land surface temperature and LAI influence the estimation of latent and sensible heat fluxes (Ghilain et al., 2012; Jarlan et al., 2008; Schwinger et al., 2010; van den Hurk, 2003; Yang et al., 1999), and therefore this study also focused on the calibration of LAI with the help of the assimilation of land surface temperature. However, there are large discrepancies between the remotely retrieved LAI and measured values, and the MODIS LAI product underestimates in situ measured LAI by 44 % on average (http: //landval.gsfc.nasa.gov/), and therefore the LAI is also calibrated by data assimilation. In summary, the novel aspects of this work are the following: (1) investigating whether data assimilation is able to correct for missing water resources management data without a priori bias correction; (2) joint assimilation of cosmic-ray neutron counts, LST and updating of LAI; and (3) application of this framework to real-world data in an irrigated area where detailed verification data were available.

2   Materials and methods

2.1   Study area and measurement

The Heihe River basin is the second-largest inland river basin of China; it is located at 97.1–102.0° E and 37.7–42.7° N and covers an area of approximately 143 000 km^2 (Li et al., 2013). In 2012, a multi-scale observation experiment of evapotranspiration with a well-equipped superstation (Daman superstation) to measure the atmospheric forcings and soil moisture at 2, 4, 10, 20, 40, 80, 120 and 160 cm depth (Xu et al., 2013) was carried out from June to September in

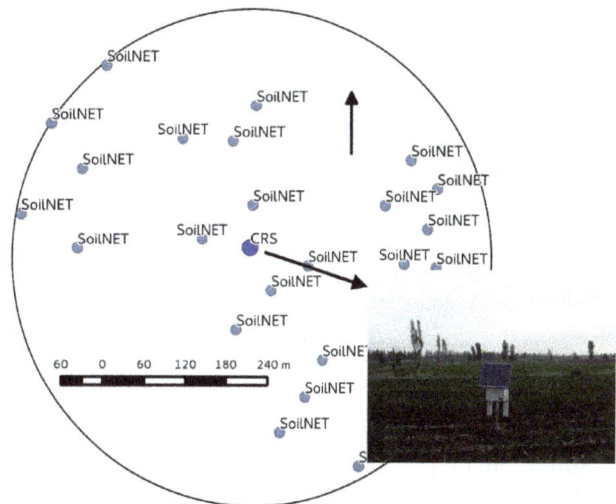

Figure 1. Map of the cosmic-ray probe and SoilNet nodes in the footprint of the CRS probe positioned at the Heihe River catchment.

the framework of the Heihe Watershed Allied Telemetry Experimental Research (HiWATER) (Li et al., 2013). SoilNet wireless network nodes (Bogena et al., 2010) were deployed to measure soil moisture content and soil temperature at four layers (4, 10, 20 and 40 cm). One cosmic-ray soil moisture probe (CRS-1000B) was installed (Han et al., 2014b) with 23 SoilNet nodes (Jin et al., 2013, 2014) in the footprint (Fig. 1). The main crop type within the footprint of the cosmic-ray probe is seed corn. The irrigation is applied through channels using the flooding irrigation method. Exact amounts of applied irrigation are therefore not available.

The measured cosmic-ray neutron count data were processed to remove the outliers according to the sensor voltage (≤ 11.8 Volt) and relative humidity (≥ 80 %) (Zreda et al., 2012). The surface fluxes were measured using the eddy covariance technique, and data were processed using EdiRe (http://www.geos.ed.ac.uk/abs/research/micromet/EdiRe) software, in which the anemometer coordinate rotation, signal lag removal, frequency response correction, density corrections and signal de-spiking were done for the raw data. The energy balance closure was not considered in this study. The LAI was measured by the LAI-2000 scanner during the field experiment; there are 17 samples collected on 14 days over 3 months.

2.2   Land surface model and data

The CLM was used to simulate the spatiotemporal distribution of soil moisture, soil temperature, land surface temperature, vegetation temperature, sensible heat flux, latent heat flux and soil heat flux of the study area. The coupled water and energy balance are modeled in CLM, and the land surface heterogeneity is represented by patched plant functional types and soil texture (Oleson et al., 2013).

The soil properties used in CLM were from the soil database of China with 1 km spatial resolution (Shangguan et al., 2013). The MODIS 500 m resolution plant functional type product (MCD12Q1) (Sun et al., 2008), which was resampled by nearest-neighbor interpolation to 1 km resolution, and the MODIS LAI product (MCD15A3) with 1 km spatial resolution (Han et al., 2012) were used as input. Due to a lack of measurement data, two atmospheric forcing data sets were used: the Global Land Data Assimilation System reanalysis data (Rodell et al., 2004) was interpolated using the National Centers for Environmental Prediction (NCEP) bilinear interpolation library iplib in spatial and temporal dimensions and used in the CLM for the spin-up period (http://www.nco.ncep.noaa.gov/pmb/docs/libs/iplib/ncep_iplib.shtml). For the 3-month data assimilation period, hourly forcing data (incident longwave radiation, incident solar radiation, precipitation, air pressure, specific humidity, air temperature and wind speed) from the Daman superstation of HiWATER were available and used.

2.3 Cosmic-ray forward model

In this study, the newly developed COsmic-ray Soil Moisture Interaction Code (COSMIC) model (Shuttleworth et al., 2013) was used as the cosmic-ray forward model to simulate the cosmic-ray neutron count rate using the soil moisture profile as input. The effective measurement depth of the cosmic-ray soil moisture probe ranges from 12 cm (wet soils) to 76 cm (dry soils) (Zreda et al., 2008), within which 86 % of the aboveground measured neutrons originate. COSMIC also calculates the effective sensor depth based on the cosmic-ray neutron intensity and the soil moisture profile values (Franz et al., 2012; Shuttleworth et al., 2013).

COSMIC makes several assumptions to calculate the number of fast neutrons reaching the cosmic-ray soil moisture probe (N_{COSMOS}) at a near-surface measurement location. The soil layer with a depth of 3 m for the complete soil profile was discretized into 300 layers for the integration of Eq. (2) in COSMIC. The number of fast neutrons reaching the cosmic-ray probe N_{COSMOS} is formulated as (Shuttleworth et al., 2013)

$$N_{COSMOS} = N \int_0^\infty \{A(z)[\alpha \rho_s(z) + \rho_w(z)] \tag{1}$$

$$\exp\left(-\left[\frac{m_s(z)}{L_1} + \frac{m_w(z)}{L_2}\right]\right)\} dz,$$

$$A(z) = \left(\frac{2}{\pi}\right) \int_0^{\pi/2} \exp\left(\frac{-1}{\cos(\theta)}\left[\frac{m_s(z)}{L_3} + \frac{m_w(z)}{L_4}\right]\right) d\theta, \tag{2}$$

$$\alpha = 0.405 - 0.102\rho_s, \tag{3}$$

$$L_3 = -31.76 + 99.38\rho_s, \tag{4}$$

where N is the high-energy neutron flux; z denotes the soil layer depth (m); ρ_s the dry soil bulk density ($g\,cm^{-3}$); ρ_w the total water density, including the lattice water ($g\,cm^{-3}$); and α denotes the ratio of fast-neutron creation factor. L_1 is the high-energy soil attenuation length with value of 162.0 $g\,cm^{-2}$ and L_2 the high-energy water attenuation length of 129.1 $g\,cm^{-2}$. In Eq. (2) θ is the angle between the vertical below the detector and the line between the detector and each point in the plane; $m_s(z)$ and $m_w(z)$ are the integrated mass per unit area of dry soil and water ($g\,cm^{-2}$), respectively. L_3 denotes the fast-neutron soil attenuation length ($g\,cm^{-2}$), and L_4 stands for the fast-neutron water attenuation length with a value of 3.16 $g\,cm^{-2}$.

The cosmic-ray neutron intensity reaching the land surface is influenced by air pressure, atmospheric water vapor content and incoming neutron flux. In order to isolate the contribution of soil moisture content to the measured neutron density, it is important to take these effects into account, and the calibrated neutron count intensity can be derived as follows

$$N_{Corr} = N_{Obs}\, f_P\, f_{wv}\, f_i, \tag{5}$$

where N_{Corr} represents corrected neutron counts and N_{Obs} the measured neutron counts. f_P is the correction factor for air pressure, f_{wv} the correction factor for atmospheric water vapor and f_i the correction factor for incoming neutron flux.

The correction factor for air pressure f_P can be calculated as (Zreda et al., 2012)

$$f_P = \exp(\frac{P - P_0}{L}), \tag{6}$$

where P (mbar) is the local air pressure, P_0 (mbar) the average air pressure during the measurement period and L ($g\,cm^{-2}$) is the mass attenuation length for high-energy neutrons; the default value of 128 $g\,cm^{-2}$ was used in this study (Zreda et al., 2012).

The correction factor f_{wv} for atmospheric water vapor is calculated as (Rosolem et al., 2013)

$$f_{wv} = 1 + 0.0054(\rho_{v0} - \rho_{v0}^{ref}), \tag{7}$$

where ρ_{v0} ($kg\,m^{-3}$) is the absolute humidity at the measurement time and ρ_{v0}^{ref} ($kg\,m^{-3}$) is the average absolute humidity during the measurement period.

Fluctuations in the incoming neutron flux should be removed because the cosmic-ray probe is designed to measure the neutron flux based on the incoming background neutron flux. The correcting factor f_i for the incoming neutron flux is calculated as

$$f_i = \frac{N_m}{N_{avg}}, \tag{8}$$

where N_m is the measured incoming neutron flux and N_{avg} is the average incoming neutron flux during the measurement period. The measured data at the Jungfraujoch station in Switzerland at 3560 m (http://cosray.unibe.ch/) were used to calculate N_m and N_{avg}. The temporal (secular or diurnal) variations caused by the sunspot cycle could be removed after this correction (Zreda et al., 2012).

In this study, the soil moisture for the CRS footprint scale was calculated from the arithmetic mean of the 23 Soil-Net soil moisture observations. The calibration of the high-energy neutron intensity parameter N in Eq. (1) was done using the measured cosmic-ray neutron counts rate and averaged soil moisture content at the CRS footprint scale. Because lattice water was unknown for this site, a value of 3 % was assumed in this study (Franz et al., 2012). Hourly soil moisture measurements for a period of 2.5 months were used for COSMIC calibration. Inside the cosmic-ray probe footprint, the amount of applied irrigation was spatially variable due to the different management practice of each farmer. The gradient search algorithm L-BFGS-B (Zhu et al., 1997) was used to minimize the root mean square error (RMSE) of the differences between simulated cosmic-ray neutron counts (using measured soil moisture by SoilNet as input to COSMIC) and the measured neutron counts N_{Corr}. The optimized parameter value of N was 615.96 counts h^{-1} in this case.

The simulated soil moisture content for 10 CLM soil layers (3.8 m depth) was used as input to COSMIC in order to simulate the corresponding neutron count intensity and compare it with the measured neutron count intensity. It should be mentioned that it is unlikely that anything beyond 1 m depth will substantially impact the results because the effective measurement depth of the cosmic-ray probe is between 12 and 76 cm. The COSMIC model assumes a more detailed soil profile. COSMIC interpolates the soil moisture information from the 10 CLM soil layers to information for 300 soil layers of 1 cm depth. The contribution of each soil layer to the measured neutron flux will change temporally depending on the soil moisture condition. Therefore the effective measurement depth of the cosmic ray probe will also change temporally. COSMIC calculates the vertically weighted soil moisture content based on the vertical distribution of soil moisture content.

2.4 Two-source formulation – TSF

The land surface temperature products of MODIS are composed of a ground temperature and vegetation temperature component, which are however unknown. CLM models the ground temperature and vegetation temperature separately, but it does not model the composed land surface temperature as seen by MODIS. The corresponding land surface temperature of CLM should therefore be modeled for data assimilation purposes. The two-source formulation (Kustas and Anderson, 2009) was used in this study to calculate the land surface temperature from the MODIS view angle using ground

temperature and vegetation temperature simulated by CLM:

$$T_s = [F_c(\Phi)T_c^4 + (1 - F_c(\Phi)T_g^4)]^{1/4}, \tag{9}$$

where T_S (K) is the composed surface temperature as seen by the MODIS sensor, $F_c(\Phi)$ is the fraction vegetation cover observed from the sensor view angle Φ (radians), T_c (K) is the vegetation temperature and T_g (K) is the ground temperature (Kustas and Anderson, 2009):

$$F_c(\Phi) = 1 - \exp\left(\frac{-0.5\Omega(\Phi)\text{LAI}}{\cos\Phi}\right), \tag{10}$$

where $\Omega(\Phi)$ is a clumping index to represent the nonrandom leaf area distributions of farmland or other heterogeneous land surfaces (Anderson et al., 2005); it is defined as

$$\Omega(\Phi) = \frac{0.49\Omega_{max}}{0.49 + (\Omega_{max} - 0.49)\exp(k\theta^{3.34})}, \tag{11}$$

$$\Omega_{max} = 0.49 + 0.51(\sin\Phi)^{0.05}, \tag{12}$$

$$k = -\{0.3 + [0.833(\sin\Phi)^{0.1}]^{14}\}. \tag{13}$$

2.5 Assimilation approach

The local ensemble transform Kalman filter (LETKF) was used as the assimilation algorithm, which is one of the square-root variants of the ensemble Kalman filter (Evensen, 2003; Hunt et al., 2007; Miyoshi and Yamane, 2007). The model uncertainties are represented using the ensemble simulation of model states, and LETKF derives the background error covariance using the model state ensemble members. LETKF uses the non-perturbed observations to update all the ensemble members of model states at each assimilation step.

In this study, $x_1^b, \ldots, x_N^b$ denote the model state ensemble members; $\bar{x}^b$ is the ensemble mean of $x_1^b, \ldots, x_N^b$; N is the ensemble size; $y_1^b, \ldots, y_N^b$ denote the mapped model state ensemble members; $\bar{y}^b$ is the ensemble mean of $y_1^b, \ldots, y_N^b$; and H is the observation operator (COSMIC for soil moisture or the two-source function for land surface temperature). The analysis step of LETKF can be summarized as follows.

Prepare the model state vector X^b:

$$X^b = [x_1^b - \bar{x}^b, \ldots, x_N^b - \bar{x}^b] \tag{14}$$

where $\bar{x}^b$ is composed of one vertically weighted soil moisture content and soil moisture content for 10 CLM layers, resulting in a state dimension equal to 11 if only the neutron count observation was assimilated; and $\bar{x}^b$ is composed of surface temperature, ground temperature, vegetation temperature and soil temperature for 15 CLM layers if only the land surface temperature observations were assimilated without soil moisture update, giving a state dimension of 18. The water and energy balance are coupled, and in CLM the energy balance is firstly solved; then the derived surface fluxes are used for updating soil moisture content. The cross correlation between the soil temperature and soil moisture can be calculated using the ensemble prediction in LETKF, and this makes the updating of soil moisture by assimilating land surface temperature possible. We also used the land surface temperature to update the soil moisture profile; in this case the soil moisture vector was augmented to the LETKF state vector of land surface temperature assimilation, resulting in a state dimension of 28.

Construct the mapped model state vector Y^b after transformation of observation operator:

$$y_i^b = H(x_i^b), \tag{15}$$

$$Y^b = \left[y_1^b - \overline{y}^b, \ldots, y_N^b - \overline{y}^b \right]. \tag{16}$$

The following analysis is looped for each model grid cell to calculate the update of model state ensemble members.

Calculate analysis error covariance matrix $\mathbf{P}^a$:

$$\mathbf{P}^a = [(N-1)\mathbf{I} + Y^{bT}\mathbf{R}^{-1}Y^b], \tag{17}$$

where $\mathbf{I}$ is the identity matrix.

The perturbations in ensemble space are calculated as

$$\mathbf{W}^a = [(N-1)\mathbf{P}^a]^{1/2}. \tag{18}$$

Calculate the analysis mean $\overline{w}^a$ in ensemble space and add to each column of W^a to get the analysis ensemble in ensemble space:

$$\overline{w}^a = \mathbf{P}^a Y^{bT}\mathbf{R}^{-1}(y^o - \overline{y}^b). \tag{19}$$

Calculate the new analysis:

$$X^a = X^b[\overline{w}^a + \mathbf{W}^a] + \overline{x}^b, \tag{20}$$

where $\mathbf{R}$ is the observation error covariance matrix, y^o is the observation vector and X^a contains the updated model ensemble members.

The LETKF method can also be extended to do parameter estimation using a state augmentation approach (Bateni and Entekhabi, 2012; Li and Ren, 2011; Moradkhani et al., 2005; Nie et al., 2011). Alternative strategies for parameter estimation are a dual approach (Moradkhani et al., 2005) with separate updating of states and parameters. Vrugt et al. (2005) also proposed a dual approach with parameter updating in an outer optimization loop using a Markov chain Monte Carlo method, and state updating in an inner loop. The a priori calibration of model parameters is also an option (Kumar et al., 2012). With the augmentation approach, the state vector of LETKF can be augmented by the parameter vector including soil properties (sand fraction, clay fraction and organic

matter density) and vegetation parameters (LAI, etc.). In a preliminary sensitivity study it was found that for this site simulation results were more sensitive to the LAI than to soil properties. Soil texture is also quite well known for this site from measurements. Therefore in this study, only the LAI was in some of the simulation scenarios calibrated. In the different scenarios of land surface temperature assimilation, the LETKF state vector was also augmented to include LAI as a calibration target. As a consequence, the augmented state vector contains surface temperature, ground temperature, vegetation temperature, 15 layers of soil temperature and LAI, making up a state dimension equal to 19 for the scenarios of land surface temperature assimilation without soil moisture update; for the scenarios of land surface temperature with soil moisture update, the state dimension is 29. The 10 layers of soil moisture and 15 layers of soil temperature are the standard CLM layout for both soil moisture and soil temperature. The hydrology calculations are done over the top 10 layers, and the bottom 5 layers are specified as bedrock. The lower 5 layers are hydrologically inactive layers. Temperature calculations are done over all layers (Oleson et al., 2013).

3 Experiment setup

First the 50 ensemble members of CLM with perturbed soil properties and atmospheric forcing data were driven from 1 January to 31 May 2012 to do the CLM spin-up; second an additional assimilation period of cosmic-ray neutron counts was done from 1 June to 30 August 2012 to reduce the spin-up error. The final CLM states on 30 August 2012 were used as the initial states for 1 June 2012 for the data assimilation scenarios. Perturbed soil properties were generated by adding a spatially uniform perturbation sampled from a uniform distribution between −10 and 10 % to the values extracted from the Soil Database of China for Land Surface Modeling (1 km spatial resolution). The LAI was perturbed with multiplicative uniform distributed random noise in the range of [0.8–1.2]. The perturbations added to the model forcings show correlations in space and time. The spatial correlation was induced by a fast Fourier transform, and the temporal correlation by a first-order auto-regressive model (Han et al., 2013; Kumar et al., 2009; Reichle et al., 2010). The statistics on the perturbation of the forcing data are summarized in Table 1. The values of standard deviations and temporal correlations in Table 1 were chosen based on previous catchment-scale and regional-scale data assimilation studies (De Lannoy et al., 2012; Kumar et al., 2012; Reichle et al., 2010).

The cosmic-ray neutron intensity was assimilated every 3 days at 12:00 Z from 1 June 2012 onwards. We found that the differences between daily assimilation and 3-day assimilation were small; therefore only the results of the 3-day assimilation are shown. The measured neutron count intensity showed large temporal fluctuations in time, and these

Table 1. Summary of perturbation parameters for atmospheric forcing data.

Variables	Noise	Standard deviation	Time correlation scale	Spatial correlation Scale	Cross correlation
Precipitation	Multiplicative	0.5	24 h	5 km	[1.0,-0.8, 0.5, 0.0,
Shortwave radiation	Multiplicative	0.3	24 h	5 km	-0.8, 1.0,-0.5, 0.4,
Longwave radiation	Additive	20 W m^{-2}	24 h	5 km	0.5, -0.5, 1.0, 0.4,
Air temperature	Additive	1 K	24 h	5 km	0.0, 0.4, 0.4, 1.0]

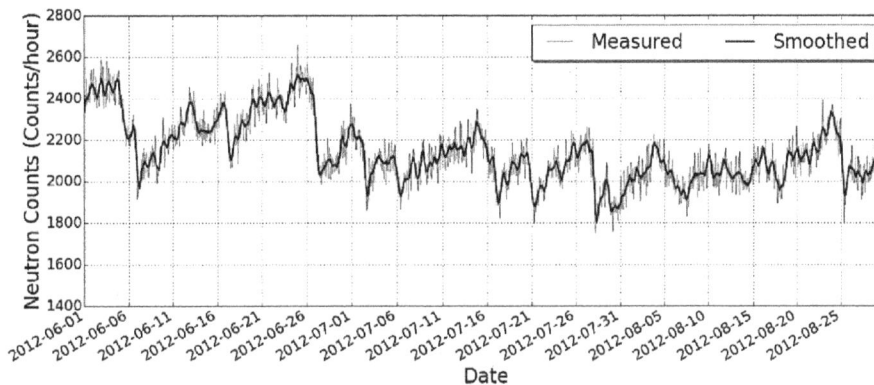

Figure 2. Measured and temporally smoothed CRS neutron counts.

fluctuations did not correspond to the temporal variations of soil moisture. Therefore the measured neutron count intensity was smoothed with the Savitzky–Golay filter using a moving average window of size 31 h and a polynomial of order 4 (Savitzky and Golay, 1964). The originally measured neutron counts and smoothed neutron counts are plotted in Fig. 2. The assimilation frequency of MODIS LST products of MOD11A1 and MYD11A1 was up to 4 times (maximum) per day depending on the data availability. There are 230 observation data (including cosmic-ray probe neutron counts, MODIS LST, MOD11A1 and MYD11A1 LST) in the whole assimilation window. The variance of the instantaneous measured neutron intensity is equal to the measured neutron count intensity (Zreda et al., 2012) and smaller for temporal averaging for daily or sub-daily applications. The instantaneous neutron intensity was assimilated in this study. The variance of MODIS LST was assumed to be 1 K (Wan and Li, 2008).

The 4-day MODIS LAI product was aggregated and used as the CLM LAI parameter. Because the LAI from MODIS is usually lower than the true value (compared with the field-measured LAI in the HiWATER experiment) and because the surface flux and surface temperature are sensitive to the LAI, two additional scenarios were investigated where LAI was calibrated to study the impact of LAI estimation on surface flux estimation within the data assimilation framework.

The following assimilation scenarios were compared:

1. CLM: open-loop simulation without assimilation.

2. Only_CRS: only the measured neutron counts were assimilated.

3. Only_LST: only the MODIS LST products were assimilated. The quality control flags of LST products were used to select the data with good quality for assimilation.

4. CRS_LST: the measured neutron counts and MODIS LST products were assimilated jointly. In the above scenarios, the neutron count data were used to update the soil moisture and the LST data were used to update the ground temperature, vegetation temperature and soil temperature.

5. LST_Feedback: we also evaluated the scenario of assimilating the LST measurements to update the soil moisture profile.

6. CRS_LST_Par_LAI: the LAI was included as variable to be calibrated; otherwise the scenario was the same as CRS_LST.

7. LST_Feedback_Par_LAI: the LAI was included as variable to be calibrated; otherwise the scenario was the same as LST_Feedback.

8. CRS_LST_True_LAI: the in situ measured LAI during the HiWATER experiment was used in the model simulation.

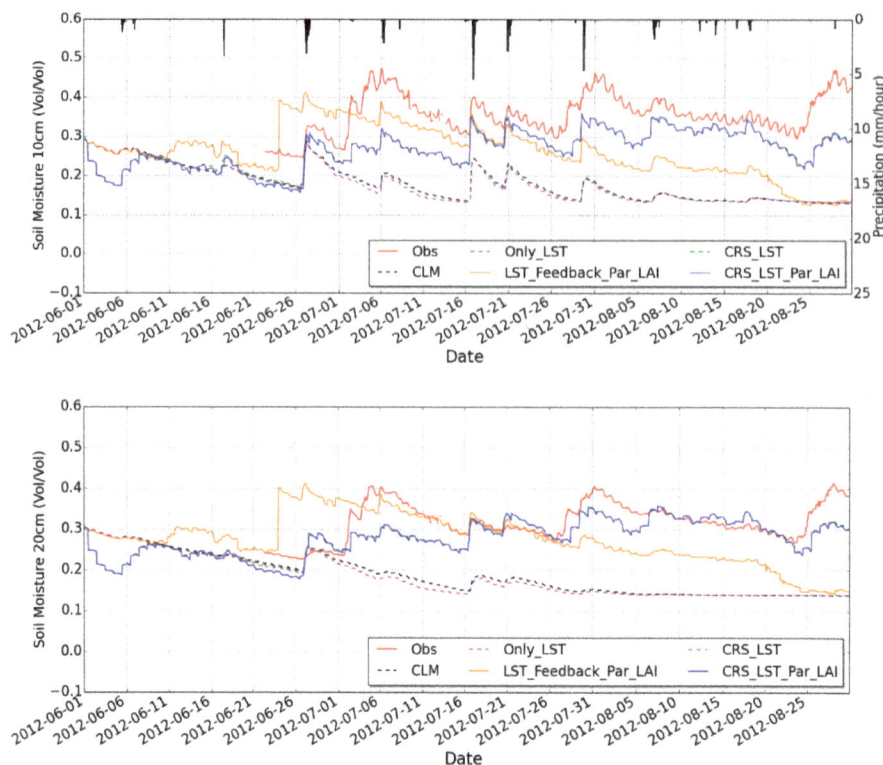

Figure 3. Soil moisture at 10 cm (upper) and 20 cm (lower) depth as obtained from an open-loop run (CLM), local sensors (Obs) and different simulation scenarios. For a description of the scenarios see Sect. 3 of the paper. The CRS neutron counts were assimilated on 1 June.

4   Results and discussion

In order to evaluate the assimilation results for the different scenarios outlined in Sect. 3, the RMSE was used:

$$\text{RMSE} = \sqrt{\frac{\sum_{n=i}^{N} (\text{estimated} - \text{measured})^2}{N}}, \qquad (21)$$

where "estimated" is the ensemble mean without assimilation or the ensemble mean after assimilation, and "measured" is measured soil moisture content evaluated at the SoilNet nodes (or latent heat flux, sensible heat flux or soil heat flux). N is the number of time steps. For the soil moisture analysis in this study, N is equal to 2184. The smaller the RMSE value is, the closer assimilation results are to measured values, which is in general considered to be desirable.

The temporal evolution of soil moisture content at 10, 20, 50 and 80 cm depth for different scenarios is plotted in Figs. 3 and 4. The RMSE values for different scenarios are summarized in Table 2. Assimilating the land surface temperature could improve the soil moisture profile estimation in the scenario of LST_Feedback_Par_LAI; the soil moisture results are better than the open-loop run at all depths. With the assimilation of CRS neutron counts, the soil moisture RMSE values at 10 and 20 cm depth (scenarios CRS_LST_Par_LAI and CRS_LST_True_LAI)

Table 2. Root mean square error (RMSE) of soil moisture profile of open-loop run (CLM), feedback assimilation of land surface temperature including LAI calibration (LST_Feedback_Par_LAI), bivariate assimilation of neutron counts and land surface temperature including LAI calibration (CRS_LST_Par_LAI) and bivariate assimilation of neutron counts and land surface temperature (CRS_LST_True_LAI).

Soil layer depth	RMSE ($m^3 \, m^{-3}$)			
	Open loop (CLM)	LST_Feedback _Par_LAI	CRS_LST _Par_LAI	CRS_LST _True_LAI
10 cm	0.202	0.137	0.085	0.086
20 cm	0.167	0.106	0.047	0.048
50 cm	0.193	0.112	0.112	0.119
80 cm	0.188	0.124	0.136	0.146

decreased significantly. The RMSE values for the scenarios Only_CRS and CRS_LST (not shown) are similar to CRS_LST_Par_LAI, which indicates that the main improvement for the soil moisture profile characterization is achieved by neutron count assimilation; and land surface temperature assimilation and LAI estimation play a minor role. Without assimilation of cosmic-ray probe neutron counts, the soil moisture simulation cannot be improved (scenario Only_LST). However, the scenarios of LST_Feedback and LST_Feedback_Par_LAI improve the soil moisture profile

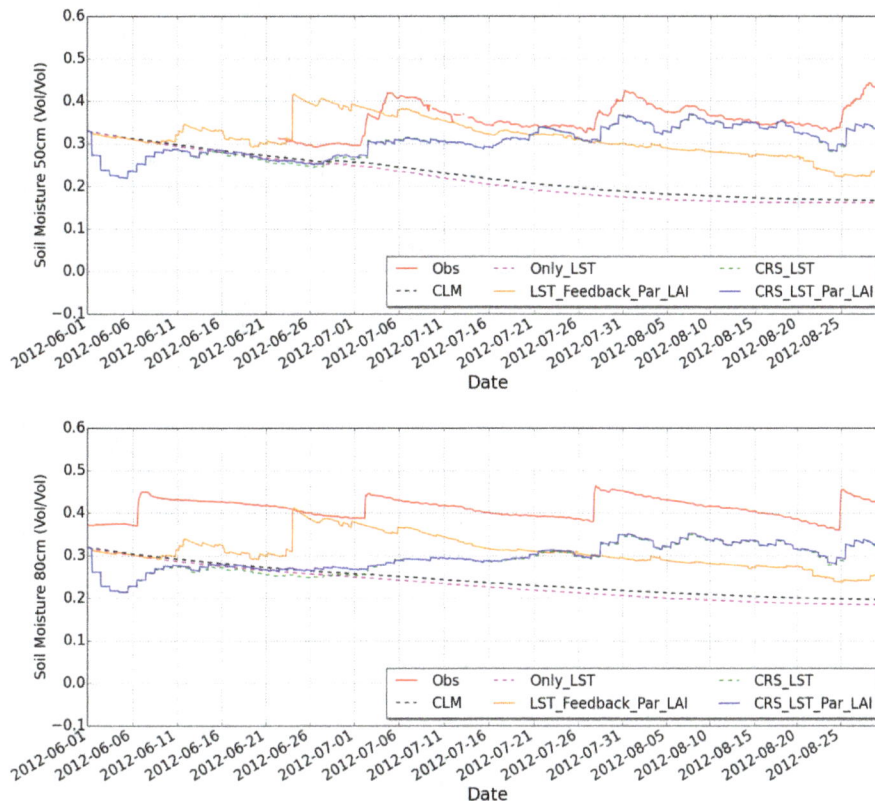

Figure 4. Same as Fig. 3 but for 50 cm (upper) and 80 cm (lower).

characterization, which shows that explicitly using LST to update soil moisture content in the data assimilation routine gives better results than using LST only to update soil moisture by the model equations. Results of LST_Feedback and LST_Feedback_Par_LAI are similar; therefore only results for LST_Feedback_Par_LAI are shown in Figs. 3 and 4. This implies that the improved soil moisture characterization due to LAI calibration is low. The results for the cosmic-ray probe neutron count assimilation proved that the cosmic-ray probe sensor can be used to improve the soil moisture profile estimation at the footprint scale.

Figure 5 depicts the scatterplots of measured ET versus modeled ET for different scenarios, and the accumulated ET for all scenarios are summarized in the lower-right corner of Fig. 5. The EC-measured ET is 384.7 mm for the assimilation period, without energy balance closure correction. The true evapotranspiration is therefore likely larger, but not much larger as the energy balance gap was limited (3.7 %). The CLM-estimated ET, without data assimilation, using only precipitation as input is 223.7 mm and is much smaller than the measured value as applied irrigation is not considered in the model. This open-loop simulated value would imply water stress and a limitation of canopy transpiration and soil evaporation due to low soil moisture content. Assimilation of land surface temperature only (Only_LST) hardly affected the estimated ET and was not able to correct

for the artificial water stress condition. However, if land surface temperature was used to update soil moisture directly, taking into account correlations between the two states in the data assimilation routine, the ET estimates improved to 336.8 and 354.8 mm for the scenarios of LST_Feedback and LST_Feedback_Par_LAI, respectively. The assimilation of land surface temperature of MODIS with soil moisture update results in significant improvements of ET.

The different neutron count assimilation scenarios also resulted in significantly improved estimates of ET. Univariate assimilation of cosmic-ray neutron data (Only_CRS) resulted in 301.9 mm ET. This shows that the impact of neutron count assimilation to correct evapotranspiration estimates is slightly smaller than the impact of land surface temperature with soil moisture update. Joint assimilation of land surface temperature data and cosmic-ray neutron data (CRS_LST) gave a slightly larger ET of 310.6 mm than Only_CRS. The scenarios of CRS_LST_Par_LAI and CRS_LST_True_LAI gave the best ET estimates (360.5 and 349.3 mm). This shows that correcting the biased LAI estimates from MODIS by in situ data or calibration helped to improve model estimates.

The RMSE values of latent heat flux, sensible heat flux and soil heat flux for all scenarios are summarized in Fig. 6. It is obvious that the RMSE values are very large for both the latent heat flux (123.9 W m^{-2}) and sensible heat flux (80.5 W m^{-2}) for the open-loop run and

Obs — 384.7 mm
CLM — 223.5 mm
Only_CRS — 301.6 mm
Only_LST — 209.6 mm
LST_Feedback — 336.6 mm
LST_Feedback_Par_LAI — 354.6 mm
CRS_LST — 310.3 mm
CRS_LST_Par_LAI — 360.2 mm
CRS_LST_True_LAI — 349.0 mm

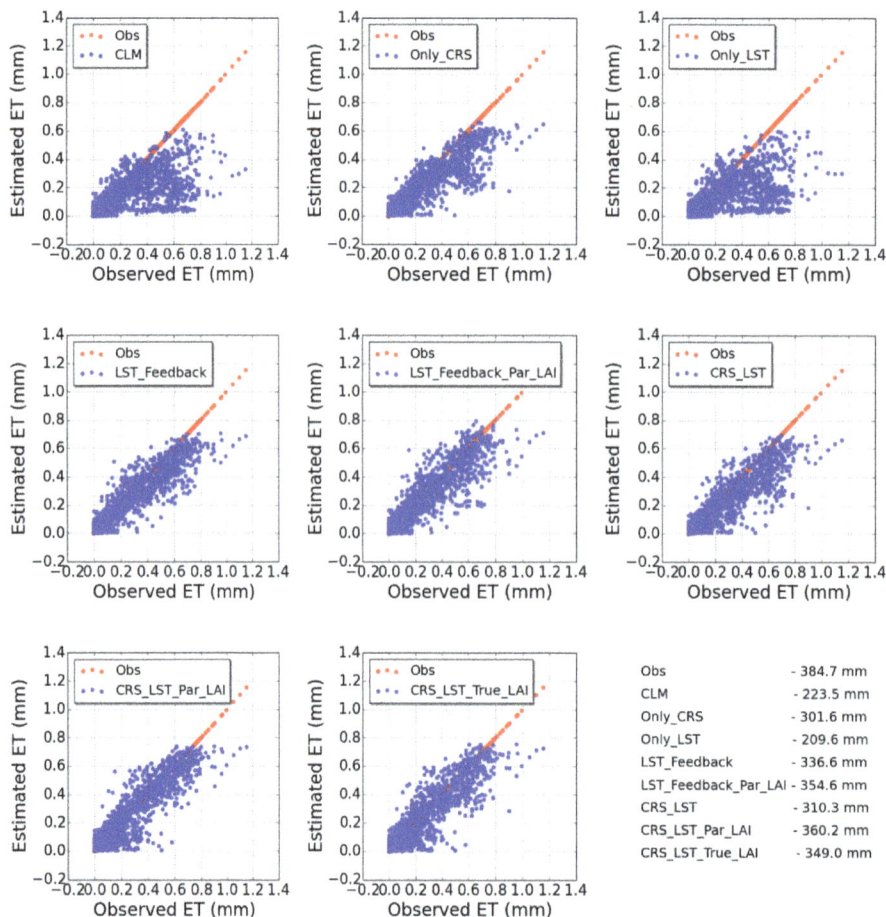

Figure 5. Evapotranspiration estimated according to different scenarios for the period June–August 2012. For a full description see Fig. 3.

all other scenarios where the soil moisture was not updated. When the land surface temperature was assimilated to update the soil moisture, the latent heat flux RMSE decreased to 60.5 (LST_Feedback) and 62.5 W m^{-2} (LST_Feedback_Par_LAI). The scenario where soil moisture and LAI are jointly updated (LST_Feedback_Par_LAI) gave worse results than the scenario of LST_Feedback. Again, the assimilation of neutron counts also resulted in a strong RMSE reduction for the latent heat flux (76.5 W m^{-2} for Only_CRS). When in addition land surface temperature was assimilated and LAI optimized, the RMSE value of latent heat flux further decreased to 56.1 W m^{-2} (70.7 W m^{-2} without LAI optimization). When the field-measured LAI was used instead in the assimilation (CRS_LST_True_LAI), the RMSE was 61.0 W m^{-2}. These results are in correspondence with the ones discussed before for soil moisture characterization. Evidently, the combined assimilation of cosmic-ray probe neutron counts and land surface temperature, and calibration of LAI (or use of field-measured LAI as model input) shows the strongest improvement for the estimation of land surface fluxes. The soil heat flux did not show a clear improvement related to assimilation and showed only some

improvement when LAI was calibrated. For the scenario of land surface temperature assimilation without soil moisture update (Only_LST), estimates of latent and sensible heat flux are not improved. It means that, under water stress conditions, the improved characterization of land surface temperature (and soil temperature) does not contribute to a better estimation of land surface fluxes.

The updated LAI for the scenarios of LST_Feedback_Par_LAI and CRS_LST_Par_LAI is shown in Fig. 7. The MODIS LAI product was used as input for CLM, and time series are plotted as blue line in Fig. 7 (Background). The LAI was also measured in the HiWATER experiment, and the measured values are shown as a green star (Observation). Ens_Mean represents the mean LAI of all ensemble members (Ensembles). It is obvious that MODIS underestimates the LAI compared with the observations. With the assimilation of land surface temperature, the LAI could be updated and be closer to the observations, but there is still a significant discrepancy between the measured LAI and the updated one. The LAI values for the scenario with LAI calibration (CRS_LST_Par_LAI) are close to the measured LAI values (CRS_LST_True_LAI), which

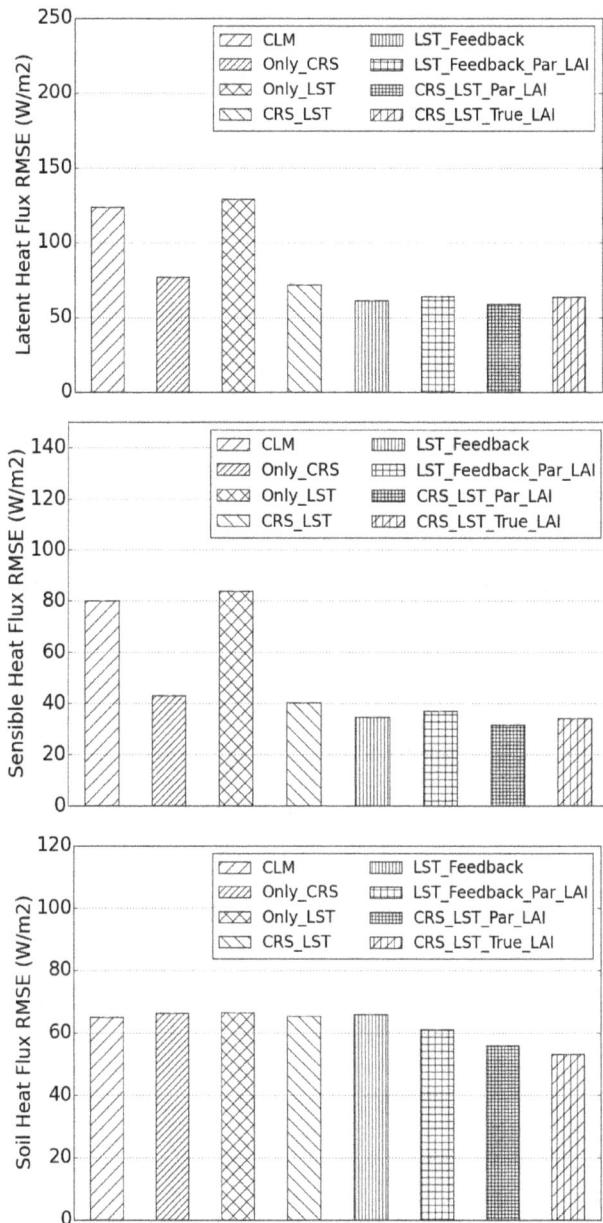

Figure 6. RMSE values of latent heat flux, sensible heat flux and soil heat flux for the period June–August 2012. For a description of the scenarios see Sect. 3 of the paper.

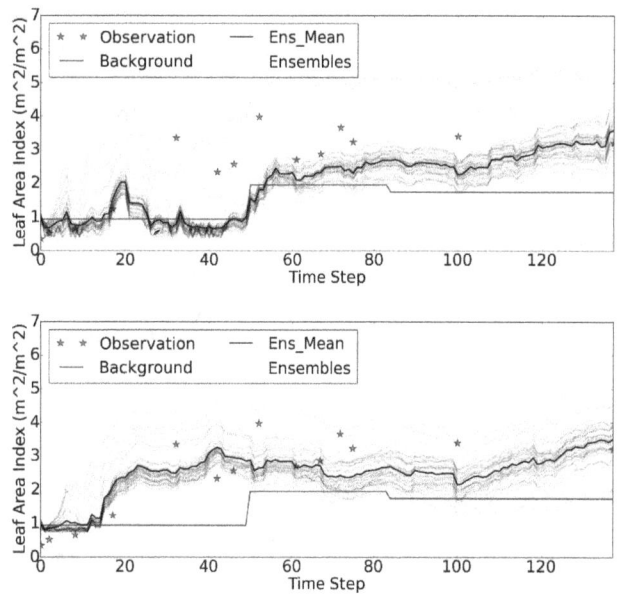

Figure 7. LAI evolution for the period June–August 2012. Displayed are the measured LAI (Observation), default values (Background), mean of ensemble members (Ens_Mean) and ensemble members (Ensembles) for the scenarios of LST_Feedback_Par_LAI (upper) and CRS_LST_Par_LAI (lower).

is an encouraging result. The calibrated LAI shows some unrealistic increases and decreases during the assimilation period, which is inherent to the data assimilation approach. A smoothed representation of the LAI might provide a more realistic picture.

This study illustrates that, for an irrigated farmland, the measured cosmic-ray probe neutron counts can be used to improve the soil moisture profile estimation significantly. Without irrigation data, CLM underestimated soil moisture content. The cosmic-ray neutron count data assimilation can be used as an alternative way to retrieve the soil moisture content profile in CLM. The improved soil moisture simulation was helpful for the characterization of the land surface fluxes. The univariate assimilation of land surface temperature without soil moisture update is not helpful for the estimation of land surface fluxes and even worsened the sensible heat flux characterization (Fig. 6). However, in a multivariate data assimilation framework where land surface temperature was assimilated together with measured cosmic-ray probe neutron counts, the land surface temperature assimilation contributed significantly to an improved ET estimation. The simulated canopy transpiration in CLM was in general too low, even when the water stress condition was corrected by assimilating neutron counts, which was related to small values of the LAI. The additional estimation of LAI through the land surface temperature assimilation resulted in an increase of the LAI, yielding an increase of estimated ET.

In general, land surface models need to be calibrated before use in land data assimilation, especially if there is an apparent large bias in the model simulation (Dee, 2005). The simulation of soil moisture and surface fluxes was biased in our study, mainly due to the lack of irrigation water as input. This bias cannot be corrected a priori without exact irrigation data, which are not available in the field. The data assimilation was proven to be an efficient way to remove the model bias in this case. We also calculated the equivalent water depth to analyze the equivalent irrigated water after each step of soil moisture update. For the scenarios of CRS_LST_Par_LAI and CRS_LST_True_LAI, the equivalent irrigation in 3 months was 693.6 and 607.6 mm, respectively. Because the irrigation method is flood irrigation, it is not easy to evaluate the true irrigation applied in the field. From the results we see, however, that the applied irrigation (in the model) is much larger than actual ET (~ 600 to 700 mm vs. ~ 400 mm). This could indicate that the amount of applied irrigation in the model is too large, but irrigation by flooding is also inefficient and results in excess runoff and infiltration to the groundwater, because it cannot be controlled as well as sprinkler irrigation or drip irrigation. Therefore, the calculated amount of irrigation could be realistic but might also be too large if soil properties are erroneous in the model.

The soil moisture content measured by the cosmic-ray probe represents the depth between 12 cm (very humid) and 76 cm (extremely dry case) depending on the amount of soil water (soil moisture content and lattice water). Therefore the effective sensor depth of the cosmic-ray probe will change over time. In order to model the variable sensor depth and the relationship between the soil moisture content and neutron counts, the new developed COSMIC model was used as the observation operator in this study. Additionally the influences of air pressure, atmospheric vapor pressure and incoming neutron counts were removed from the originally measured neutron counts. Because there is still some water in the crop which also affects the cosmic-ray probe sensor, the COSMIC observation operator could be improved to include vegetation effects. Several default parameters proposed by Shuttleworth et al. (2013) were used in the COSMIC model, and these parameters probably need further calibration following the development of the COSMIC model.

The spatial distribution of soil moisture for the study area was very heterogeneous due to the small farmland patches and different irrigation periods for the different farmlands. Therefore the soil moisture content inferred by SoilNet may not represent the true soil moisture content of the cosmic-ray probe footprint, which is a further limitation of this study. Although the Cosmic-ray Soil Moisture Observing System (COSMOS) has been designed as a continental-scale network by installing 500 COSMOS probes across the USA (Zreda et al., 2012), there are still some disadvantages of COSMOS compared with remote sensing. COSMOS is also expensive for extensive deployment to measure continental/regional-scale soil moisture.

5 Summary and conclusions

In this paper, we studied the univariate assimilation of MODIS land surface temperature products, the univariate assimilation of measured neutron counts by the cosmic-ray probe, the bivariate assimilation of land surface temperature and neutron count data, and the additional calibration of LAI for an irrigated farmland at the Heihe Catchment in China, where data on the amount of applied irrigation were lacking. The most important objective of this study was to test whether data assimilation is able to correct for the absence of information on water resources management as model input, a situation commonly encountered in large-scale land surface modeling. For the specific case of lacking irrigation data, no prior bias correction is possible. The bias-blind assimilation without explicit bias estimation was used. We focused on the model bias introduced by the forcing data and the LAI, and neglected the other sources of bias. When LAI was calibrated, this was done at each data assimilation step of land surface temperature. The data assimilation experiments were carried out with the CLM, and the data assimilation algorithm used was the LETKF. A likely further model bias, besides missing information on irrigation, is the underestimation of LAI by MODIS, which was used to force the model.

The results show that the direct assimilation of measured comic-ray neutron counts improves the estimation of soil moisture significantly, whereas univariate assimilation of land surface temperature without soil moisture update does not improve soil moisture estimation. However, if the land surface temperature was assimilated to update the soil moisture profile directly with the help of the state augmentation method, the evapotranspiration and soil moisture could be improved significantly. This result suggests that the land surface temperature remote sensing products are needed to correct the characterization of the soil moisture profile and the evapotranspiration. The improved soil moisture estimation after the assimilation of neutron counts resulted in a better ET estimation during the irrigation season, correcting the too-low ET of the open-loop simulation. The joint assimilation of neutron counts and MODIS land surface temperature improved the ET estimation further compared to neutron count assimilation only. The best ET estimation was obtained for the joint assimilation of cosmic-ray neutron counts, MODIS land surface temperature including calibration of the LAI (or if field-measured LAI was used as input). This shows that bias due to neglected information on water resources management can be corrected by data assimilation if a combination of soil moisture and land surface temperature data is available.

We can conclude that data assimilation of neutron counts and land surface temperature is useful for ET and soil moisture estimation of an irrigated farmland, even if irrigation data are not available and excluded from model input. The land surface temperature measurements are an alternative data source to improve the soil moisture and land surface flux estimation under water stress conditions. This shows the potential of data assimilation to correct also a systematic model bias. LAI optimization further improves simulation results, which is also likely related to a systematic underestimation of LAI by the MODIS remote sensing product. The results of using the calibrated LAI are comparable to the results of using field-measured LAI as model input.

Acknowledgements. This work is supported by the NSFC (National Science Foundation of China) project (grant nos. 41271357, 91125001), the Knowledge Innovation Program of the Chinese Academy of Sciences (grant no. KZCX2-EW-312) and the Transregional Collaborative Research Centre 32, financed by the German Science Foundation. Jungfraujoch neutron monitor data were kindly provided by the Cosmic-ray Group, Physikalisches Institut, University of Bern, Switzerland. We acknowledge computing resources and time on the Supercomputing Center of Cold and Arid Region Environment and Engineering Research Institute of Chinese Academy of Sciences.

Edited by: H. Cloke

References

Anderson, M. C., Norman, J. M., Kustas, W. P., Li, F., Prueger, J. H., and Mecikalski, J. R.: Effects of Vegetation Clumping on Two–Source Model Estimates of Surface Energy Fluxes from an Agricultural Landscape during SMACEX, J. Hydrometeorol., 6, 892–909, 2005.

Barrett, D. J. and Renzullo, L. J.: On the Efficacy of Combining Thermal and Microwave Satellite Data as Observational Constraints for Root-Zone Soil Moisture Estimation C-7972-2009, J. Hydrometeorol., 10, 1109–1127, 2009.

Bateni, S. M. and Entekhabi, D.: Surface heat flux estimation with the ensemble Kalman smoother: Joint estimation of state and parameters, Water Resour. Res., 48, W08521, doi:10.1029/2011wr011542, 2012.

Bogena, H. R., Herbst, M., Huisman, J. A., Rosenbaum, U., Weuthen, A., and Vereecken, H.: Potential of Wireless Sensor Networks for Measuring Soil Water Content Variability, Vadose Zone J., 9, 1002–1013, 2010.

Bogena, H. R., Huisman, J. A., Baatz, R., Hendricks Franssen, H. J., and Vereecken, H.: Accuracy of the cosmic-ray soil water content probe in humid forest ecosystems: The worst case scenario, Water Resour. Res.., 49, 5778–5791, 2013.

Bosilovich, M. G., Radakovich, J. D., da Silva, A., Todling, R., and Verter, F.: Skin temperature analysis and bias correction in a coupled land-atmosphere data assimilation system, J. Meteorol. Soc. Jpn., 85, 205–228, 2007.

Collatz, G. J., Ball, J. T., Grivet, C., and Berry, J. A.: Physiological and Environmental-Regulation of Stomatal Conductance, Photosynthesis and Transpiration – a Model That Includes a Laminar Boundary-Layer, Agr. Forest Meteorol., 54, 107–136, 1991.

Crow, W. T., Kustas, W. P., and Prueger, J. H.: Monitoring root-zone soil moisture through the assimilation of a thermal remote sensing-based soil moisture proxy into a water balance model, Remote Sens. Environ., 112, 1268–1281, 2008.

Crow, W. T., van den Berg, M. J., Huffman, G. J., and Pellarin, T.: Correcting rainfall using satellite-based surface soil moisture retrievals: The Soil Moisture Analysis Rainfall Tool (SMART), Water Resour. Res., 47, W08521, doi:10.1029/2011wr010576, 2011.

Das, N. N., Mohanty, B. P., Cosh, M. H., and Jackson, T. J.: Modeling and assimilation of root zone soil moisture using remote sensing observations in Walnut Gulch Watershed during SMEX04, Remote Sens. Environ., 112, 415–429, 2008.

De Lannoy, G. J. M., Houser, P. R., Pauwels, V. R. N., and Verhoest, N. E. C.: State and bias estimation for soil moisture profiles by an ensemble Kalman filter: Effect of assimilation depth and frequency, Water Resour. Res., 43, W06401, doi:10.1029/2006WR005100, 2007.

De Lannoy, G. J. M., Reichle, R. H., Arsenault, K. R., Houser, P. R., Kumar, S., Verhoest, N. E. C., and Pauwels, V. R. N.: Multiscale assimilation of Advanced Microwave Scanning Radiometer-EOS snow water equivalent and Moderate Resolution Imaging Spectroradiometer snow cover fraction observations in northern Colorado, Water Resour. Res., 48, W01522, doi:10.1029/2011WR010588, 2012.

Dee, D. P.: Bias and data assimilation, Q. J. Roy. Meteorol. Soc., 131, 3323–3343, 2005.

Desilets, D. and Zreda, M.: Footprint diameter for a cosmic-ray soil moisture probe: Theory and Monte Carlo simulations, Water Resour. Res., 49, 3566–3575, 2013.

Desilets, D., Zreda, M., and Ferré, T. P. A.: Nature's neutron probe: Land surface hydrology at an elusive scale with cosmic rays, Water Resour. Res., 46, W11505, doi:10.1029/2009WR008726, 2010.

Draper, C. S., Mahfouf, J. F., and Walker, J. P.: Root zone soil moisture from the assimilation of screen-level variables and remotely sensed soil moisture, J. Geophys. Res.-Atmos., 116, D02127, doi:10.1029/2010JD013829, 2011.

Entekhabi, D., Njoku, E. G., O'Neill, P. E., Kellogg, K. H., Crow, W. T., Edelstein, W. N., Entin, J. K., Goodman, S. D., Jackson, T. J., Johnson, J., Kimball, J., Piepmeier, J. R., Koster, R. D., Martin, N., McDonald, K. C., Moghaddam, M., Moran, S., Reichle, R., Shi, J. C., Spencer, M. W., Thurman, S. W., Tsang, L., and Van Zyl, J.: The Soil Moisture Active Passive (SMAP) Mission, Proc. IEEE, 98, 704–716, doi:10.1109/jproc.2010.2043918, 2010.

Evensen, G.: The ensemble Kalman filter: Theoretical formulation and practical implementation, Ocean Dynam., 53, 343–367, 2003.

Franz, T. E., Zreda, M., Ferre, T. P. A., Rosolem, R., Zweck, C., Stillman, S., Zeng, X., and Shuttleworth, W. J.: Measurement depth of the cosmic ray soil moisture probe affected by hydrogen from various sources, Water Resour. Res., 48, W08515, doi:10.1029/2012WR011871, 2012.

Franz, T. E., Zreda, M., Rosolem, R., and Ferre, T. P. A.: A universal calibration function for determination of soil moisture with cosmic-ray neutrons, Hydrol. Earth Syst. Sci., 17, 453–460, doi:10.5194/hess-17-453-2013, 2013.

Ghent, D., Kaduk, J., Remedios, J., Ardo, J., and Balzter, H.: Assimilation of land surface temperature into the land surface model JULES with an ensemble Kalman filter, J. Geophys. Res.-Atmos., 115, D19112, doi:10.1029/2010JD014392, 2010.

Ghilain, N., Arboleda, A., Sepulcre-Cantò, G., Batelaan, O., Ardö, J., and Gellens-Meulenberghs, F.: Improving evapotranspiration in a land surface model using biophysical variables derived from MSG/SEVIRI satellite, Hydrol. Earth Syst. Sci., 16, 2567–2583, doi:10.5194/hess-16-2567-2012, 2012.

Han, X., Li, X., Hendricks Franssen, H. J., Vereecken, H., and Montzka, C.: Spatial horizontal correlation characteristics in the land data assimilation of soil moisture, Hydrol. Earth Syst. Sci., 16, 1349–1363, doi:10.5194/hess-16-1349-2012, 2012.

Han, X. J., Franssen, H. J. H., Li, X., Zhang, Y. L., Montzka, C., and Vereecken, H.: Joint Assimilation of Surface Temperature and L-Band Microwave Brightness Temperature in Land Data Assimilation, Vadose Zone J., 12, doi:10.2136/Vzj2012.0072, 2013.

Han, X. J., Franssen, H. J. H., Montzka, C., and Vereecken, H.: Soil moisture and soil properties estimation in the Community Land Model with synthetic brightness temperature observations, Water Resour. Res., 50, 6081–6105, 2014a.

Han, X. J., Jin, R., Li, X., and Wang, S. G.: Soil Moisture Estimation Using Cosmic-Ray Soil Moisture Sensing at Heterogeneous Farmland, IEEE Geosci. Remote S., 11, 1659–1663, 2014b.

Hunt, B. R., Kostelich, E. J., and Szunyogh, I.: Efficient data assimilation for spatiotemporal chaos: A local ensemble transform Kalman filter, Physica D: Nonlinear Phenomena, 230, 112–126, 2007.

Jarlan, L., Balsamo, G., Lafont, S., Beljaars, A., Calvet, J. C., and Mougin, E.: Analysis of leaf area index in the ECMWF land surface model and impact on latent heat and carbon fluxes: Application to West Africa, J. Geophys. Res.-Atmos., 113, D24117, doi:10.1029/2007JD009370, 2008.

Jin, R., Wang, X., Kang, J., Wang, Z., Dong, C., and Li, D.: HiWATER:SoilNET observation dataset in the middle reaches of the Heihe river basin, Heihe Plan Science Data Center, doi:10.3972/hiwater.120.2013.db, 2013.

Jin, R., Li, X., Yan, B., Li, X., Luo, W., Ma, M., Guo, J., Kang, J., Zhu, Z., and Zhao, S.: A Nested Ecohydrological Wireless Sensor Network for Capturing the Surface Heterogeneity in the Midstream Areas of the Heihe River Basin, China, IEEE Geosci. Remote S., 1–5, 2014.

Kerr, Y. H., Waldteufel, P., Wigneron, J. P., Delwart, S., Cabot, F., Boutin, J., Escorihuela, M. J., Font, J., Reul, N., Gruhier, C., and Others: The SMOS Mission: New Tool for Monitoring Key Elements of the Global Water Cycle, Proc. IEEE, 98, 666–687, 2010.

Kumar, S. V., Reichle, R. H., Koster, R. D., Crow, W. T., and Peters-Lidard, C. D.: Role of Subsurface Physics in the Assimilation of Surface Soil Moisture Observations, J. Hydrometeorol., 10, 1534–1547, 2009.

Kumar, S. V., Reichle, R. H., Harrison, K. W., Peters-Lidard, C. D., Yatheendradas, S., and Santanello, J. A.: A comparison of methods for a priori bias correction in soil moisture data assimilation, Water Resour. Res., 48, W03515, doi:10.1029/2010WR010261, 2012.

Kustas, W. and Anderson, M.: Advances in thermal infrared remote sensing for land surface modeling, Agr. Forest Meteorol., 149, 2071–2081, 2009.

Li, C. and Ren, L.: Estimation of Unsaturated Soil Hydraulic Parameters Using the Ensemble Kalman Filter, Vadose Zone J., 10, 1205, doi:10.2136/vzj2010.0159, 2011.

Li, F. Q., Crow, W. T., and Kustas, W. P.: Towards the estimation root-zone soil moisture via the simultaneous assimilation of thermal and microwave soil moisture retrievals, Adv. Water Resour., 33, 201–214, 2010.

Li, X., Cheng, G. D., Liu, S. M., Xiao, Q., Ma, M. G., Jin, R., Che, T., Liu, Q. H., Wang, W. Z., Qi, Y., Wen, J. G., Li, H. Y., Zhu, G. F., Guo, J. W., Ran, Y. H., Wang, S. G., Zhu, Z. L., Zhou, J., Hu, X. L., and Xu, Z. W.: Heihe Watershed Allied Telemetry Experimental Research (HiWATER): Scientific Objectives and Experimental Design, B. Am. Meteorol. Soc., 94, 1145–1160, 2013.

Miyoshi, T. and Yamane, S.: Local Ensemble Transform Kalman Filtering with an AGCM at a T159/L48 Resolution, Mon. Weather Rev., 135, 3841–3861, 2007.

Montzka, C., Grant, J. P., Moradkhani, H., Franssen, H. J. H., Weihermuller, L., Drusch, M., and Vereecken, H.: Estimation of Radiative Transfer Parameters from L-Band Passive Microwave Brightness Temperatures Using Advanced Data Assimilation, Vadose Zone J., 12, doi:10.2136/Vzj2012.0040, 2013.

Moradkhani, H., Sorooshian, S., Gupta, H. V., and Houser, P. R.: Dual state–parameter estimation of hydrological models using ensemble Kalman filter, Adv. Water Resour., 28, 135–147, 2005.

Nie, S., Zhu, J., and Luo, Y.: Simultaneous estimation of land surface scheme states and parameters using the ensemble Kalman filter: identical twin experiments, Hydrol. Earth Syst. Sci., 15, 2437–2457, doi:10.5194/hess-15-2437-2011, 2011.

Oleson, K., Lawrence, D. M., Bonan, G., Drewniak, B., Huang, M., Koven, C. D., Levis, S., Li, F., Riley, W. J., Subin, Z. M., Swenson, S. C., Thornton, P. E., Bozbiyik, A., Fisher, B. E. A., Kluzek, E., Lamarque, J. F., Lawrence, P. J., Leung, L. R., Lipscomb, W., Muszala, S., Ricciuto, D. M., Sacks, W., Sun, Y., Tang, J., and Yang, Z.-L.: Technical Description of version 4.5 of the Community Land Model (CLM), Ncar Technical Note NCAR/TN-503+STR, National Center for Atmospheric Research, Boulder, CO, USA, 422 pp., 2013.

Pauwels, V. R. N., Balenzano, A., Satalino, G., Skriver, H., Verhoest, N. E. C., and Mattia, F.: Optimization of Soil Hydraulic Model Parameters Using Synthetic Aperture Radar Data: An Integrated Multidisciplinary Approach, IEEE T. Geosci. Remote, 47, 455–467, 2009.

Reichle, R. H.: Data assimilation methods in the Earth sciences, Adv. Water Resour., 31, 1411–1418, 2008.

Reichle, R. H., Kumar, S. V., Mahanama, S. P. P., Koster, R. D., and Liu, Q.: Assimilation of Satellite-Derived Skin Temperature Observations into Land Surface Models, J. Hydrometeorol., 11, 1103–1122, 2010.

Robinson, D. A., Jones, S. B., Wraith, J. M., Or, D., and Friedman, S. P.: A Review of Advances in Dielectric and Electrical Conductivity Measurement in Soils Using Time Domain Reflectometry, Vadose Zone J., 2, 444–475, 2003.

Rodell, M., Houser, P. R., Jambor, U., Gottschalck, J., Mitchell, K., Meng, C. J., Arsenault, K., Cosgrove, B., Radakovich, J., Bosilovich, M., Entin, J. K., Walker, J. P., Lohmann, D., and Toll, D.: The global land data assimilation system, B. Am. Meteorol. Soc., 85, 381–394, doi:10.1175/Bams-85-3-381, 2004.

Rosolem, R., Shuttleworth, W. J., Zreda, M., Franz, T. E., Zeng, X., and Kurc, S. A.: The Effect of Atmospheric Water Vapor on Neutron Count in the Cosmic-Ray Soil Moisture Observing System, J. Hydrometeorol., 14, 1659–1671, 2013.

Rosolem, R., Hoar, T., Arellano, A., Anderson, J. L., Shuttleworth, W. J., Zeng, X., and Franz, T. E.: Translating aboveground cosmic-ray neutron intensity to high-frequency soil moisture profiles at sub-kilometer scale, Hydrol. Earth Syst. Sci., 18, 4363–4379, doi:10.5194/hess-18-4363-2014, 2014.

Savitzky, A. and Golay, M. J. E.: Smoothing and Differentiation of Data by Simplified Least Squares Procedures, Anal. Chem., 36, 1627–1639, 1964.

Schwinger, J., Kollet, S. J., Hoppe, C. M., and Elbern, H.: Sensitivity of Latent Heat Fluxes to Initial Values and Parameters of a Land-Surface Model, Vadose Zone J., 9, 984–1001, 2010.

Shangguan, W., Dai, Y., Liu, B., Zhu, A., Duan, Q., Wu, L., Ji, D., Ye, A., Yuan, H., Zhang, Q., Chen, D., Chen, M., Chu, J., Dou, Y., Guo, J., Li, H., Li, J., Liang, L., Liang, X., Liu, H., Liu, S., Miao, C., and Zhang, Y.: A China data set of soil properties for land surface modeling, J. Adv. Model Earth Sys., 5, 212–224, 2013.

Shuttleworth, J., Rosolem, R., Zreda, M., and Franz, T.: The COsmic-ray Soil Moisture Interaction Code (COSMIC) for use in data assimilation, Hydrol. Earth Syst. Sci., 17, 3205–3217, doi:10.5194/hess-17-3205-2013, 2013.

Sun, W. X., Liang, S. L., Xu, G., Fang, H. L., and Dickinson, R.: Mapping plant functional types from MODIS data using multi-source evidential reasoning, Remote Sens. Environ., 112, 1010–1024, 2008.

van den Hurk, B. J. J. M.: Impact of leaf area index seasonality on the annual land surface evaporation in a global circulation model, J. Geophys. Res., 108, 4191, doi:10.1029/2002jd002846, 2003.

Vrugt, J. A., Diks, C. G. H., Gupta, H. V., Bouten, W., and Verstraten, J. M.: Improved treatment of uncertainty in hydrologic modeling: Combining the strengths of global optimization and data assimilation, Water Resour. Res., 41, W01017, doi:10.1029/2004wr003059, 2005.

Wan, Z. and Li, Z. L.: Radiance-based validation of the V5 MODIS land-surface temperature product, Int. J. Remote Sens., 29, 5373–5395, 2008.

Xu, T. R., Liang, S. L., and Liu, S. M.: Estimating turbulent fluxes through assimilation of geostationary operational environmental satellites data using ensemble Kalman filter, J. Geophys. Res.-Atmos., 116, D09109, doi:10.1029/2010JD015150, 2011.

Xu, Z. W., Liu, S. M., Li, X., Shi, S. J., Wang, J. M., Zhu, Z. L., Xu, T. R., Wang, W. Z., and Ma, M. G.: Intercomparison of surface energy flux measurement systems used during the HiWATER-MUSOEXE, J. Geophys. Res.-Atmos., 118, 13140–13157, 2013.

Yang, K., Koike, T., Kaihotsu, I., and Qin, J.: Validation of a Dual-Pass Microwave Land Data Assimilation System for Estimating Surface Soil Moisture in Semiarid Regions, J. Hydrometeorol., 10, 780–793, 2009.

Yang, Z. L., Dai, Y., Dickinson, R. E., and Shuttleworth, W. J.: Sensitivity of ground heat flux to vegetation cover fraction and leaf area index, J. Geophys. Res.-Atmos., 104, 19505–19514, 1999.

Zhu, C. Y., Byrd, R. H., Lu, P. H., and Nocedal, J.: Algorithm 778: L-BFGS-B: Fortran subroutines for large-scale bound-constrained optimization, Acm Transactions on Mathematical Software, 23, 550–560, 1997.

Zreda, M., Desilets, D., Ferre, T. P. A., and Scott, R. L.: Measuring soil moisture content non-invasively at intermediate spatial scale using cosmic-ray neutrons, Geophys. Res. Lett., 35, L21402, doi:10.1029/2008GL035655, 2008.

Zreda, M., Shuttleworth, W. J., Zeng, X., Zweck, C., Desilets, D., Franz, T., and Rosolem, R.: COSMOS: the COsmic-ray Soil Moisture Observing System, Hydrol. Earth Syst. Sci., 16, 4079–4099, doi:10.5194/hess-16-4079-2012, 2012.

7

Divergence of actual and reference evapotranspiration observations for irrigated sugarcane with windy tropical conditions

R. G. Anderson[1,*]**, D. Wang**[1]**, R. Tirado-Corbalá**[1,**]**, H. Zhang**[1,***]**, and J. E. Ayars**[1]

[1]USDA, Agricultural Research Service, San Joaquin Valley Agricultural Sciences Center, Water Management Research Unit, Parlier, California, USA
[*]USDA, Agricultural Research Service, U.S. Salinity Laboratory, Contaminant Fate and Transport Unit, Riverside, California, USA
[**]Crops and Agro-Environmental Science Department,University of Puerto Rico, Mayagüez, Puerto Rico, USA
[***]USDA, Agricultural Research Service, Water Management Research Unit, Fort Collins, Colorado, USA

Correspondence to: R. G. Anderson (ray.anderson@ars.usda.gov)

Abstract. Standardized reference evapotranspiration (ET) and ecosystem-specific vegetation coefficients are frequently used to estimate actual ET. However, equations for calculating reference ET have not been well validated in tropical environments. We measured ET (ET_{EC}) using eddy covariance (EC) towers at two irrigated sugarcane fields on the leeward (dry) side of Maui, Hawaii, USA in contrasting climates. We calculated reference ET at the fields using the short (ET_0) and tall (ET_r) vegetation versions of the American Society for Civil Engineers (ASCE) equation. The ASCE equations were compared to the Priestley–Taylor ET (ET_{PT}) and ET_{EC}. Reference ET from the ASCE approaches exceeded ET_{EC} during the mid-period (when vegetation coefficients suggest ET_{EC} should exceed reference ET). At the windier tower site, cumulative ET_r exceeded ET_{EC} by 854 mm over the course of the mid-period (267 days). At the less windy site, mid-period ET_r still exceeded ET_{EC}, but the difference was smaller (443 mm). At both sites, ET_{PT} approximated mid-period ET_{EC} more closely than the ASCE equations ((ET_{PT}-ET_{EC}) < 170 mm). Analysis of applied water and precipitation, soil moisture, leaf stomatal resistance, and canopy cover suggest that the lower observed ET_{EC} was not the result of water stress or reduced vegetation cover. Use of a custom-calibrated bulk canopy resistance improved the reference ET estimate and reduced seasonal ET discrepancy relative to ET_{PT} and ET_{EC} in the less windy field and had mixed performance in the windier field. These divergences suggest that modifications to reference ET equations may be warranted in some tropical regions.

1 Introduction

Accurate estimates of evapotranspiration (ET) are needed for numerous purposes, including efficient irrigation scheduling (Davis and Dukes, 2010), parameterizing and running different classes of biogeochemical and hydrologic models (Fisher et al., 2005; Zhao et al., 2013), assessing changes in regional hydrology under different cultivation systems (Ferguson and Maxwell, 2011; Holwerda et al., 2013; Waterloo et al., 1999), and evaluating the impacts of agricultural production on regional and global climate (Kueppers et al., 2007; Lo and Famiglietti, 2013; Puma and Cook, 2010) and hydrology (Anderson et al., 2012; Vörösmarty et al., 1998). In irrigated agriculture, underestimation of required ET can lead to suboptimal yield due to water stress (Kang et al., 2002), whereas overestimation of ET can lead to excessive applied water, thus reducing water available for other uses or additional acreage (Perry, 2005), degrading water quality (Smith, 2000), and decreasing economic competitiveness (Hargreaves and Samani, 1984).

While accurate ET estimates are essential, ET can be challenging to measure. Numerous approaches have been developed to measure or estimate ET, including lysimeters (Meissner et al., 2010), micrometeorological methods (Anderson and Goulden, 2009; Baldocchi, 2003; Hemakumara et al., 2003), satellite remote sensing (Bastiaanssen et al., 2005; Tang et al., 2009), and water balance methods. While these approaches vary in their spatial/temporal scale and methodological assumptions and accuracy, most require significant observational costs, technical expertise, or have operational difficulties that are too high for most farmers.

Because of the difficulties in actual ET measurement, the vegetation coefficient/reference ET approach (Jensen, 1968) has gained widespread acceptance for estimating actual ET for varied applications (e.g., Arnold et al., 1998; Cristea et al., 2013). This approach involves calculating a reference ET for a standard land surface, usually grass or alfalfa, using meteorological data and relating the reference surface to the ecosystem/land cover of interest with empirical coefficient(s):

$$ET_A = K_C * ET_0, \qquad (1)$$

where ET_A is actual ET, ET_0 is reference ET, and K_c is the coefficient for the specific land cover type. Two of the most commonly used standard methods include the Food and Agricultural Organization (FAO) approach presented in Irrigation and Drainage paper 56, hereafter referred to as FAO-56 (Allen et al., 1998), and the American Society of Civil Engineers approach, hereafter referred to as ASCE (Allen et al., 2005). Both approaches are based on the combination Penman–Monteith (PM) formula (Monteith, 1965) and account for ET from both solar irradiation and advectively driven ET due to wind and vapor pressure deficit (VPD). Both the FAO-56 and ASCE approaches assume standard measurement conditions and surface parameters (canopy height, surface resistance, albedo, etc.), thus allowing canopy and atmospheric resistance terms to be condensed into constants. Both methods also provide scaling procedures to account for variation in meteorological measurements as well as missing or erroneous data.

Validation work of standardized reference ET equations against large weighing lysimeters with reference surfaces has been done primarily in the western continental USA with low atmospheric humidity (Evett et al., 2000; Jensen et al., 1990). Internationally, most other reference ET validation has been done in Mediterranean climates with similar low humidity (Lecina et al., 2003; Ventura et al., 1999). Relatively little evaluation of these equations has been done in areas with higher relative humidity, presumably because of the perceived lack of use for reference ET equations in these areas. However, reference ET equations are used in more humid regions for applications such as watershed modeling (Rao et al., 2011), forecasting water demand (Tian and Martinez, 2012), and determining irrigation needs (Suleiman and Hoogenboom, 2007). As such, it is necessary to test these reference ET equations in regions with high relative humidity to ensure accurate ET parameterization.

One major tropical and subtropical crop that has generally high ET is sugarcane. Sugarcane is a good crop to test reference ET parameterizations because of its longer full-canopy period, when actual crop ET should be at its maximum relative to reference ET equations, and its high crop coefficient (K_c) that generally exceeds 1. Previous research in irrigated sugarcane has found full-canopy ET rates that equal or exceed evaporation rates from open-water pans (Campbell et al., 1960; Thompson and Boyce, 1967). Since the development and implementation of reference ET equations, researchers have generally found irrigated sugarcane to have a K_c greater than 1 in Australia and Swaziland (Inman-Bamber and McGlinchey, 2003), Brazil (Da Silva et al., 2012), and Texas (Salinas and Namken, 1977). However, all of these studies found variable and differing K_c values, with Inman-Bamber and McGlinchey noting a correspondence between meteorological events and outlying daily K_c values. Sugarcane's high water use, the potential for expanded irrigation to reduce yield deficits and increase production in tropical regions (Inman-Bamber et al., 1999), and the potential for sugarcane irrigation to stress water resources during dry periods in tropical areas (Ramjeawon, 1994) make it a good case study for evaluating reference ET equations in tropical regions.

To evaluate the performance of standardized reference ET equations, we established two eddy covariance (EC) towers over irrigated sugarcane fields in Hawaii, USA to measure ET (ET_{EC}). We calculated reference ET using the ASCE approach for short (ET_0) and tall (ET_r) reference vegetation. The FAO-56 ET_0 was not used as it is identical to ASCE ET_0 for calculations on a daily time step (Irmak et al., 2006; Suleiman and Hoogenboom, 2009). We also compared ET_{EC} to the Priestley–Taylor (PT) ET equation (ET_{PT}). Our objectives were to (1) determine if standardized reference ET equations adequately parameterized actual ET across differing microclimates, (2) determine the meteorological conditions that contribute to discrepancies in the standardized equations, and (3) examine corrections to improve estimates of reference ET under relatively more humid conditions.

2 Methods

2.1 Study region

We evaluated reference ET approaches in two sugarcane (*Saccharum officinarum* L.) fields with identical cultivars (Heinz et al., 1981) at a commercial farm on Maui, Hawaii (Fig. 1 and Table 1). Climatic conditions vary across the farm, with changes in precipitation, wind, solar irradiation, and air temperature due to orographic effects. Normal annual precipitation ranges from 275 to 1275 mm year^{-1} from the leeward (south) side to the windward (northeast) side of the plantation (Giambelluca et al., 2013). Elevations on the plantation range from near sea level to ~ 340 m. The western side of the plantation is generally windier (Table 1). Drip irrigation is used to maximize limited surface and ground water resources (Moore and Fitschen, 1990). Drip tape spacing is 2.70 m with sugarcane rows planted 45 cm away from the tape on both sides; the tape irrigates at 1.58 L^{-1} h^{-1} m^{-1} and is regulated to 83 kPa of pressure at the head of the row. Irrigation amounts were recorded by the farm; rainfall was

Table 1. Eddy covariance field site information.

Micrometeorological site information		
Field	Lee	Windy
Latitude (°N)	20.784664	20.824633
Longitude (°W)	156.403869	156.491278
Elevation (m)	203	44
Date field planted	28 March 2011	11 May 2011
Date tower established	21 July 2011	23 July 2011
Begin of mid-period (cover > 80 %)	3 November 2011	5 December 2011
End of analysis	26 July 2012	27 August 2012
Natural Resource Conservation Service (NRCS) soil series	Waiakoa very stony, silty clay loam	Pulehu cobbly silt loam
Bulk density*	1.22	1.35
Porosity (%)	54	49
Soil texture classification**	Clay	Sandy clay loam
Soil texture – sand (%)	31	51
Soil texture – silt (%)	15	16
Soil texture – clay (%)	54	33
Soil volumetric water content (VWC) at saturation (mm/40 cm depth)	216	196
Soil water storage (water content at 30 % VWC-wilting point) (mm)	60	72
Wilting point (% VWC)	15	12
Matric potential at 30 % VWC (MPa)	NA***	−0.01
Matric potential at 24 % VWC (MPa)	NA	−0.033
Field size (ha)	99.1	62.6
Field length (m) (predominant wind)	> 500	415
Field length (m) (shortest direction)	220	150
Mean meteorological observations (1 August 2011–31 July 2012)		
Mean daily air temperature (°C)	22.3	23.4
Mean minimum daily air temperature (°C)	17.8	20.4
Mean maximum daily air temperature (°C)	27.3	26.9
Mean daily wind speed (m s^{-1})	2.0	4.6
Mean daily net radiation (MJ m^{-2} day^{-1})	10.7	11.3
Mean daily relative humidity (%)	65	62

* All reported soil properties averaged/summed over the first 40 cm of soil depth. ** Soil texture was determined in the lab using the hydrometer method. ** Matric potential not available for Lee because of extreme logistical difficulty in obtaining intact Tempe cell samples at depth for determination of water retention characteristics.

recorded at nearby weather stations (Supplement S1). As is typical for Hawaii (Heinz and Osgood, 2009), sugarcane is grown on a 24-month rotation with planting and harvesting throughout most of the year. Peak ET, as determined by the length of the mid-season period, lasts significantly longer (330 days) than for sugarcane in other regions (190–220 days) (Doorenbos and Pruitt, 1977; Inman-Bamber and McGlinchey, 2003).

2.2 EC measurements and data analysis

We installed two micrometeorological towers in contrasting micro-climates (Fig. 1 and Table 1). These towers are at the "Windy" site (lower elevation, higher wind velocity, more constant wind direction, and sandy clay loam soil) and the "Lee" site (higher elevation, lower wind velocity, and clay

soil). Field fetch in the prevailing wind directions was over 200 m for both towers. The slope in both fields, as determined using the 1/3 arcsec ($\sim$ 10 m) digital elevation model from the US Geological Survey's National Elevation Dataset (http://ned.usgs.gov/index.html), is less than 3 %. Beyond the edge of each field, Windy was surrounded by sugarcane fields on all sides for over 1500 m; Lee was bordered by non-irrigated rangeland in the non-prevailing wind directions (east and south) and contiguous sugarcane fields to the north and east.

Tower instrumentation included an integrated EC system (EC150 – Campbell Scientific, Logan, Utah, USA[1]) with an open-path infrared gas analyzer, aspirated temperature probe,

[1] Mention of trade names or commercial products in this publication is solely for the purpose of providing specific information

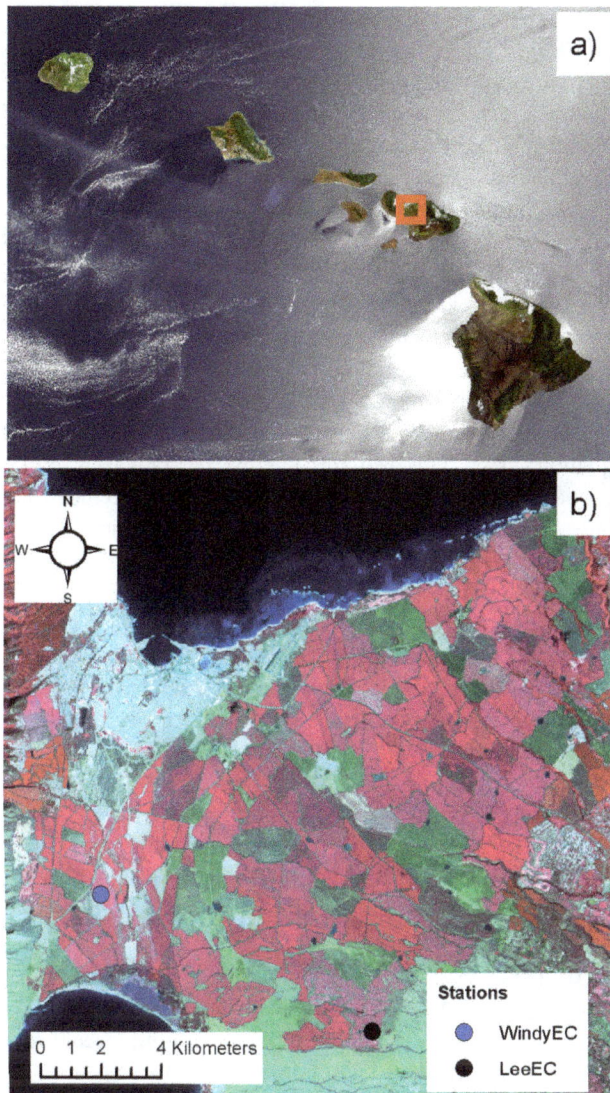

Figure 1. (a) True color image of the main Hawaiian islands from the moderate resolution imaging spectroradiometer (250 m resolution – image date: 27 May 2003). Study region is outlined in red box. (b) The study region on central Maui showing the location of the eddy covariance towers (Windy and Lee) used in this study. Image is false color Landsat 7 (30 m resolution – image date: 5 February 2000).

attached 3-D sonic anemometer head (CSAT3A – Campbell Scientific), and enhanced barometer (PTB110 – Vaisala, Vantaa, Finland). Relative humidity and air temperature were measured by a combined temperature and relative humidity probe (HMP45C – Vaisala). Net radiation was measured with a single component net radiometer (NR-Lite2 – Kipp and Zonen, Delft, Netherlands). We corrected the single component net radiometer for the effect of wind following Cobos and Baker (2003). Ground heat flux was measured as the average

---

and does not imply recommendation or endorsement by the US Department of Agriculture.

of four self-calibrating heat flux plates (HFP01SC – Huskeflux, Delft, Netherlands). The plates were installed at 5 cm depth at four lateral locations perpendicular to the irrigation drip line (Sect. 2.1): 0 cm (drip line), 45 cm (sugarcane row), 75 cm, and 135 cm (mid-point between drip lines). All instruments were factory calibrated to ISO 9001:2008 standards prior to deployment; data were recorded and processed on solid-state data loggers (CR3000, Campbell Scientific).

Two water content reflectometry probes (CS616 – Campbell Scientific) were installed at 20 cm depth at two locations perpendicular to the drip line (45 and 135 cm) to measure soil volumetric water content (VWC). These locations were chosen to correspond with the sugarcane row (center of root zone) and halfway between sugarcane rows. VWC was measured to independently assess potential water stress in both fields. VWC was calculated using a quadratic equation with empirically determined coefficients specific to each field following the manufacturer's recommendation. Soil water retention and permanent wilting point were also determined for Windy but, due to rockiness at the Lee site, could not be determined for Lee because of the logistical difficulty and equipment risk in obtaining intact Tempe cell samples below the surface. More technical details on soil calibrations are provided in Supplement S1.

The EC150 system measured CO_2, H_2O, wind velocity, and sonic temperature at 10 Hz. Other variables were averaged to 30 min fluxes. We processed raw covariances on the data logger and post-processed high-frequency time series data with commercial software (Eddy Pro Advanced V 3.0 and 4.0 – LI-COR, Lincoln, Nebraska USA). Data logger flux calculations were downloaded daily via a cellular modem. High-frequency (10 Hz) data and half-hourly fluxes were transferred monthly via a data card. Raw time series data were checked following the tests of Vickers and Mahrt (1997). Sonic anemometer tilt was corrected using double rotation (Kaimal and Finnigan, 1994); lags between the infrared gas analyzer and sonic anemometer were determined using maximum covariance. We corrected for density fluctuations (Webb et al., 1980), low pass filtering (Moncrieff et al., 1997), and high pass filtering (Moncrieff et al., 2004). Flux footprint lengths were calculated following Kljun et al. (2004), and quality flags were assigned following the CarboEurope standard (Mauder and Foken, 2011). We independently calculated stability (Obukhov, 1971). After installation, tower heights were periodically adjusted to keep meteorological instrumentation ~ 3.0–3.3 m above the zeroplane displacement height, which was assumed to be 67 % of canopy height (Arya, 2001). Canopy height was measured biweekly, concurrent with the vegetation cover observations (Sect. 2.4). Additional detailed EC cross-validation activities are described in Supplement S1.

Half-hourly fluxes with instrumentation errors flagged by the EC150 system, rainfall, or lack of turbulence (friction velocity $< 0.1\,\mathrm{m\,s^{-1}}$) were excluded. Excluded fluxes were gap-filled as a function of fluxes measured from similar meteorological periods using the Max Planck Institute tool (http://www.bgc-jena.mpg.de/~MDIwork/eddyproc/index.php) (Reichstein et al., 2005). Gap-filled fluxes were used to calculate daily and cumulative fluxes but were excluded from half-hourly analyses. We corrected fluxes for energy budget closure by regressing daily EC-observed available energy against measured available energy (net radiation minus ground heat flux) and forcing the regression through the origin, preserving the daily mean Bowen ratio and adjusting each day's ET by the regression slope for the entire study period (Anderson and Wang, 2014; Leuning et al., 2012).

2.3 Reference ET equations, corrections, and evaluation of controls

At each tower, daily and hourly reference ET was calculated using the ASCE short (ET_0) and tall (ET_r) reference equations, where short and tall refer to parameterized surfaces similar to well-watered fescue grass (short) and alfalfa (tall), with differences in the equations due to assumed leaf area index (LAI) and bulk canopy resistance to ET:

$$ET_{sz} = \frac{0.408\Delta\,(R_n - G) + \gamma\frac{C_n}{T+273}u_2\,(e_s - e_a)}{\Delta + \gamma\,(1 + C_d u_2)}. \qquad (2)$$

As shown in Eq. (2), ET_{sz} is the reference ET type (ET_r or ET_0 in mm day^{-1} or mm h^{-1} depending on time step); R_n and G are net radiation and ground heat flux (MJ m^{-2} day^{-1} or MJ m^{-2} h^{-1}), respectively; γ is the psychrometric constant (kPa °C^{-1}); T is mean daily or hourly air temperature (°C); u_2 is mean daily or hourly wind speed measured at or scaled to 2 m height; e_s and e_a are mean saturation and actual vapor pressure (kPa), respectively; and C_n and C_d are empirical numerator and denominator constants that change with the reference surface and time step (Table 1 in Allen et al., 2005). We scaled all meteorological variables from 3 m above the zero-plane displacement to 2 m, following the ASCE procedure for adjusting meteorological measurements at non-standard heights. Following ASCE, mean daily meteorological values were calculated as an average of daily minimum and maximum values as opposed to averaging all 24 h of measurements. Differences between these averaging approaches were small (mean T difference of 0.26 and 0.27 °C in Windy and Lee, respectively). Measured net radiation and ground heat fluxes were used for all calculations.

We also calculated another reference using the PT equation (Priestley and Taylor, 1972). PT was chosen as a comparison because of its different treatment of advection in comparison to the PM-type equations, its wide usage, and the relative simplicity of its meteorological inputs compared to PM. The PT equation is

$$ET_{PT} = \frac{\alpha}{\lambda} * \frac{\Delta\,(R_n - G)}{\Delta + \gamma}. \qquad (3)$$

ET_{PT} is the PT ET (mm day^{-1}); Δ, γ, R_n, and G are the same as in Eq. (2); λ is the latent heat of vaporization; and α is an empirical constant. We assumed that λ is 2.45 MJ mm^{-1}, which is the same as the ASCE/FAO-56 approach. We used an α of 1.26, which is widely, but not universally, representative of a well-watered surface across a variety of climates (e.g., Eichinger et al., 1996; McAneney and Itier, 1996).

To examine the discrepancies between the ASCE equations (ET_0 and ET_r), the PT equation (ET_{PT}), and measured ET_{EC} we inverted the PM equation to calculate bulk canopy resistance (r_c) from ET_{EC} and ET_{PT} and compared the calculated r_c to the constant r_c used to calculate ET_0 and ET_r during the mid-period. The ASCE parameterization to calculate atmospheric resistance (r_a) was used in the inverted PM equation. Days with available energy (net radiation (R_n) − ground heat flux (G)) of < 5 MJ day^{-1} were excluded because low radiation values would result in extreme r_c values and to avoid including days with precipitation, which would bias the net radiation measurement of the NR-Lite2.

Once the discrepancies between reference and measured ET became apparent (see Sects. 3.2 and 3.3), we attempted two corrections to the ASCE reference ET approach to better parameterize sugarcane water use. One was a climatological correction to the ET coefficient (K_{C-adj}). Following the FAO-56 approach (Allen et al., 1998), an adjustment term (K_{adj}) was calculated:

$$K_{adj} = 0.04 * (U_{2avg} - 2) - 0.004 * (RH_{avg} - 45) * h_{avg}^{0.3}, \qquad (4)$$

$$K_{C-adj} = K_{C-FAO} + K_{adj}. \qquad (5)$$

In Eqs. (4) and (5), K_{C-FAO} is the literature mid-canopy K_C value, U_{2avg} is mean location wind speed (m s^{-1}) at 2 m height, RH_{avg} is mean location relative humidity, and h_{avg} is average vegetation height. For our study we used average wind speed, relative humidity, and vegetation height over the mid-period to calculate these parameters in the absence of longer-term climate data. The FAO-56 provides a range of mid-period K_C values for sugarcane (1.25–1.40) for short reference ET. For adjustment, we chose the lowest end of the range (1.25) for K_{C-FAO} to enable the most conservative estimate of parameterized ET.

The second correction was to parameterize the ASCE–PM equation with a custom constant r_c. To estimate a r_c value, an intermediate bulk canopy resistance of 165 s m^{-1} was used, which was chosen as the weighted average of the r_c calculated by inverting the ET_{PT} at Windy and Lee. We then ran the full-form PM equation to calculate a new reference ET (ET_{r-cane}).

Along with corrections to the reference ET equations, we examined potential controls on the discrepancies between reference and measured ET values. Daytime and nighttime r_c were investigated by inverting the full PM equation with measured ET to see if there was a systematic time-of-day difference between the fields and to see if errors in daytime- or nighttime-parameterized r_c were disproportionally contributing to discrepancies in reference ET. Daily daytime and nighttime r_c were calculated for days that had at least eight (daytime) and four (nighttime) non-gap-filled half-hourly flux measurements. For these calculations, daytime was defined as $R_n > 50\,\mathrm{Wm^{-2}}$ and nighttime as $R_n < -10\,\mathrm{Wm^{-2}}$. We used this definition to avoid including periods with near zero Rn that would blow up the inverted PM equation. Finally, we evaluated the correlation between meteorological observations and discrepancies between the ASCE tall reference ET equation (ET_r) and ET_{EC} to assess the importance of the advective and radiation terms in the PM equation.

2.4　Canopy cover and determination of mid-period

We measured fractional canopy cover with an optical camera to obtain an independent, conservative determination of the mid-season period (mid-period) for intercomparison of measured and reference ET. The mid-period is one of the growth/ET stages in the FAO/ASCE methodology and corresponds to maximum plant transpiration and the highest ecosystem coefficient (K_{c-mid}). In unstressed sugarcane, the mid-period coefficient should exceed 1 (Allen et al., 1998), thus measured ET should exceed reference ET. The camera (TetraCam ADC multispectral camera, TetraCam Inc., Chatsworth, California, USA) contains a single precision 3.2 megapixel image sensor optimized for capturing green, red, and near-infrared wavebands of reflected light. A telescoping pole tripod system (GeoData Systems Management Inc., Berea, Ohio, USA) was used to suspend the camera directly above the plant at a height of 7 m and aim vertically downward at nadir view. Each field was photographed every $\sim 16 \pm 2$ days. Ten images were taken in two lines perpendicular to the irrigation line at pre-selected sampling locations in each field at solar noon ± 2 hours; sampling locations were identical throughout the study. Each image was preprocessed with image processing software (LView Pro 2006 – CoolMoom Corp., Hallandale, Florida, USA) to paint out the pixels of soil, grass, shadow, and other background. The preprocessed image was then analyzed using proprietary software (PixelWrench, TetraCam Inc.) to classify fractional vegetation cover based on threshold analysis, and the cover readings from the 10 locations were averaged to determine mean and standard error of field vegetation cover. We considered the beginning of the mid-period to be the latter of the beginning date of mid-period from the FAO-56 K_C curve (Allen et al., 1998) or the date when canopy cover clearly exceeded 80 %, which has been shown to coincide with the start of mid-period (Carr and Knox, 2011). The end of the K_{C-mid} period was set to 27 August 2012, which was the last date of irrigation data prior to the end of the FAO-56 mid-period. Finally, we further restricted the end of the mid-period in the earlier planted field (Lee) to ensure that the length of the mid-period was identical in both fields for intercomparison purposes.

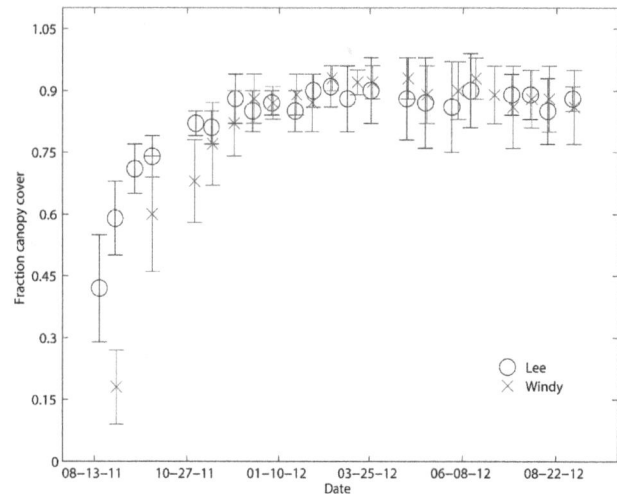

Figure 2. Measured mean and standard deviation of fractional vegetation cover from TetraCam for Windy and Lee fields.

2.5　LAI and stomatal resistance measurements

We measured LAI and leaf stomatal resistance in a field campaign during the mid-period for both EC fields (July 2012). LAI was measured using a nondestructive, optical plant canopy analyzer (LAI 2200, LI-COR Inc.) on 13 July in the Lee field and 16 July in the Windy field. At each of the 10 TetraCam sampling locations in each field (Sect. 2.4), we made 10 below-canopy and 5 above-canopy measurements with the optical canopy analyzer; we then used the manufacturer's software (FV2200, LI-COR Inc.) to determine mean and standard error of LAI for both fields. To observe leaf level stomatal resistance, we used a steady-state diffusion porometer (SC-1, Decagon Devices Inc.), which has been used to observe response to different irrigation regimes in multiple agronomic crops (e.g., Ballester et al., 2013; Hirich et al., 2014; Mabhaudhi et al., 2013; Mendez-Costabel et al., 2014). At each TetraCam point nine leaves were measured: three fully sunlit upper-canopy leaves near (< 20 vertical cm away from) the top visible dewlap (TVD) point (Glaz et al., 2008), three mid-level leaves that were attached to the cane stalk below the TVD height but were still mostly sunlit, and three lower canopy leaves that were partially to mostly shaded. Porometry measurements were made in a 30 s measurement window using the porometer's automatic mode. We also repeated the stomatal resistance measurements at five of the TetraCam points in the Windy field to evaluate the larger discrepancies in reference ET observed in that field.

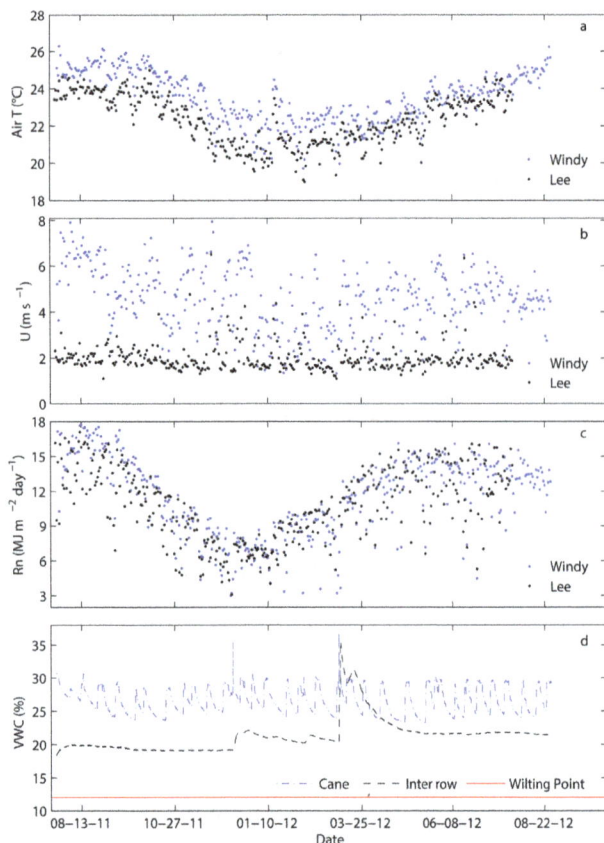

Figure 3. Meteorological and soil observations during the study period: (**a**) mean daily air temperature; (**b**) mean 24 h wind velocity; (**c**) cumulative daily net radiation; and (**d**) soil volumetric water content (VWC) data from Windy field at 20 cm depth underneath cane row (45 cm away from drip line) and inter-row or midway between drip lines (137 cm away from drip line). Wilting point noted as solid red line (12 % VWC).

3   Results

3.1   Fractional vegetation cover, LAI, and leaf stomatal resistance

Fractional vegetation cover increased rapidly in both fields after the beginning of the EC measurements (Fig. 2). Initial cover was < 20 % in Windy and < 45 % in Lee (112 and 142 days after planting (DAP), respectively). Some early TetraCam sampling dates were missed due to initial equipment failures. Vegetation cover exceeded 80 % in Lee on 3 November 2011 and 5 December 2011 in Windy (220 and 208 DAP, respectively), which we considered the onset of the mid-period. Both of these dates are later than the onset of mid-period according to the FAO-56 curve (180 DAP). Variation in cover was largest at the beginning of the study period (standard deviation of ∼ 10%) (Fig. 2). Vegetation cover was least variable near the onset of the mid-period (standard de-

Table 2. A summary of cumulative irrigation, rain, actual measured evapotranspiration-ET_{EC}, and reference evapotranspiration values (ASCE short-ET_0 and tall-ET_r, Priestley–Taylor ET_{PT}, and a custom cane reference ET-ET_{r-cane}) for the entire study period and the mid-period. All values are in mm.

	Lee		Windy	
	Whole study	Mid-period	Whole study	Mid-period
Irrigation	1599	1348	1928	1221
Rain	58	58	140	122
ET_{EC}	1191	843	1389	1001
ET_0	1487	1042	2099	1367
ET_r	1828	1292	2861	1861
ET_{PT}	1470	1008	1707	1096
ET_{r-cane}	1317	947	1662	1128

viation < 5 %). Mean canopy height reached 3.97 m in Lee and 4.09 m in Windy by the end of the study.

Mean ± standard error of measured LAI was 4.9 ± 0.2 in Windy on 13 July 2012 and 4.7 ± 0.3 in Lee on 16 July 2012. Midday leaf stomatal resistance (r_s) observations of fully sunlit leaves in Windy ($n = 32$) and Lee ($n = 21$) showed substantial variation, ranging from 45 to 259 s m^{-1} in Windy and 40 to 640 s m^{-1} in Lee. Median r_s in Windy and Lee were 112 and 114 s m^{-1}, respectively. Mean ± standard deviations of r_s in Windy and Lee were 125 ± 57 s m^{-1} and 161 ± 157 s m^{-1}, respectively. There were two observations in Lee of sunlit stomatal resistance of > 500 s m^{-1}. Excluding these two observations resulted in a revised mean and median r_s in Lee of 114 and 104 s m^{-1}, respectively. Mean sunlit stomatal resistance was not significantly different ($p < 0.01$) from 100 s m^{-1} in either Windy ($p = 0.02$) or Lee ($p = 0.09$).

3.2   Meteorological observations

Air temperature and net radiation were similar in both Windy and Lee (Fig. 3a and c; Table 1). In Windy, mean daily air temperature ranged from 19.0 to 25.0 °C over the study period, whereas in Lee, mean daily air temperature ranged from 19.7 to 26.3 °C. Mean air temperature was higher in Windy than Lee (23.5 and 22.3 °C, respectively) with a similar, low day-to-day variability (standard deviation of 1.3 °C for both fields). Daily net radiation (Rn) was also similar between fields; Rn was slightly higher in Windy versus Lee (11.5 and 10.9 MJ m^{-2} day^{-1}; Fig. 3c and Table 1). Both fields showed larger relative variations in R_n (∼ 10 MJ m^{-2} day^{-1}) than in other meteorological observations. Wind velocities were sharply divergent between the two fields. Mean wind velocity was more than twice as high (4.6 m s^{-1} versus 2.0 m s^{-1}) in Windy compared to Lee (Fig. 3b; Table 1). Wind velocities were also more variable in Windy than Lee (standard deviation of 1.4 and 0.7 m s^{-1}, respectively).

VWC observations in the Windy field underneath the sugar cane row/line varied from 23 to 30 % during the mid-

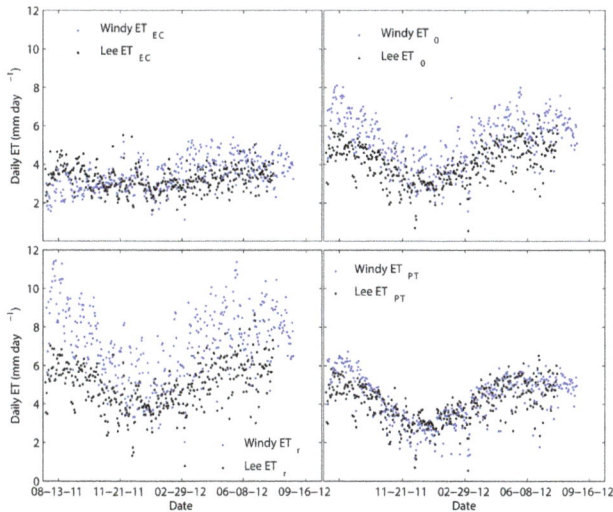

Figure 4. Daily measured and reference ETs for EC tower fields from the tower's establishment until the end of the study period for each field.

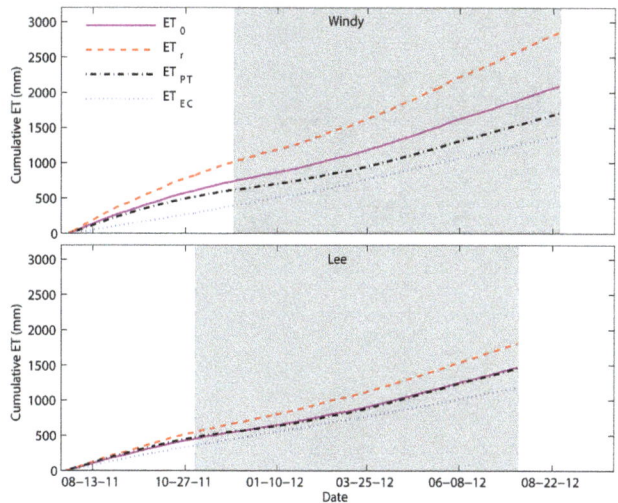

Figure 5. Cumulative measured and reference ET for Windy and Lee. Shaded background indicates mid-period when ground canopy cover exceeded 80 %.

period except after major rain events in December 2011 and March 2012 when they spiked to 36–37 % (Fig. 3d). At all times, VWC remained well above wilting point (12 %) for both sensors (Table 1). Available plant water in the top 40 cm of the soil at minimum VWC was ∼ 40 mm. Soil matric potentials in Windy near typical maximum (30 %) and minimum (24 %) soil VWC were −0.01 and −0.033 MPa, respectively (Table 1). Shallow VWC observations underneath the cane row are likely indicative of plant water stress due to the majority of drip-irrigated Hawaiian sugarcane roots being at less than 50 cm depth (Evensen et al., 1997). VWC observations between drip lines showed relatively little periodicity compared to underneath the cane row, indicating that neither irrigation events nor root depletion was impacting VWC at this location. Due to difficulties with instrument installation and instrument failure we were not able to obtain a reliable time series of soil VWC observations in the Lee field. Precipitation at both fields was less than 150 mm over the course of the study, with irrigation providing more than 90 % of the water input (Table 2).

From the tower's establishment until the end of the study period, daily EC measured ET (ET_{EC}) ranged from 1.6 to 5.5 mm day^{-1} with a mean of 3.2 mm day^{-1} in Lee and 1.6 to 5.5 mm day^{-1} with a mean 3.8 mm day^{-1} in Windy (Fig. 4). ET_{EC} showed relatively little seasonal variation (< 3 mm day^{-1} from summer maxima to winter minima) and greater day-to-day variations of 1–2 mm day^{-1}. Cumulatively, mid-period ET_{EC} was 158 mm higher in Windy than in Lee (Fig. 5; Table 2). Factors contributing to higher ET_{EC} in Windy include higher wind speed, slightly higher Rn, a higher mean air temperature, and lower mean daily relative humidity. However, maximum daily air temperature is higher near Lee than Windy. Ground heat flux was minimal (< 3 %

of R_n during daytime periods) at both sites during the mid-period.

Quality control checks on the EC data indicated no significant issues with ET measurements. Energy closure varied significantly between the sites, with daily energy closure of the turbulent fluxes of 75 % at Lee and 97 % at Windy. As data processing and instrumentation were identical between sites, the difference in energy closure is very likely due to the differences in topography and turbulence between the two fields, particularly nighttime turbulence (Anderson and Wang, 2014). Friction velocity at Windy rarely dropped below the critical threshold (0.1 m s^{-1}) at night (2.5 % of the half-hourly fluxes). Mean 90 % footprint lengths during the study period, determined following Kljun et al. (2004), were 158 m in Windy and 124 m in Lee, which indicate that our EC towers were observing the field of interest even during the rare periods (∼ 7 % of record) when we were observing in the short fetch direction (Table 1), such as during Kona winds (winds from the south and west). During the predominant trade wind flows (prevailing winds from the northeast), our fetch in both fields was > 200 m.

3.3 Reference ET at EC tower sites

Daily short (ET_0) and tall (ET_r) ASCE reference ET were significantly different between the two sites (Fig. 4). In Windy, ET_0 ranged from 1.6 to 8.1 mm day^{-1} over the study period with a mean of 5.2 mm day^{-1} (5.1 mm day^{-1} over the mid-period). ET_r ranged from 2.0 to 12.3 mm day^{-1} with a mean of 7.14 mm day^{-1} (7.0 mm day^{-1} for mid-period). For Lee, ET_0 varied from 0.6 to 6.5 mm day^{-1} with a mean of 4.0 mm day^{-1} (3.9 mm day^{-1} for mid-period). ET_r ranged from 0.8 to 8.6 mm day^{-1} with a mean of 5.0 mm day^{-1} (4.8 mm day^{-1} mid-period). The PT ET (ET_{PT}) showed less

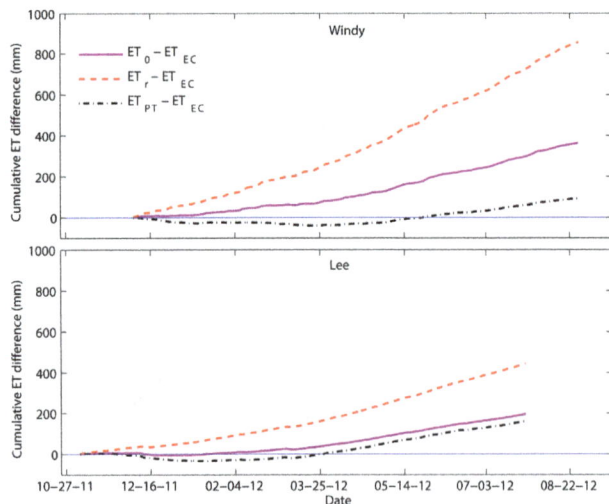

Figure 6. Cumulative difference between reference and measured ET since the beginning of the mid-period in each EC tower field.

difference between the two fields. Mean ET_{PT} was slightly higher at Windy (4.3 and 4.1 mm day^{-1} mid-period) than at Lee (4.0 and 3.8 mm day^{-1} mid-period).

Over the course of the study, the cumulative ET_0 in Windy was 612 mm higher than in Lee, and cumulative ET_r was 1032 mm higher (Fig. 5; Table 2). Similar to the daily values, cumulative ET_{PT} values were considerably closer, with Windy exceeding Lee by 237 mm. As expected, the cumulative difference between reference equations and ET_{EC} grew in the early portion of the study period prior to the mid-period (Fig. 5). During the mid-period, the difference between ET_r and ET_{EC} grew significantly larger in both EC fields. Windy also saw increasing differences between ET_0, ET_{PT}, and ET_{EC}, whereas in Lee, cumulative ET_0 and ET_{PT} tracked quite closely with each other.

To further evaluate these discrepancies between reference and ET_{EC}, we calculated the cumulative difference between the three reference ET equations and ET_{EC} during the mid-period (Fig. 6). ET_{PT} was the only equation with near zero cumulative difference for a substantial amount of the mid-period for both fields; ET_0 was near 0 for the Lee field from October 2011 to February 2012 but not for the Windy field. Over the mid-period in Windy, the difference between cumulative ET_{EC} and ET_{PT} ranged from -40 mm in March 2012 to 92 mm at the end of the study period (August 2012) with cumulative differences of < 40 mm until July 2012. In Lee, the differences were greater, varying between -33 and 161 mm. The difference with ET_0 ranged from 0 (at the beginning of mid-period) to 362 and 195 mm in Windy and Lee, respectively. ET_r showed the greatest cumulative differences of 854 and 443 mm in Windy and Lee.

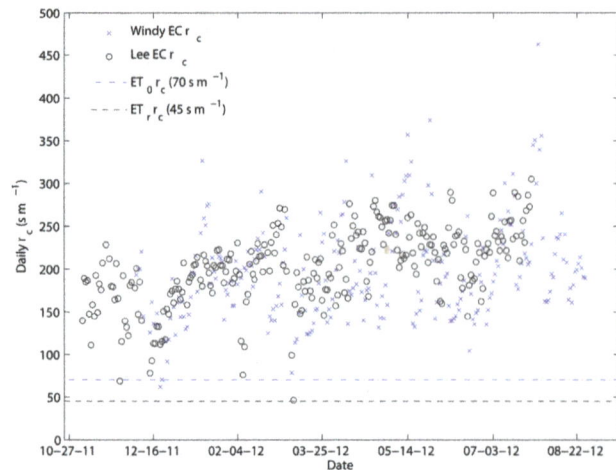

Figure 7. Calculated daily bulk canopy resistance at Windy and Lee from the EC towers for the mid-period. Dotted lines show daily time step resistances from short-canopy ($ET_0 - 70$ s m^{-1}) and tall-canopy ($ET_r - 50$ s m^{-1}) reference surfaces.

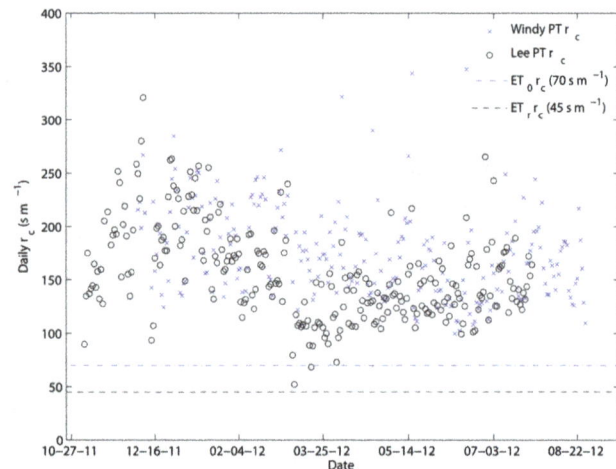

Figure 8. Calculated daily bulk canopy resistances at Windy and Lee from inverting the Priestley–Taylor (PT) ET for the mid-period. Dotted lines again show daily time step resistances from short and tall canopy for comparison.

3.4 Bulk canopy resistances at EC towers, soil observations, and patterns in ET discrepancies

r_c varied considerably between Windy and Lee for ET_{EC}. For the mid-period, mean $\pm$ standard deviations of daily r_c at Lee and Windy were 201 ± 47 and 145 ± 36 s m^{-1}, respectively (Fig. 7). With respect to ET_{PT}, mean $\pm$ standard deviations of daily r_c at Lee and Windy during the mid-period were 146 ± 28 s m^{-1} and 175 ± 42 s m^{-1}, respectively (Fig. 8). In all cases, mean r_c values were significantly higher (> 75 s m^{-1}) than the daily r_c values used to parameterize the ET_0 and ET_r equations.

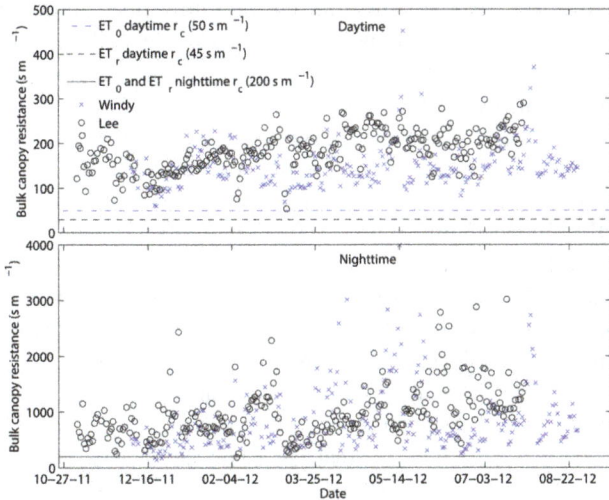

Figure 9. Calculated mean nighttime and daytime bulk canopy resistances (following Fig. 6) compared to assumed resistances.

Daily daytime and nighttime r_c are shown in Fig. 9. Nighttime r_c shows greater difference between towers, with mean $\pm$ standard deviation in Windy and Lee of 675 ± 289 and 808 ± 445 s m^{-1} and substantially larger absolute and relative standard deviation in r_c. For both fields, daytime and nighttime r_c were larger than the ASCE r_c parameterizations for almost all days. One other notable feature of the resistance terms was the low atmospheric resistance (r_a); in Windy and Lee, mean daily r_a were 17.7 and 38.6 s m^{-1}, respectively, over the study period.

With respect to meteorological controls on the discrepancies between ET_r and ET_{EC}, the only parameter that was highly correlated to ET discrepancy (ET_r-ET_{EC}) was VPD with a coefficient of determination (r^2) of 0.66 (Fig. 10a). VPD showed a much stronger correlation with ET discrepancy than ET_{EC} ($r^2 = 0.19$) (Fig. 10b). Available energy was moderately correlated with ET discrepancy ($r^2 = 0.37$) while all other tested parameters (daily minimum, mean and maximum wind speed and temperature) had weak or no correlation with ET discrepancy ($r^2 < 0.1$).

3.5   Corrections to better parameterize sugarcane water use

The climatological K_C adjustment (K_{adj}) had relatively little impact on calculated water use. In the Windy field K_{adj} was -0.0126, and in Lee K_{adj} was -0.0359. For both fields, the wind adjustment offset the relative humidity/vegetation height adjustment as all three parameters were greater than zero. The magnitude of the K_{adj} term was insufficient to account for the observed discrepancies between reference ET and ET_{EC}.

Cumulative differences between ET_{r-cane} and ET_{EC} are shown in Fig. 11 along with the differences between ET_{PT} and ET_{EC}. ET_{r-cane} showed some improvements over ET_{PT}

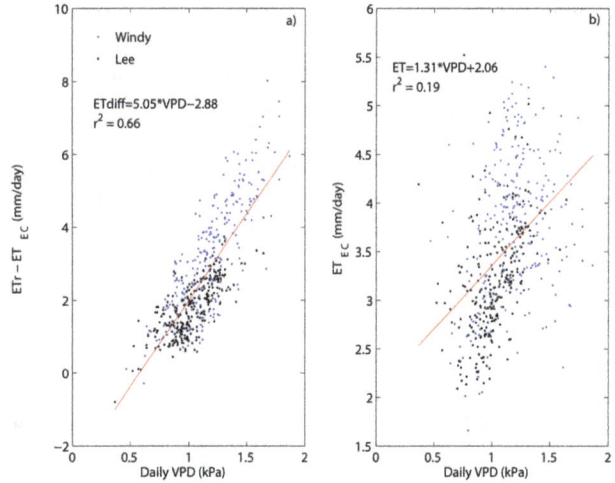

Figure 10. (a) Relationship between daily ET discrepancy (ET_r-ET_{EC}) and daily vapor pressure deficit (VPD) from the beginning of the mid-period to the end of the study period. Regression equation is fitted to the entire pool of data from Lee and Windy. **(b)** Relationship between measured ET and daily VPD; time period and regression approach are the same as in **(a)**.

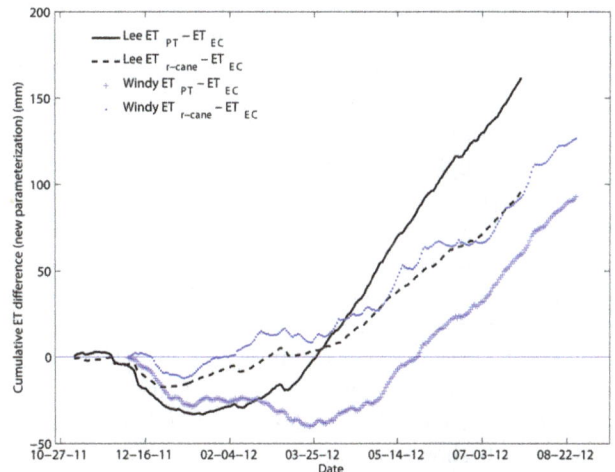

Figure 11. Cumulative difference between new reference ET (custom bulk canopy resistance of 165 s m^{-1}) and measured ET for both EC tower fields during the mid-period.

in predicting measured ET between October 2011 and March 2012; in particular ET_{r-cane} had less underestimation of ET (15 to 27 mm improvement) in winter and spring for both fields and had consistently better performance in the Lee field. ET_{r-cane} had worse performance than ET_{PT} during the summer in the Windy field (40 mm). The minimum cumulative difference between ET_{r-cane} and ET_{EC} was -12 and -18 mm in Windy and Lee, respectively. The maximum cumulative difference between ET_{r-cane} and ET_{EC} was 132 and 164 mm at the end of the study period in Windy and Lee, respectively.

4 Discussion

4.1 Is Hawaiian sugarcane representative of a fully transpiring reference ET surface?

Well-irrigated, full-canopy sugarcane has generally been reported to have an ET rate 1.1 to 1.4 times the ASCE/FAO-56 reference ET_0 equation (Da Silva et al., 2012; Inman-Bamber and McGlinchey, 2003), and rain-fed sugarcane has been reported to have an ET rate approaching ET_0 (Cabral et al., 2012). Furthermore, a reference PM–ET equation designed specifically for sugarcane, created by McGlinchey and Inman-Bamber (1996), has a bulk canopy resistance that is slightly lower than the daily ASCE ET_r equation (40 versus $45\,\mathrm{s\,m^{-1}}$ for ASCE ET_r). Therefore, the significant overestimation of measured ET (ET_{EC}) by the ET_0 and ET_r equations found in this study was quite surprising. Although Windy and Lee fields had slight differences in planting dates, available soil water capacity, and fetch (Table 1), we do not believe these account for the observed ET/reference ET differences between the fields. Seasonal variation in temperature in Hawaii is quite small; wind speeds appeared to be uncorrelated to seasonality. Wind fields in central Maui are generally very strong, and our separate calculations of reference ET using independent farm weather station observations (Supplement S1) and publicly available airport weather data from Kahului airport (http://mesonet.agron.iastate. edu/request/download.phtml?networkHI_ASOS – station ID PHOG) show higher-than-typical values of reference ET for a tropical region.

The quality of EC observations was good, especially at the Windy tower where high turbulence, flux footprints that were well within field boundaries, low proportion of time periods requiring gap filling, and excellent energy budget closure ($H + \mathrm{LE}$ was $> 95\,\%$ of daily R_n-G) indicated that the methodological requirements of the EC method were well satisfied (Anderson and Wang, 2014). At the Lee tower, EC measurements showed a more typical pattern, with a larger number of gaps during still nighttime periods when ET is low. Furthermore, seasonal and annual totals of ET have been shown to be relatively insensitive to gap-filling methodologies (Alavi et al., 2006). Finally, while the gap-filling method of Reichstein et al. (2005) may systematically underestimate wet canopy evaporation due to exclusion of all EC periods during and immediately after rain, this bias is likely to be insignificant at our sites due to the low precipitation (Table 2) and drip irrigation that would minimize wetting of the leaves.

One hypothesis is that portions of the fields measured by our EC towers were under significant water stress or had less-than-optimal cover and thus were not representative of a reference ET type surface. Uniformity of irrigation is a major concern with drip irrigation, particularly with sub- and near-surface drip lines where root development can plug or pinch drip lines, leading to insufficient irrigation (e.g., Soopramanien et al., 1990). At our field with higher ET (Windy),

visible dry lines arising from pinched drip tubes appeared in parts of the field at and after the end of the study period. However, there are multiple independent lines of evidence against this hypothesis.

With respect to canopy cover, the TetraCam observations of cover (Fig. 2) show that fractional cover remained above $80\,\%$, a threshold for the mid-period K_C (Carr and Knox, 2011; Inman-Bamber and McGlinchey, 2003). More evidence for full canopy comes from the LAI measurements made in July 2012 toward the end of the mid-period. In both Lee and Windy, mean LAI (4.7 and 4.9) were slightly higher than the LAI (4.5) parameterized in the ET_r equation (Allen et al., 2005). These two types of data indicate that incomplete cover is not an issue with our study sites.

Another possibility is that the sugarcane leaves are under significant water stress and thus are transpiring at a lower rate. Four factors show that the sugarcane is unlikely to be water stressed. First, porometer measurements from the July 2012 campaign of midday, sunlit leaf stomatal resistance were not significantly $> 100\,\mathrm{s\,m^{-1}}$. The $100\,\mathrm{s\,m^{-1}}$ comes from the mean leaf level stomatal resistance of a sunlit leaf on a well-watered plant, as measured by Szeicz and Long (1969), and which is used as a basis for scaling bulk canopy resistance in the ASCE and FAO-56 approaches (Allen et al., 1998, 2005). Second, we compared the daily observed ET coefficient (K_C) from the day immediately preceding a substantial irrigation or rain event (defined as $> 8\,\mathrm{mm\,day^{-1}}$) during the mid-period with daily K_C 2 and 3 days after the irrigation event using a paired t test ($n = 106$ in Windy and $n = 98$ in Lee). We reasoned that stressed full-canopy sugarcane would respond to irrigation within 3 days, but that 3 days were short enough to avoid confounding changes due to variations in field water budgets. Neither field showed significantly greater daily ET_{EC} following an irrigation during the mid-period ($p > 0.40$ for all tests). Third, the soil VWC data from the Windy field indicate relatively high soil moisture content; available soil water underneath the cane row in the middle of the root zone always remained at $> 50\,\%$ of available capacity. Windy's soils were also near field capacity (and far above permanent wilting point) based on matric potential at typical maximum and minimum soil VWC (Table 1). The VWC content also argues against severe water stress that might persist after irrigation relieves the soil moisture deficit; thus if the ASCE reference ET equations and coefficients were applicable to this situation, we should see at least some days with ET_{EC} in the range of ET_0 and ET_r (6–$10\,\mathrm{mm\,day^{-1}}$ in Windy) when soil moisture was near or above field capacity. Fourth, measured irrigation plus precipitation as recorded by the plantation was compared to measured cumulative ET_{EC}, with cumulative mid-period irrigation and precipitation exceeding ET_{EC} by 342 mm in Windy (Table 2). At all times in the Windy field, cumulative ET_{EC} was significantly less than irrigation plus precipitation. In Lee, by early January 2012 cumulative precipitation and irrigation exceeded ET_{EC}; by the end of the mid-period (July 2012), cumulative irrigation and precipitation exceeded cumulative ET_{EC} by > 500 mm (Table 2). In summary, the evidence of full canopy and the lack of evidence of water stress indicated that the mid-period sugarcane at our study fields should be fully transpiring.

4.2 Why do the standardized ASCE reference ET equations differ between similar sites?

Without clear evidence of water stress or lack of canopy cover over the study sites, we examine some explanations for the overestimation of the ASCE ET_0 and ET_r compared to ET_{EC} and ET_{PT}. Four hypotheses include (1) scaling of leaf level stomatal resistance to whole canopy bulk resistance, (2) incorrect parameterization of daytime leaf level resistance, (3) underestimation of nighttime bulk canopy resistance, and (4) underestimation of atmospheric resistance. Scaling up leaf level resistance measurements has long been recognized as a major challenge (Bailey and Davies, 1981; Furon et al., 2007; Sprintsin et al., 2012) due to heterogeneity of environmental variables. The ASCE/FAO reference ET methods take a single layer "big leaf" approach to scaling to convert non-stressed leaf resistances (r_s) into whole canopy bulk resistances (r_c) by using an "effective LAI" where r_c is calculated by dividing r_s by effective LAI. ASCE assumes that effective LAI is equivalent to 0.5 times measured LAI, which is assumed to be 2.9 for ET_0 and 4.5 for ET_r, thus resulting in effective LAIs of 1.4 and 2.3, respectively. Studies of well-watered crops have found effective LAIs which vary quite significantly from those assumed for the reference surface. Tolk et al. (1996) found an effective LAI of 1.3 for irrigated maize in Texas that was only 30 % of the maximum measured LAI. Other studies (Alfieri et al., 2008; Mehrez et al., 1992) have assumed effective LAI as a linear function of LAI, with effective LAI equaling 50 % of LAI when LAI is 6. Ultimately, the effective LAI concept is only a presumed distribution of leaves with differing r_s (Bailey and Davies, 1981); there is a possibility that the relatively unique production system in our study fields results in a different, distinctive leaf distribution with a lower effective LAI. Along with effective LAI, another leaf parameter that could be different is leaf level resistance (r_s). Although we did not find a highly significant difference between measured r_s and the r_s assumed in the ASCE parameterizations (100 s m^{-1}), we were able to measure r_s in only one field campaign during the mid-period, where r_s observations were limited by clouds and other logistical limitations. A large number of r_s observations are needed to accurately characterize r_c (Denmead, 1984); more than we could feasibly measure during our field campaign. We also note that other researchers (e.g., Zhang et al., 2008) have found non-stressed r_s values greater than 100 s m^{-1}.

Two other nonbiological factors could help explain the discrepancy between ASCE reference and mid-period ET_{EC}. One is nighttime r_c. Both ASCE approaches assume a night-time r_c of 200 s m^{-1}, which is based on measurements of damp soil beneath a grass lysimeter (Allen et al., 2006). Measured nighttime r_c at our fields was significantly higher. We suspect that the taller sugarcane canopy and substantial layer of trash and lodged cane minimizes bare soil water evaporation, thus increasing nighttime r_c. Oliver and Singels (2012) found a significant decrease in soil evaporation in sugarcane with surfaces covered by crop residue. Furthermore, the minimal daytime ground heat flux (< 5 %) further reduces nighttime ET. Another factor is the canopy energy storage that is considerable in high-biomass systems (Anderson and Wang, 2014). Finally, we note that nighttime r_c is likely to be a locally specific value; 200 s m^{-1} is too low for our study region but too high for other regions with significant advection (Evett et al., 2012).

Along with nighttime r_c, we examined the role of atmospheric resistance (r_a) in parameterizing ET, given the low observed mean r_a at Windy (< 20 s m^{-1}) and the demonstrated importance of atmospheric resistance/conductance parameterizations in coastal tropical regions for accurate ET parameterization (e.g., Holwerda et al., 2012). Given the canopy architecture of mid-period sugarcane in our study fields, we were not certain about the equations that are commonly used to parameterize zero plane displacement height and roughness lengths, which are also used in the ASCE reference ET equations. To test the effect of r_a uncertainty, a sensitivity analysis was conducted. We used r_a that was 200 and 50 % of the original r_a and recalculated r_c for both EC towers. In all cases, the new r_a changed the r_c values by < 10 s m^{-1}, with most r_c values changed by < 5 s m^{-1}. These values are too small to explain the discrepancy between observed and parameterized r_c. The presence of r_a in both the numerator and denominator of the PM equation limits the impact of variation in r_a on r_c.

Finally, we note that the ASCE and FAO reference ET and PT ET equations show varying sensitivity to meteorological variables depending upon climate. Multiple studies have shown spatial, seasonal, and interannual variation in the sensitivity of reference ET to meteorological inputs, with the most sensitive input (air temperature, wind velocity, relative humidity, etc.) changing depending upon season and location (e.g., Bandyopadhyay et al., 2009; Estévez et al., 2009; Gong et al., 2006; Huo et al., 2013; Irmak et al., 2006; Liang et al., 2008; Liu et al., 2014). Irmak et al. (2006) and Estévez et al. (2009) found increased sensitivity to reference ET parameterization at locations with higher wind velocities in the United States and Spain, respectively. Bandyopadhyay et al. (2009) and Huo et al. (2013) reported that decreased wind velocities accounted for the largest proportion of decreased reference ET in climatically differing regions in India and China. Across a large river basin in China (Chiang Jiang), Gong et al. (2006) showed that sensitivities of reference ET to other meteorological variables (air temperature and relative humidity) depended significantly on the spatial pattern of wind sensitivity. With respect to the PT equation, variability in the PT coefficient (α) has been found at lower to middle LAI (LAI less than 3) depending upon the soil wetness and covering (Ding et al., 2013). This may be particularly relevant for our system in early growth stages with fractional soil wetness and partial cover from sugarcane detritus (trash). Conversely, at mid- to full canopy (LAI greater than 3) or when soil moisture was greater than 50 % of the available field capacity, α showed little sensitivity.

5   Summary and conclusion

We investigated discrepancies between two standardized reference ET equations and EC-measured ET at two field sites over irrigated sugarcane on Maui. At both fields, measured daily ET during the mid-period should have approached the tall reference ET equation and exceeded the short reference ET equation. At both fields, both ASCE reference ET equations significantly overestimated mid-period ET compared to EC observations of ET. The PT equation performed substantially better at the Windy field than the short reference ET, while the short reference ET equation and PT were more closely matched at the Lee field. We used a custom bulk canopy resistance derived from inverting PT ET; the custom cane reference ET equation had less seasonal variation in ET discrepancy. Multiple independent field observations did not indicate insufficient canopy cover or plant water stress reducing ET_{EC} significantly.

This study indicated nighttime bulk canopy resistance, leaf stomatal resistance, and effective LAI as possible causes for the discrepancy in bulk canopy resistance (and reference ET estimates) between the ASCE reference equations and mid-period ET_{EC}. The higher bulk canopy resistances and relationship between ET discrepancies and vapor pressure deficit indicated that the ASCE equations overestimated the advective component of ET. Ultimately, validation with field methods, including micrometeorology and water balance methods, is needed to establish the accuracy of the ASCE equations in a region where they have not been tested previously. Adjusting the bulk canopy resistance to local climate to reduce the advective component of ET may make the full ASCE PM equation a more appropriate equation in this region.

The PT equation performs better than ET_r or ET_0 in our study region. The PT equation likely provides a more robust estimation of reference ET in regions with high humidity. The simplicity of the PT equation also makes it attractive for use in larger scale project planning as it has been parameterized in satellite-based ET models (e.g., Choi et al., 2011; Jin et al., 2011) and can be used in regions with a relative paucity of surface meteorological data, unlike the ASCE/FAO equations that require near-surface wind speed and humidity data, which are currently supplied by surface meteorological stations and interpolated in satellite-based approaches (Allen et al., 2007; Hart et al., 2009).

The results illustrate the importance of the careful use of reference ET equations and coefficients for assessing actual ET in hydrologic applications. Our finding of high bulk canopy resistance and low atmospheric resistance supports Widmoser's (2009) recommendation for research of the canopy resistance/atmospheric resistance ratio. Many areas with changing hydrology (Elison Timm et al., 2011) and areas that currently and may soon use irrigation in previously non-irrigated fields (Baker et al., 2012; Salazar et al., 2012) are outside of the semi-arid areas where reference ET methods have been primarily developed and tested. As such, it will be important to ensure that the appropriate reference equation is used to parameterize evaporative demand.

Acknowledgements. We thank Ilja van Meerveld, Lixin Wang, Maarten Waterloo, and two anonymous reviewers for their constructive feedback on this manuscript. Don Schukraft discussed previous meteorological investigations and observations at the farm. Jim Gartung, ARS-Parlier, assisted with the establishment of the EC tower and TetraCam measurements. David Grantz provided insight into the historical evaluation of ET data for Hawaiian sugarcane. Adel Youkhana, Neil Abranyi, Jason Drogowski, and the farm crew assisted with data collection and field logistical support. This research was supported by USDA Agricultural Research Service, national program 211: Water Availability and Watershed Management and by the US Navy, Office of Naval Research. Note: the US Department of Agriculture (USDA) prohibits discrimination in all its programs and activities on the basis of race, color, national origin, age, disability, and where applicable, sex, marital status, familial status, parental status, religion, sexual orientation, genetic information, political beliefs, reprisal, or because all or part of an individual's income is derived from any public assistance program (not all prohibited bases apply to all programs.) Persons with disabilities who require alternative means of communication of program information (Braille, large print, audiotape, etc.) should contact USDA's TARGET Center at (202) 720-2600 (voice and TDD). To file a complaint of discrimination, write to USDA, Director, Office of Civil Rights, 1400 Independence Avenue, S.W., Washington, D.C. 20250-9410, or call (800) 795-3272 (voice) or (202) 720-6382 (TDD). USDA is an equal opportunity provider and employer.

Edited by: L. Wang

References

Alavi, N., Warland, J. S., and Berg, A. A.: Filling gaps in evapotranspiration measurements for water budget studies: Evaluation of a Kalman filtering approach, Agr. Forest Meteorol., 141, 57–66, doi:10.1016/j.agrformet.2006.09.011, 2006.

Alfieri, J. G., Niyogi, D., Blanken, P. D., Chen, F., LeMone, M. A., Mitchell, K. E., Ek, M. B., and Kumar, A.: Estimation of the Minimum Canopy Resistance for Croplands and Grasslands Using Data from the 2002 International H2O Project, Mon. Weather Rev., 136, 4452–4469, doi:10.1175/2008MWR2524.1, 2008.

Allen, R. G., Pereira, L. S., Raes, D., and Smith, M.: Crop evapotranspiration?: guidelines for computing crop water requirements, Food and Agriculture Organization of the United Nations, Rome, 1998.

Allen, R. G., Walter, I. A., Elliott, R. L., Howell, T. A., Itenfisu, D., Jensen, M. E., and Snyder, R. L: The ASCE standardized reference evapotranspiration equation, American Society of Civil Engineers, Reston, Va., 2005.

Allen, R. G., Pruitt, W. O., Wright, J. L., Howell, T. A., Ventura, F., Snyder, R., Itenfisu, D., Steduto, P., Berengena, J., Yrisarry, J. B., Smith, M., Pereira, L. S., Raes, D., Perrier, A., Alves, I., Walter, I., and Elliott, R.: A recommendation on standardized surface resistance for hourly calculation of reference ETo by the FAO56 Penman-Monteith method, Agr. Water Manage., 81, 1–22, doi:10.1016/j.agwat.2005.03.007, 2006.

Allen, R. G., Tasumi, M., and Trezza, R.: Satellite-Based Energy Balance for Mapping Evapotranspiration with Internalized Calibration (METRIC)-Model, J. Irrig. Drain. Eng., 133, 380–394, doi:10.1061/(ASCE)0733-9437(2007)133:4(380), 2007.

Anderson, R. G. and Goulden, M. L.: A mobile platform to constrain regional estimates of evapotranspiration, Agr. Forest Meteorol., 149, 771–782, doi:10.1016/j.agrformet.2008.10.022, 2009.

Anderson, R. G. and Wang, D.: Energy budget closure observed in paired Eddy Covariance towers with increased and continuous daily turbulence, Agr. Forest Meteorol., 184, 204–209, doi:10.1016/j.agrformet.2013.09.012, 2014.

Anderson, R. G., Lo, M.-H., and Famiglietti, J. S.: Assessing surface water consumption using remotely-sensed groundwater, evapotranspiration, and precipitation, Geophys. Res. Lett., 39, L16401, doi:10.1029/2012GL052400, 2012.

Arnold, J. G., Srinivasan, R., Muttiah, R. S., and Williams, J. R.: Large Area Hydrologic Modeling and Assessment Part I: Model Development, J. Am. Water Resour. Assoc., 34, 73–89, doi:10.1111/j.1752-1688.1998.tb05961.x, 1998.

Arya, S. P.: Introduction to micrometeorology, Academic Press, San Diego, 2001.

Bailey, W. G. and Davies, J. A.: Bulk stomatal resistance control on evaporation, Boundary-Lay. Meteorol., 20, 401–415, doi:10.1007/BF00122291, 1981.

Baker, J. M., Griffis, T. J., and Ochsner, T. E.: Coupling landscape water storage and supplemental irrigation to increase productivity and improve environmental stewardship in the US Midwest, Water Resources Res., 48, W05301, doi:10.1029/2011WR011780, 2012.

Baldocchi, D. D.: Assessing the eddy covariance technique for evaluating carbon dioxide exchange rates of ecosystems: past, present and future, Glob. Change Biol., 9, 479–492, doi:10.1046/j.1365-2486.2003.00629.x, 2003.

Ballester, C., Jiménez-Bello, M. A., Castel, J. R., and Intrigliolo, D. S.: Usefulness of thermography for plant water stress detection in citrus and persimmon trees, Agr. Forest Meteorol., 168, 120–129, doi:10.1016/j.agrformet.2012.08.005, 2013.

Bandyopadhyay, A., Bhadra, A., Raghuwanshi, N. S., and Singh, R.: Temporal Trends in Estimates of Reference Evapotranspiration over India, J. Hydrol. Eng., 14, 508–515, doi:10.1061/(ASCE)HE.1943-5584.0000006, 2009.

Bastiaanssen, W. G. M., Noordman, E. J. M., Pelgrum, H., Davids, G., Thoreson, B. P., and Allen, R. G.: SEBAL Model with Remotely Sensed Data to Improve Water-Resources Management under Actual Field Conditions, J. Irrig. and Drain. Eng., 131, 85–93, doi:10.1061/(ASCE)0733-9437(2005)131:1(85), 2005.

Cabral, O. M. R., Rocha, H. R., Gash, J. H., Ligo, M. A. V., Tatsch, J. D., Freitas, H. C., and Brasilio, E.: Water use in a sugarcane plantation, GCB Bioenergy, 4, 555–565, doi:10.1111/j.1757-1707.2011.01155.x, 2012.

Campbell, R., Chang, J.-H., and Cox, D.: Evapotranspiration of Sugar Cane in Hawaii as Measured by In-Field Lysimeters in Relation to Climate, in: Proceedings of the International Society of Sugarcane Technologists; 10th Congress, Hawaii, 3–22 May, 1959, 637–645, 1960.

Carr, M. K. V. and Knox, J. W.: The Water Relations and Irrigation Requirements of Sugar Cane (Saccharum Officinarum): A Review, Exp. Agric., 47, 1–25, doi:10.1017/S0014479710000645, 2011.

Choi, M., Woong Kim, T., and Kustas, W. P.: Reliable estimation of evapotranspiration on agricultural fields predicted by the Priestley–Taylor model using soil moisture data from ground and remote sensing observations compared with the Common Land Model, Int. J. Remote Sens., 32, 4571–4587, doi:10.1080/01431161.2010.489065, 2011.

Cobos, D. R. and Baker, J. M.: Evaluation and Modification of a Domeless Net Radiometer, Agron. J., 95, 177–183, 2003.

Cristea, N. C., Kampf, S. K., and Burges, S. J.: Revised Coefficients for Priestley-Taylor and Makkink-Hansen Equations for Estimating Daily Reference Evapotranspiration, J. Hydrol. Eng., 18, 1289–1300, doi:10.1061/(ASCE)HE.1943-5584.0000679, 2013.

Da Silva, T. G. F., de Moura, M. S. B., Zolnier, S., Soares, J. M., Vieira, V. J. S., and Júnior, W. G. F.: Water requirement and crop coefficient of irrigated sugarcane in a semi-arid region, Revista Brasileira de Engenharia Agricola e Ambiental, 16, 64–71, 2012.

Davis, S. L. and Dukes, M. D.: Irrigation scheduling performance by evapotranspiration-based controllers, Agr. Water Manage., 98, 19–28, doi:10.1016/j.agwat.2010.07.006, 2010.

Denmead, O. T.: Plant physiological methods for studying evapotranspiration: Problems of telling the forest from the trees, Agr. Water Manage., 8, 167–189, doi:10.1016/0378-3774(84)90052-0, 1984.

Ding, R., Kang, S., Li, F., Zhang, Y., and Tong, L.: Evapotranspiration measurement and estimation using modified Priestley–Taylor model in an irrigated maize field with mulching, Agr. Forest Meteorol., 168, 140–148, doi:10.1016/j.agrformet.2012.08.003, 2013.

Doorenbos, J. and Pruitt, W.: Crop water requirements. FAO irrigation and drainage paper 24, Land and Water Development Division, FAO, Rome, 1977.

Eichinger, W. E., Parlange, M. B., and Stricker, H.: On the Concept of Equilibrium Evaporation and the Value of the

Priestley-Taylor Coefficient, Water Resour. Res., 32, 161–164, doi:10.1029/95WR02920, 1996.

Elison Timm, O., Diaz, H. F., Giambelluca, T. W., and Takahashi, M.: Projection of changes in the frequency of heavy rain events over Hawaii based on leading Pacific climate modes, J. Geophys. Res., 116, D04109, doi:10.1029/2010JD014923, 2011.

Estévez, J., Gavilán, P., and Berengena, J.: Sensitivity analysis of a Penman-Monteith type equation to estimate reference evapotranspiration in southern Spain, Hydrol. Process., 23, 3342–3353, doi:10.1002/hyp.7439, 2009.

Evensen, C. I., Muchow, R. C., El-Swaify, S. A., and Osgood, R. V.: Yield Accumulation in Irrigated Sugarcane: I. Effect of Crop Age and Cultivar, Agron. J., 89, 638–646, doi:10.2134/agronj1997.00021962008900040016x, 1997.

Evett, S. R., Howell, T. A., Todd, R. W., Schneider, A. D., and Tolk, J. A.: Alfalfa reference ET measurement and prediction, in National irrigation symposium. Proceedings of the 4th Decennial Symposium, Phoenix, Arizona, USA, 14–16 November 2000, 266–272, American Society of Agricultural Engineers, St. Joseph Mich., 2000.

Evett, S. R., Lascano, R. J., Howell, T. A., Tolk, J. A., O'Shaughnessy, S. A., and Colaizzi, P. D.: Single- and dual-surface iterative energy balance solutions for reference ET, Trans. ASABE, 55, 533–541, doi:10.13031/2013.41388, 2012.

Ferguson, I. M. and Maxwell, R. M.: Hydrologic and land–energy feedbacks of agricultural water management practices, Environ. Res. Lett., 6, 014006, doi:10.1088/1748-9326/6/1/014006, 2011.

Fisher, J. B., DeBiase, T. A., Qi, Y., Xu, M., and Goldstein, A. H.: Evapotranspiration models compared on a Sierra Nevada forest ecosystem, Environ. Model. Soft., 20, 783–796, doi:10.1016/j.envsoft.2004.04.009, 2005.

Furon, A. C., Warland, J. S., and Wagner-Riddle, C.: Analysis of Scaling-Up Resistances from Leaf to Canopy Using Numerical Simulations, Agron. J., 99, 1483, doi:10.2134/agronj2006.0335, 2007.

Giambelluca, T. W., Chen, Q., Frazier, A. G., Price, J. P., Chen, Y.-L., Chu, P.-S., Eischeid, J. K., and Delparte, D. M.: Online Rainfall Atlas of Hawai'i, B. Am. Meteorol. Soc., 94, 313–316, doi:10.1175/BAMS-D-11-00228.1, 2013.

Glaz, B., Reed, S. T., and Albano, J. P.: Sugarcane Response to Nitrogen Fertilization on a Histosol with Shallow Water Table and Periodic Flooding, J. Agron. Crop Sci., 194, 369–379, doi:10.1111/j.1439-037X.2008.00329.x, 2008.

Gong, L., Xu, C., Chen, D., Halldin, S., and Chen, Y. D.: Sensitivity of the Penman–Monteith reference evapotranspiration to key climatic variables in the Changjiang (Yangtze River) basin, J. Hydrol., 329, 620–629, doi:10.1016/j.jhydrol.2006.03.027, 2006.

Hargreaves, G. H. and Samani, Z. A.: Economic Considerations of Deficit Irrigation, J. Irrig. Drain. Eng., 110, 343–358, doi:10.1061/(ASCE)0733-9437(1984)110:4(343), 1984.

Hart, Q. J., Brugnach, M., Temesgen, B., Rueda, C., Ustin, S. L., and Frame, K.: Daily reference evapotranspiration for California using satellite imagery and weather station measurement interpolation, Civil Eng. Environ. Systems, 26, 19–33, 2009.

Heinz, D. J. and Osgood, R. V.: A History of the Experiment Station: Hawaiian Sugar Planters' Association, Hawaiian Planters' Record, 61, 1–108, 2009.

Heinz, D. J., Tew, T. L., Meyer, H. K., and Wu, K. K.: Registration of H65-7052 Sugarcane (Reg. No. 51), Crop Sci., 21, 634–635, doi:10.2135/cropsci1981.0011183X002100040050x, 1981.

Hemakumara, H., Chandrapala, L., and Moene, A. F.: Evapotranspiration fluxes over mixed vegetation areas measured from large aperture scintillometer, Agr. Water Manage., 58, 109–122, doi:10.1016/S0378-3774(02)00131-2, 2003.

Hirich, A., Choukr-Allah, R., and Jacobsen, S.-E.: Deficit Irrigation and Organic Compost Improve Growth and Yield of Quinoa and Pea, J. Agronom. Crop Sci., 200, 390–398, doi:10.1111/jac.12073, 2014.

Holwerda, F., Bruijnzeel, L. A., Scatena, F. N., Vugts, H. F., and Meesters, A. G. C. A.: Wet canopy evaporation from a Puerto Rican lower montane rain forest: The importance of realistically estimated aerodynamic conductance, J. Hydrol., 414–415, 1–15, doi:10.1016/j.jhydrol.2011.07.033, 2012.

Holwerda, F., Bruijnzeel, L. A., Barradas, V. L., and Cervantes, J.: The water and energy exchange of a shaded coffee plantation in the lower montane cloud forest zone of central Veracruz, Mexico, Agr. Forest Meteorol., 173, 1–13, doi:10.1016/j.agrformet.2012.12.015, 2013.

Huo, Z., Dai, X., Feng, S., Kang, S., and Huang, G.: Effect of climate change on reference evapotranspiration and aridity index in arid region of China, J. Hydrol., 492, 24–34, doi:10.1016/j.jhydrol.2013.04.011, 2013.

Inman-Bamber, N. G. and McGlinchey, M. G.: Crop coefficients and water-use estimates for sugarcane based on long-term Bowen ratio energy balance measurements, Field Crop Res., 83, 125–138, doi:10.1016/S0378-4290(03)00069-8, 2003.

Inman-Bamber, N. G., Robertson, M. J., Muchow, R. C., Wood, A. W., Pace, R., and Spillman, M. F.: Boosting yields with limited irrigation water, Proc. Aust. Soc. Sugar Cane Techol., 21, 203–211, 1999.

Irmak, S., Payero, J. O., Martin, D. L., Irmak, A., and Howell, T. A.: Sensitivity Analyses and Sensitivity Coefficients of Standardized Daily ASCE-Penman-Monteith Equation, J. Irrig. Drain. Eng., 132, 564–578, doi:10.1061/(ASCE)0733-9437(2006)132:6(564), 2006.

Jensen, M. E.: Water consumption by agricultural plants, in: Water deficit and plant growth, Vol. 1. Development, control and measurement, Vol. II, 1–22, New York, London: Academic Press, 1968.

Jensen, M. E., Burman, R. D., and Allen, R. G.: Evapotranspiration and irrigation water requirements: a manual, American Society of Civil Engineers, New York, N.Y., 1990.

Jin, Y., Randerson, J. T., and Goulden, M. L.: Continental-scale net radiation and evapotranspiration estimated using MODIS satellite observations, Remote Sens. Environ., 115, 2302–2319, doi:10.1016/j.rse.2011.04.031, 2011.

Kaimal, J. C. and Finnigan, J. J.: Atmospheric boundary layer flows: their structure and measurement, Oxford University Press, USA, 1994.

Kang, S., Zhang, L., Liang, Y., Hu, X., Cai, H., and Gu, B.: Effects of limited irrigation on yield and water use efficiency of winter wheat in the Loess Plateau of China, Agr. Water Manage., 55, 203–216, doi:10.1016/S0378-3774(01)00180-9, 2002.

Kljun, N., Calanca, P., Rotach, M. W., and Schmid, H. P.: A Simple Parameterisation for Flux Footprint

Predictions, Boundary-Lay. Meteorol., 112, 503–523, doi:10.1023/B:BOUN.0000030653.71031.96, 2004.

Kueppers, L. M., Snyder, M. A., and Sloan, L. C.: Irrigation cooling effect: Regional climate forcing by land-use change, Geophys. Res. Lett., 34, L03703, doi:10.1029/2006GL028679, 2007.

Lecina, S., Martínez-Cob, A., Pérez, P. J., Villalobos, F. J., and Baselga, J. J.: Fixed versus variable bulk canopy resistance for reference evapotranspiration estimation using the Penman–Monteith equation under semiarid conditions, Agr. Water Manage., 60, 181–198, doi:10.1016/S0378-3774(02)00174-9, 2003.

Leuning, R., van Gorsel, E., Massman, W. J., and Isaac, P. R.: Reflections on the surface energy imbalance problem, Agr. Forest Meteorol., 156, 65–74, doi:10.1016/j.agrformet.2011.12.002, 2012.

Liang, L., Li, L., Zhang, L., Li, J., and Li, B.: Sensitivity of penman-monteith reference crop evapotranspiration in Tao'er River Basin of northeastern China, Chinese Geogr. Sci., 18, 340–347, doi:10.1007/s11769-008-0340-x, 2008.

Liu, H., Zhang, R., and Li, Y.: Sensitivity analysis of reference evapotranspiration (ETo) to climate change in Beijing, China, Desal. Water Treat., 52, 2799–2804, doi:10.1080/19443994.2013.862030, 2014.

Lo, M.-H. and Famiglietti, J. S.: Irrigation in California's Central Valley Strengthens the Southwestern U.S. Water Cycle, Geophys. Res. Lett., 6, 301–306, doi:10.1002/grl.50108, 2013.

Mabhaudhi, T., Modi, A. T., and Beletse, Y. G.: Response of taro (Colocasia esculenta L. Schott) landraces to varying water regimes under a rainshelter, Agr. Water Manage., 121, 102–112, doi:10.1016/j.agwat.2013.01.009, 2013.

Mauder, M. and Foken, T.: Documentation and Instruction Manual of the Eddy Covariance Software Package TK2, Universitätsbibliothek Bayreuth, Bayreuth, available at: https://epub.uni-bayreuth.de/342/1/ARBERG046.pdf (last access: 25 October 2014), 2011.

McAneney, K. J. and Itier, B.: Operational limits to the Priestley-Taylor formula, Irrig. Sci., 17, 37–43, doi:10.1007/s002710050020, 1996.

McGlinchey, M. G. and Inman-Bamber, N.G.: Predicting sugarcane water use with the Penman–Monteith equation, in Proceedings of the International Conference on Evapotranspiration and Irrigation Scheduling, San Antonio, 592–598, ASAE, St. Joseph Mich., 1996.

Mehrez, M. B., Taconet, O., Vidal-Madjar, D., and Valencogne, C.: Estimation of stomatal resistance and canopy evaporation during the HAPEX-MOBILHY experiment, Agr. Forest Meteorol., 58, 285–313, doi:10.1016/0168-1923(92)90066-D, 1992.

Meissner, R., Rupp, H., Seeger, J., Ollesch, G., and Gee, G. W.: A comparison of water flux measurements: passive wick-samplers versus drainage lysimeters, Eur. J. Soil Sci., 61, 609–621, doi:10.1111/j.1365-2389.2010.01255.x, 2010.

Mendez-Costabel, M. P., Wilkinson, K. L., Bastian, S. E. P., Jordans, C., McCarthy, M., Ford, C. M., and Dokoozlian, N. K.: Effect of increased irrigation and additional nitrogen fertilisation on the concentration of green aroma compounds in V itis vinifera L. Merlot fruit and wine: Green aroma compounds in Merlot, Aust. J. Grape Wine Res., 20, 80–90, doi:10.1111/ajgw.12062, 2014.

Moncrieff, J. B., Massheder, J. M., de Bruin, H., Elbers, J., Friborg, T., Heusinkveld, B., Kabat, P., Scott, S., Soegaard, H., and Verhoef, A.: A system to measure surface fluxes of momentum, sensible heat, water vapour and carbon dioxide, J. Hydrol., 188–189, 589–611, doi:10.1016/S0022-1694(96)03194-0, 1997.

Moncrieff, J., Clement, R., Finnigan, J., and Meyers, T.: Averaging, Detrending, and Filtering of Eddy Covariance Time Series, in: Handbook of Micrometeorology, Vol. 29, edited by: Lee, X., Massman, W., and Law, B., 7–31, Kluwer Academic Publishers, Dordrecht, 2004.

Monteith, J. L.: Evaporation and environment, Symp. Soc. Exp. Biol., 19, 205–234, 1965.

Moore, R. C. and Fitschen, J. C.: The drip irrigation revolution in the Hawaiian sugarcane industry, in Visions of the future?: proceedings of the 3. National Irrigation Symposium, held in conjunction with the 11. Annual International Irrigation Exposition, 1990, Phoenix Civic Plaza, Phoenix, Arizona, ASAE, St. Joseph, Mich., 1990.

Obukhov, A. M.: Turbulence in an atmosphere with a nonuniform temperature, Boundary-Lay. Meteorol., 2, 7–29, doi:10.1007/BF00718085, 1971.

Olivier, F. and Singels, A.: The effect of crop residue layers on evapotranspiration, growth and yield of irrigated sugarcane, Water SA, 38, 78–86, doi:10.4314/wsa.v38i1.10, 2012.

Perry, C.: Irrigation reliability and the productivity of water: A proposed methodology using evapotranspiration mapping, Irrig. Drain. Systems, 19, 211–221, doi:10.1007/s10795-005-8135-z, 2005.

Priestley, C. H. B. and Taylor, R. J.: On the Assessment of Surface Heat Flux and Evaporation Using Large-Scale Parameters, Mon. Weather Rev., 100, 81–92, doi:10.1175/1520-0493(1972)100<0081:OTAOSH>2.3.CO;2, 1972.

Puma, M. J. and Cook, B. I.: Effects of irrigation on global climate during the 20th century, J. Geophys. Res., 115, D16120, doi:10.1029/2010JD014122, 2010.

Ramjeawon, T.: Water resources management on the small Island of Mauritius, Int. J. Water Resour. D., 10, 143–155, doi:10.1080/07900629408722619, 1994.

Rao, L., Sun, G., Ford, C., and Vose, J.: Modeling potential evapotranspiration of two forested watersheds in the southern Appalachians, Trans. ASABE, 54, 2067–2078, doi:10.13031/2013.40666, 2011.

Reichstein, M., Falge, E., Baldocchi, D., Papale, D., Aubinet, M., Berbigier, P., Bernhofer, C., Buchmann, N., Gilmanov, T., Granier, A., Grunwald, T., Havrankova, K., Ilvesniemi, H., Janous, D., Knohl, A., Laurila, T., Lohila, A., Loustau, D., Matteucci, G., Meyers, T., Miglietta, F., Ourcival, J.-M., Pumpanen, J., Rambal, S., Rotenberg, E., Sanz, M., Tenhunen, J., Seufert, G., Vaccari, F., Vesala, T., Yakir, D., and Valentini, R.: On the separation of net ecosystem exchange into assimilation and ecosystem respiration: review and improved algorithm, Glob. Change Biol., 11, 1424–1439, doi:10.1111/j.1365-2486.2005.001002.x, 2005.

Salazar, M. R., Hook, J. E., Garcia y Garcia, A., Paz, J. O., Chaves, B., and Hoogenboom, G.: Estimating irrigation water use for maize in the Southeastern USA: A modeling approach, Agr. Water Manage., 107, 104–111, doi:10.1016/j.agwat.2012.01.015, 2012.

Salinas, F. and Namken, L. N.: Irrigation scheduling for sugarcane in the Lower Rio Grande Valley of Texas, Proc. Am. Soc. Sugar Cane Technol., 6, 186–191, 1977.

Smith, M.: The application of climatic data for planning and management of sustainable rainfed and irrigated crop production, Agr. Forest Meteorol., 103, 99–108, doi:10.1016/S0168-1923(00)00121-0, 2000.

Soopramanien, G. C., Berthelot, B., and Batchelor, C. H.: Irrigation research, development and practice in Mauritius, Agr. Water Manage., 17, 129–139, doi:10.1016/0378-3774(90)90060-C, 1990.

Sprintsin, M., Chen, J. M., Desai, A., and Gough, C. M.: Evaluation of leaf-to-canopy upscaling methodologies against carbon flux data in North America, J. Geophys. Res., 117, G011023, doi:10.1029/2010JG001407, 2012.

Suleiman, A. A. and Hoogenboom, G.: Comparison of Priestley-Taylor and FAO-56 Penman-Monteith for Daily Reference Evapotranspiration Estimation in Georgia, J. Irrig. Drain. Eng., 133, 175–182, doi:10.1061/(ASCE)0733-9437(2007)133:2(175), 2007.

Suleiman, A. A. and Hoogenboom, G.: A comparison of ASCE and FAO-56 reference evapotranspiration for a 15-min time step in humid climate conditions, J. Hydrol., 375, 326–333, doi:10.1016/j.jhydrol.2009.06.020, 2009.

Szeicz, G. and Long, I. F.: Surface Resistance of Crop Canopies, Water Resources Res., 5, 622–633, doi:10.1029/WR005i003p00622, 1969.

Tang, Q., Peterson, S., Cuenca, R. H., Hagimoto, Y., and Lettenmaier, D. P.: Satellite-based near-real-time estimation of irrigated crop water consumption, J. Geophys. Res., 114, D05114, doi:10.1029/2008JD010854, 2009.

Thompson, G. D. and Boyce, J. P.: Daily measurements of potential evapotranspiration from fully canopied sugarcane, Agric. Meteorol., 4, 267–279, doi:10.1016/0002-1571(67)90027-1, 1967.

Tian, D. and Martinez, C. J.: Forecasting Reference Evapotranspiration Using Retrospective Forecast Analogs in the Southeastern United States, J. Hydrolmeteorol., 13, 1874–1892, doi:10.1175/JHM-D-12-037.1, 2012.

Tolk, J. A., Howell, T. A., Steiner, J. L., and Krieg, D. R.: Corn canopy resistance determined from whole plant transpiration, in: Proceedings of the International Conference on Evapotranspiration and Irrigation Scheduling, San Antonio, 347–351, ASAE, St. Joseph Mich., 1996.

Ventura, F., Spano, D., Duce, P., and Snyder, R. L.: An evaluation of common evapotranspiration equations, Irrig. Sci., 18, 163–170, doi:10.1007/s002710050058, 1999.

Vickers, D. and Mahrt, L.: Quality Control and Flux Sampling Problems for Tower and Aircraft Data, J. Atmos. Ocean. Technol., 14, 512–526, doi:10.1175/1520-0426(1997)014<0512:QCAFSP>2.0.CO;2, 1997.

Vörösmarty, C. J., Federer, C. A., and Schloss, A. L.: Potential evaporation functions compared on US watersheds: Possible implications for global-scale water balance and terrestrial ecosystem modeling, J. Hydrol., 207, 147–169, doi:10.1016/S0022-1694(98)00109-7, 1998.

Waterloo, M. J., Bruijnzeel, L. A., Vugts, H. F., and Rawaqa, T. T.: Evaporation from Pinus caribaea plantations on former grassland soils under maritime tropical conditions, Water Resour. Res., 35, 2133–2144, doi:10.1029/1999WR900006, 1999.

Webb, E. K., Pearman, G. I., and Leuning, R.: Correction of flux measurements for density effects due to heat and water vapour transfer, Q. J. Roy. Meteorol. Soc., 106, 85–100, doi:10.1002/qj.49710644707, 1980.

Widmoser, P.: A discussion on and alternative to the Penman–Monteith equation, Agr. Water Manage., 96, 711–721, doi:10.1016/j.agwat.2008.10.003, 2009.

Zhang, B., Kang, S., Li, F., and Zhang, L.: Comparison of three evapotranspiration models to Bowen ratio-energy balance method for a vineyard in an arid desert region of northwest China, Agr. Forest Meteorol., 148, 1629–1640, doi:10.1016/j.agrformet.2008.05.016, 2008.

Zhao, L., Xia, J., Xu, C., Wang, Z., Sobkowiak, L., and Long, C.: Evapotranspiration estimation methods in hydrological models, J. Geograph. Sci., 23, 359–369, doi:10.1007/s11442-013-1015-9, 2013.

8

Using measured soil water contents to estimate evapotranspiration and root water uptake profiles – a comparative study

M. Guderle[1,2,3] **and A. Hildebrandt**[1,2]

[1]Friedrich Schiller University, Institute for Geosciences, Burgweg 11, 07749 Jena, Germany
[2]Max Planck Institute for Biogeochemistry, Biogeochemical Processes, Hans-Knöll-Str. 10, 07745 Jena, Germany
[3]International Max Planck Research School for Global Biogeochemical Cycles, Hans-Knöll-Str. 10, 07745 Jena, Germany

Correspondence to: M. Guderle (marcus.guderle@uni-jena.de)

Abstract. Understanding the role of plants in soil water relations, and thus ecosystem functioning, requires information about root water uptake. We evaluated four different complex water balance methods to estimate sink term patterns and evapotranspiration directly from soil moisture measurements. We tested four methods. The first two take the difference between two measurement intervals as evapotranspiration, thus neglecting vertical flow. The third uses regression on the soil water content time series and differences between day and night to account for vertical flow. The fourth accounts for vertical flow using a numerical model and iteratively solves for the sink term. None of these methods requires any a priori information of root distribution parameters or evapotranspiration, which is an advantage compared to common root water uptake models. To test the methods, a synthetic experiment with numerical simulations for a grassland ecosystem was conducted. Additionally, the time series were perturbed to simulate common sensor errors, like those due to measurement precision and inaccurate sensor calibration. We tested each method for a range of measurement frequencies and applied performance criteria to evaluate the suitability of each method. In general, we show that methods accounting for vertical flow predict evapotranspiration and the sink term distribution more accurately than the simpler approaches. Under consideration of possible measurement uncertainties, the method based on regression and differentiating between day and night cycles leads to the best and most robust estimation of sink term patterns. It is thus an alternative to more complex inverse numerical methods. This study demonstrates that highly resolved (temporally and spatially)

soil water content measurements may be used to estimate the sink term profiles when the appropriate approach is used.

1 Introduction

Plants play a key role in the Earth system by linking the water and the carbon cycle between soil and atmosphere (Feddes et al., 2001; Chapin et al., 2002; Feddes and Raats, 2004; Teuling et al., 2006b; Schneider et al., 2009; Seneviratne et al., 2010; Asbjornsen et al., 2011). Knowledge of evapotranspiration and especially root water uptake profiles is key to understanding plant–soil-water relations and thus ecosystem functioning, in particular efficient plant water use, storage keeping and competition in ecosystems (Davis and Mooney, 1986; Le Roux et al., 1995; Jackson et al., 1996; Hildebrandt and Eltahir, 2007; Arnold et al., 2009; Schwendenmann et al., 2014).

For estimation of root water uptake, models are prevalent in many disciplines. Most commonly, root water uptake is applied as a sink term S, incorporated in the 1-D soil water flow equation (Richards equation, Eq. 1)

$$\frac{\partial \theta}{\partial t} = \frac{\partial}{\partial z}\left[K(h)\left(\frac{\partial h}{\partial z} + 1\right)\right] - S(z, t), \tag{1}$$

where θ is the volumetric soil water content, t is time, z is the vertical coordinate, h is the soil matric potential, $K(h)$ is the unsaturated soil hydraulic conductivity and $S(z, t)$ is the sink term (water extraction by roots, evaporation, etc.). The sink term profile $S(z, t)$ depends on root activity, which has to be known previously. Often root activity is assumed to be related to rooting profiles, represented by power laws (Gale and Grigal, 1987; Jackson et al., 1996; Schenk, 2008; Kuhlmann et al., 2012). The parameters of those rooting profile functions are cumbersome to measure in the field, and the relevance for root water uptake distribution is also uncertain (Hamblin and Tennant, 1987; Lai and Katul, 2000; Li et al., 2002; Doussan et al., 2006; Garrigues et al., 2006; Schneider et al., 2009). Therefore, assumptions have to be made in order to determine the sink term for root water uptake in soil water flow models. The lack of an adequate description of root water uptake parameters was mentioned by Gardner (1983) and is currently still an issue (Lai and Katul, 2000; Hupet et al., 2002; Teuling et al., 2006a, b). For those reasons, methods for estimating root water uptake are a paramount requirement.

Standard measurements, for instance of soil water content profiles, are recommended to be used for estimation of evapotranspiration and root water uptake at low cost, since the evolution of soil moisture in space and time is expected to contain information on root water uptake (Musters and Bouten, 2009; Hupet et al., 2002; Zuo and Zhang, 2002; Teuling et al., 2006a). Methods using these measurements are, for instance, simple water balance approaches, which estimate evapotranspiration (Wilson et al., 2001; Schume et al., 2005; Kosugi and Katsuyama, 2007; Breña Naranjo et al., 2011) and root water uptake (Clothier and Green, 1995; Coelho and Or, 1996; Hupet et al., 2002) by calculating the difference in soil water storage between two different observation times. The advantages of these simple water balance methods are the small amount of information required and the simple methodology. However, a disadvantage is that the depletion of soil water is assumed to occur only by root water uptake and soil evaporation, and soil water fluxes are negligible. This is only the case during long dry periods with high atmospheric demand (Hupet et al., 2002).

A possible alternative which allows for the consideration of vertical soil water fluxes is the inverse use of numerical soil water flow models (Musters and Bouten, 1999; Musters et al, 2000; Vrugt et al., 2001; Hupet et al., 2002; Zuo and Zhang, 2002). Root water uptake or parameters on the root water uptake function are estimated by minimizing the differences between measured soil water contents and the corresponding model results by an objective function (Hupet et al., 2002). However, the quality of the estimation depends, on the one hand, strongly on system boundary conditions (e.g., incoming flux, drainage flux or location of the groundwater table) and soil parameters (e.g., hydraulic conductivity), which are, on the other hand, notoriously uncertain under natural conditions (Musters and Bouten, 2000; Kollet, 2009). Another problem is that the applied models for soil water flow potentially ignore biotic processes. For example, Musters et al. (2000) and Hupet et al. (2002) attempted to fit parameters for root distributions in a model determining uptake profiles from water availability, whereas empirical and modeling

studies suggest that adjustment of root water uptake distribution may also be from physiological adaptations (Jackson et al., 2000; Zwieniecki et al., 2003; Bechmann et al., 2014). In order to avoid this problem, Zuo and Zhang (2002) coupled a water balance approach to a soil water model, which enabled them to estimate root water uptake without the a priori estimation of root water uptake parameters.

A second option for accounting for vertical soil water flow in a water balance approach is to analyze the soil moisture fluctuation between day and night (Li et al., 2002). In comparatively dry soil, Li et al. (2002) fitted third-order polynomials to the daytime- and nighttime-measured soil water content time series and calculated vertical soil water flow using the first derivative of the fitted polynomials during nighttime.

Up to now, little effort has been made to compare those different data-driven methods for estimating evapotranspiration and root water uptake profiles in temperate climates. In this paper, we compare those water balance methods we are aware of that do not require any a priori information of root distribution parameters. We used artificial data of soil moisture and sink term profiles to compare the quality of the estimates of the different methods. Furthermore, we investigated the influence of sensor errors on the outcomes, as these uncertainties can have a significant impact on both data-driven approaches and soil hydrological models (Spank et al., 2013). For this, we artificially introduced measurement errors to the synthetic soil moisture time series that are typical for soil water content measurements: sensor calibration error and limited precision.

Our results indicate that highly resolved soil water content measurements can provide reliable predictions of the sink term or root water uptake profile when the appropriate approach is used.

2 Material and methods

Table A1 summarizes the variable names used in this section together with their units.

2.1 Target variable and general procedure

The evapotranspiration E consists of soil evaporation E_s and the plant transpiration E_t (Eq. 2):

$$E = (E_s + E_t). \tag{2}$$

The distinction between soil evaporation and combined transpiration is not possible for any of the applied water balance methods. Therefore, the water extraction from soil by plant roots and soil evaporation is referred to as the sink term profile in the rest of the paper. The integrated sink term over

Table 1. The abbreviation and full name of the methods for further use, overviews of the four applied data-driven methods, and the required input data.

Abbreviation	Method	Method short description	Input data
sssl	Single-step, single-layer water balance	Water balance (Breña Naranjo et al., 2011)	Volumetric soil water content at a single depth, precipitation
ssml	Single-step, multi-layer water balance	Water balance over entire soil profile (Clothier and Green, 1995; Coelho and Or, 1996; Hupet et al., 2002)	Volumetric soil water content at several depths, precipitation
msml	Multi-step, multi-layer regression	Approach to use the short-term fluctuations of soil moisture (Li et al., 2002)	Volumetric soil water content at several depths, precipitation
im	Inverse model	Water balance solved iteratively with a numerical soil water flow model (Zuo and Zhang, 2002; Ross, 2003)	Soil hydraulic parameters Volumetric soil water content at several depths, precipitation

the entire soil profile results in the total evapotranspiration (Eq. 3):

$$E(t) = \int_{z=z_r}^{0} S(t,z)\mathrm{d}z \rightarrow E_j = \sum_{i=1}^{n} S_{i,j} \cdot d_{z,i}, \qquad (3)$$

where z is the soil depth, $d_{z,i}$ is the thickness of the soil layer i, t is time and j is the time step. For matters of simplicity we will drop the index j when introducing the estimation methods in the following.

In this study, synthetic time series of volumetric soil water content generated by a soil water flow model coupled with a root water uptake model (Sect. 2.3) were treated as measured data and are used as the basis for all methods (Sect. 2.2) estimating the sink term $\widetilde{S}(z)$, and total evapotranspiration $\widetilde{E}$. In order to investigate the influence of sensor errors, the generated time series were systematically disturbed, as shown in Sect. 2.4. Based on these estimations, we evaluate the data-driven methods on predicting evapotranspiration $\widetilde{E}$ and sink term profiles using the quality criteria given in Sect. 2.5. As the depth at which a given fraction of root water uptake occurred is often of interest in ecohydrological studies (e.g., Clothier and Green, 1999; Plamboeck et al., 1999; Ogle et al., 2004), estimated sink term profiles were compared accordingly. Specifically, we determined up to which depths 25, 50 and 90 % ($z_{25\%}$, $z_{50\%}$ and $z_{90\%}$) of water extraction takes place.

2.2 Investigated data-driven methods for estimation of the sink term profile

In the following we introduce the four investigated methods. They are summarized in Table 1.

2.2.1 Single-step, single-layer (sssl) water balance

Breña Naranjo et al. (2011) derived the sink term using time series of rainfall and changes of soil water content between two observation times (single step), based on measurements at one single soil depth (single layer). The complete water balance equation for this single-layer method is

$$\widetilde{E}_{\mathrm{sssl}} = P - q - z_r \frac{\Delta\theta}{\Delta t}, \qquad (4)$$

where z_r is the active rooting depth, which is also the depth of the single soil layer, and is taken equal to the measurement depth of volumetric soil water content, θ. Δt indicates the length of the considered single time step. P is the rainfall and q the percolation out of the soil layer during the same time step. When rainfall occurs, infiltration as well as soil water flow takes place. It is assumed that percolation occurs only during this time and persists only up to several hours after the rainfall event (Breña Naranjo et al., 2011). Since the percolation flux is unknown, the methods cannot be applied during these wet times. During dry periods, q is set to zero and Eq. (4) simplifies to Eq. (5) (Breña Naranjo et al., 2011):

$$\widetilde{E}_{\mathrm{sssl}} = z_r \frac{\Delta\theta}{\Delta t}. \qquad (5)$$

We applied Eq. (5) to estimate evaporation (in the single-layer method equal to the sink term) from artificial soil water contents at 30 cm. Required input information is thus only time series of soil water content and active rooting depth z_r. Additionally, rainfall measurements are required to select dry periods, where no percolation occurs. These could start several hours up to several days after a rainfall event (Breña Naranjo et al., 2011), and the exact timing depends on the amount of rainfall and the site-location parameters like soil type and vegetation. In this study we waited until 24 h after the end of the precipitation event before applying the model.

2.2.2 Single-step, multi-layer (ssml) water balance

This method is similar to the sssl method introduced above. It calculates the sink term based on two observation times (single step), but is extended to several measurement depths (multi-layer). The water balance during dry periods of each layer is the same as in Eq. (5), and uptake in individual layers is calculated by neglecting vertical soil water fluxes and therefore assuming that the change in soil water content is only caused by root water uptake (Hupet et al., 2002):

$$\widetilde{S}_{\mathrm{ssml},\,i} = d_{z,\,i}\frac{\Delta\theta_i}{\Delta t}, \qquad (6)$$

where $\widetilde{S}_{\mathrm{ssml},i}$ is the estimated sink term in soil layer i, $\Delta\theta_i$ is the change in soil water content in the soil layer i over the single time step (Δt) and $d_{z,\,i}$ is the thickness of the soil layer i. Actual evapotranspiration (E_{ssml}) is calculated by summing up $\widetilde{S}_{\mathrm{ssml},i}$ over all depths in accordance with (Eq. 3). The application of the ssml method is restricted to dry periods. It requires time series of volumetric soil water content and rainfall measurements as input to select dry periods.

2.2.3 Multi-step, multi-layer (msml) regression

The third method derives actual evapotranspiration and sink term profiles from diurnal fluctuation of soil water contents (Li et al., 2002). It uses a regression over multiple time steps (multi-step) and can be applied at several measurement depths (multi-layer).

During daytime, evapotranspiration leads to a decrease in volumetric soil water content. This extraction of soil water extends over the entire active rooting depth. Additionally, soil water flow occurs both at night and during the daytime (Khalil et al., 2003; Verhoef et al., 2006; Chanzy et al., 2012), following potential gradients in the soil profile. Thus, during dry weather conditions, the time series of soil water content shows a clear day–night signal (Fig. 1). We split up the time series by fitting a linear function to each day and night branch of the time series. The onset of transpiration is mainly defined by opening and closure of plant stomata, which is according to the supply of solar energy (Loheide, 2008; Maruyama and Kuwagata, 2008; Sánchez et al., 2013), usually 1 or 2 h after sunrise or before sunset (Lee, 2009).

Here, the basic assumption is that the soil water flow does not change significantly between day and night (Fig. S1 in the Supplement). The slope of the fitted linear functions gives the rate of root water extraction and vertical flow. This can also be shown mathematically by disassembling the Richards equation (Eq. 1) in vertical flow (subscript flow) and sink

term (subscript extr) (Eq. 7), whereas the change in soil water content over time ($\partial\theta/\partial t$) integrates both fluxes:

$$\frac{\partial\theta}{\partial t} = \left.\frac{\partial\theta}{\partial t}\right|_{\mathrm{flow}} + \left.\frac{\partial\theta}{\partial t}\right|_{\mathrm{extr}} = m_{\mathrm{tot}}, \qquad (7)$$

where m_{tot} corresponds to the slope of the fitted linear function for the day or night branch. Assuming that evapotranspiration during the night is negligible, the slope for the night branch is entirely due to soil water flow. During the day, uptake processes and soil water flow act in parallel:

$$\mathrm{day}: \quad m_{\mathrm{tot}} = m_{\mathrm{flow}} + m_{\mathrm{extr}}, \qquad (8a)$$

$$\mathrm{night}: \quad m_{\mathrm{tot}} = m_{\mathrm{flow}}. \qquad (8b)$$

The sink term can be calculated from Eq. (8a), assuming that m_{flow} can be estimated from Eq. (8b) and using the average of the antecedent and the preceding night. A similar procedure has previously been applied in diurnal groundwater table fluctuations (Loheide, 2008). Also the extraction will be overestimated if day and night fluxes are not separately considered. With the soil layer thickness of the respective layer i ($d_{z,i}$) taken into account, the mean daily sink term of soil layer i ($\widetilde{S}_{\mathrm{msml},i}$) is obtained:

$$\widetilde{S}_{\mathrm{msm},\,i} = (m_{\mathrm{tot},\,i} - \bar{m}_{\mathrm{flow},\,i})\cdot d_{z,\,i}. \qquad (9)$$

Since a diurnal cycle of soil moisture is only identifiable up to a time interval of 12 h, the regression method is limited to a minimum measurement frequency of 12 h. Furthermore, as rainfall causes changes of soil water content and blurs the diurnal signal, the msml regression is only applicable during dry periods. Time series of soil water content and rainfall measurements to select dry periods are required as input.

2.2.4 Inverse model (im)

The fourth approach is the most complex. The inverse model (im) estimates the average root water uptake by solving the Richards equation (Eq. 1) and iteratively searching for the sink term profile which produces the best fit between the numerical solution and measured values of soil moisture content (Zuo and Zhang, 2002). The advantage of this method is the estimation of root water uptake without the a priori estimation of rooting profile function parameters, since they are highly uncertain, as elucidated in the Introduction. We implemented the inverse water balance approach after Zuo and Zhang (2002) with the Fast Richards Solver (Ross, 2003), which is available as Fortran 90 code. We modified the original method by changing the convergence criterion. In the following section, we first introduce the iterative procedure as proposed by Zuo and Zhang (2002) and then explain the modification which we made.

The iterative procedure by Zuo and Zhang (2002) runs the numerical model over a given time step (Δt) in order to estimate the soil water content profile $\widetilde{\theta}_i^{(v=0)}$ at the end of the time step, and assuming that the sink term ($\widetilde{S}_{\mathrm{im},i}^{(v=0)}$) is zero over the entire profile. Here $\sim$ depicts the estimated values at the respective soil layer i, and v indicates the iteration step. Next, the sink term profile $\widetilde{S}_{\mathrm{im},i}^{(v=1)}$ is set equal to the difference between previous approximation $\widetilde{\theta}_i^{(v=0)}$ and measurements θ_i while accounting for soil layer thickness and the length of the time step for units.

In the following iterations, $\widetilde{S}_{\mathrm{im},i}^{(v)}$ is used with the Richards equation to calculate the new soil water contents $\widetilde{\theta}_i^{(v)}$. The new average sink term $\widetilde{S}_{\mathrm{im},i}^{(v+1)}$ is then determined with Eq. (10):

$$\widetilde{S}_{\mathrm{im},i}^{(v+1)} = \widetilde{S}_{\mathrm{im},i}^{(v)} + \frac{\widetilde{\theta}_i^{(v)} - \theta_i}{\Delta t} \cdot d_{z,i}. \tag{10}$$

This iteration process continues until a specified decision criterion ε_{ZZ} is reached:

$$\varepsilon_{ZZ} \geq \frac{1}{n} \sum_{i=1}^{n} \left[\frac{\widetilde{\theta}_i^{(v)} - \theta_i}{\theta_i} \right]^2, \tag{11}$$

where n is the number of soil layers in the soil column.

Since ε_{ZZ} is a normalized root-mean-square error over depth, good and poor estimations cancel between layers. This leads to termination of the iterative procedure even if the estimation of the sink term is very poor in several layers. We therefore propose a slightly adapted termination process which applies to separate soil layers as follows. The estimation of the sink term in general is applied as proposed by Zuo and Zhang (2002):

1. Calculate the difference between the estimated and measured soil water content (Eq. 12) and compare the change in this difference to the difference of the previous iteration (Eq. 13):

$$e_i^{(v)} = \left| \theta_i - \widetilde{\theta}_i^{(v)} \right|, \tag{12}$$

$$\varepsilon_{\mathrm{GH},i}^{(v)} = e_i^{(v-1)} - e_i^{(v)}. \tag{13}$$

2. In soil layers where $\varepsilon_{\mathrm{GH}}^{(v)} < 0$, set the root water uptake rate back to the value of the previous iteration ($\widetilde{S}_{\mathrm{im},i}^{(v+1)} = \widetilde{S}_{\mathrm{im},i}^{(v-1)}$), since the current iteration was no improvement. Only if $\varepsilon_{\mathrm{GH},i}^{(v)} \geq 0$, go to step (3). This prevents acceptance of the estimated sink term $\widetilde{S}_{\mathrm{im},i}^{(v)}$ even if it leads to a worse fit than the previous iteration.

3. If $e_i^{(v)} > 1 \times 10^{-4}$, calculate $\widetilde{S}_{\mathrm{im},i}^{(v+1)}$ according Eq. (10); otherwise the current iteration sink term ($\widetilde{S}_{\mathrm{im},i}^{(v+1)} = \widetilde{S}_{\mathrm{im},i}^{(v)}$) is retained, as it results in a good fit between estimated and measured soil water contents.

The iteration process continues until the convergence criterion $\varepsilon_{\mathrm{GH}}^{(v)}$ (Eq. 13) no longer changes between iterations (i.e., all layers have reached a satisfactory fit), or after a specified number of iterations (we chose 3000).

Besides the soil water content measurements and the rainfall, the input information required is the soil hydraulic parameters.

2.3 Generation of synthetic reference data

We used synthetic time series of volumetric soil water content with a measurement frequency of 1, 3, 6, 12 and 24 h. The time series of soil water content as well as the sink term profiles were generated with a Soil water flow model (Fast Richards Solver (Ross, 2003), same as used in Sect. 2.2 for the im). These were treated as measured data and are used as the basis for all methods. The synthetic data are based on meteorological and soil data from the Jena Biodiversity Experiment (Roscher et al., 2011). Root water uptake was calculated using a simple macroscopic root water uptake model which uses an exponential root distribution with water stress compensation (Li et al., 2001). Soil evaporation is taken as 20 % of total evapotranspiration.

The soil profile is based on the Jena Experiment, both in terms of measurement design and soil properties. The model was set up for a one-dimensional homogeneous soil profile 220 cm deep. Measurement points were set at depths of 15, 30, 60, 100, 140, 180 and 220 cm. The spatial resolution of the soil model is according to the measurement points 15, 15, 30, 40, 40, 40 and 40 cm. The advantage of the applied soil water flow model is that the water fluxes are calculated with the matrix flux potential (Kirchhoff transformation), which allows for spatial discretization with large nodal spacing (Ross, 2006). We used a maximum rooting depth of 140 cm, with 60 % of root length density located in the top 15 cm of the root zone, which corresponds to mean values measured on the field site (Ravenek et al., 2014). We used van Genuchten soil hydraulic parameters (van Genuchten, 1980) derived from the program ROSETTA (Schaap et al., 2001) based on the texture of a silty loam: $\theta_{\mathrm{s}} = 0.409$ ($\mathrm{cm^3\,cm^{-3}}$), $\theta_{\mathrm{r}} = 0.069$ ($\mathrm{cm^3\,cm^{-3}}$), $K_{\mathrm{sat}} = 1.43 \times 10^{-6}$ ($\mathrm{m\,s^{-1}}$), $\alpha = 0.6$ ($\mathrm{m^{-1}}$) and $n_{\mathrm{vG}} = 1.619$ (–).

Upper boundary conditions are derived from measured precipitation and potential evapotranspiration calculated after Penman–Monteith (Allen et al., 1998) from measurements of the climate station at the experimental site (Weather Station Saaleaue, Max Planck Institute for Biogeochemistry, http://www.bgc-jena.mpg.de/wetter/). The weather data used have a measurement resolution of 10 min. Before applying evapotranspiration and rainfall as input data to generate the

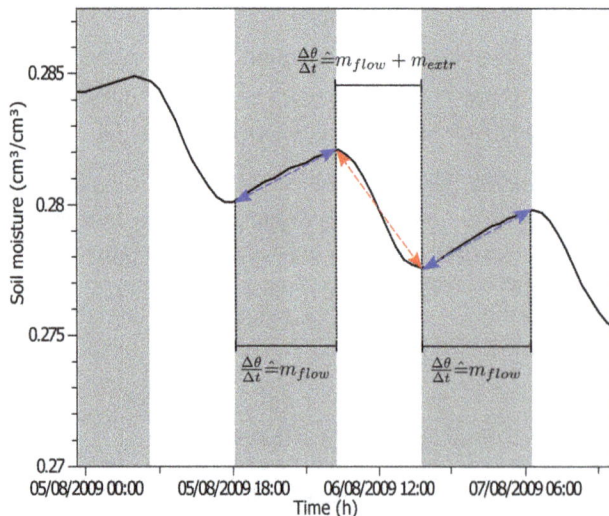

Figure 1. Short-term fluctuations in soil moisture in 15 cm depth during August 2009, showing the rewetting of soil at nighttime (blue line) and the water extraction during the day (red line); dashed lines depict the change between times with soil water extraction (white) and rewetting of soil (grey).

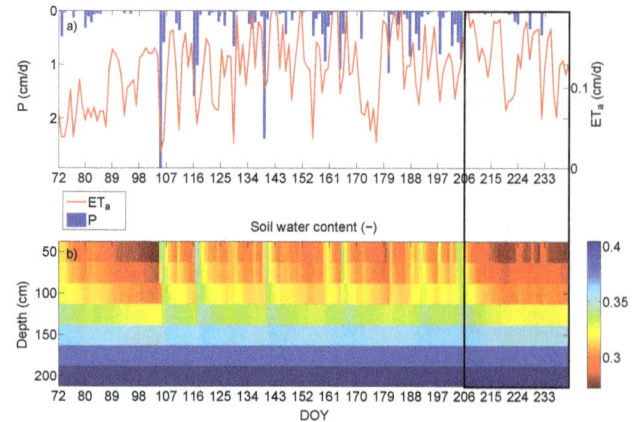

Figure 2. Actual evapotranspiration (ET_a) and precipitation (P) ($cm\,day^{-1}$) in the growing season (from March 2009 to September 2009) (**a**) and synthetic time series of soil water content (**b**) with daily resolution.

synthetic reference soil moisture and root water uptake data, both data sets were aggregated to the temporal resolutions applied for the reference run (1 h). Soil moisture and root water uptake were generated with the same temporal resolution. When translating the evapotranspiration into sink term profiles (four-digit precision), rounding errors introduce a small inaccuracy. Thus, the sum of the sink term in the reference run deviates by 0.02 % compared to the original evapotranspiration.

The lower boundary is given by the groundwater table, which fluctuates around -200 cm at the field site, but was set to constant head for simplification. Initial conditions are taken as the equilibrium (no flow) hydraulic potential profile in the soil.

We run the model with precipitation data from the field site for the year 2009, starting on 1 January to calculate time series of soil water content and the root water uptake up to September 2009. The atmospheric boundary conditions during the growing season are shown in Fig. 2a as daily values. For testing the methods, we used the period from 26 July to 28 August 2009, which covers a dry period with little rainfall (Fig. 2, black-outlined area). The times were chosen to cover a representative but dry period during the growing season and to guarantee a warm-up phase for the soil model.

The described forward simulation produces time series of soil water contents and root water uptake. Soil water content time series were used instead of measured data (synthetic measurements) as input for the investigated methods, while evapotranspiration and sink term profiles were used to evaluate them, based on the quality criteria described in Sect. 2.5.

2.4 Influence of soil moisture sensor uncertainty

Data-driven methods are as good as their input data. Therefore, we investigate and quantify the influence of common uncertainties of soil moisture sensor measurements on the estimation of sink term profiles. Sensor performance is usually characterized by three criteria, namely the accuracy, the precision and the resolution. The correctness of a measurement is described by the accuracy and for water content sensors depends greatly on the soil-specific calibration. Repeatability of many single measurements is referred to as precision, while the resolution describes the fineness of a measurement.

In this paper, we investigated the uncertainty of the applied methods stemming from calibration error (accuracy) and precision. For this we superimposed the original synthetic soil water content measurements generated in Sect. 2.3 with artificial errors. Three types of errors were implemented, as follows. (i) Precision error: the time series for each soil layer were perturbed with Gaussian noise of zero mean and standard deviation of 0.067 vol. % corresponding to a precision of 0.2 vol. %; (ii) calibration error: the perturbed time series were realigned along a new slope, which pivoted around a random point within the measurement range and a random intercept within ± 1.0 vol. %; (iii) calibration and precision: perturbed series were created as a random combination of (i) and (ii), which is a common case in field studies (Spank et al., 2013). Errors were applied independently to all soil depths, and 100 new time series were created for each of the error types. We determined the quality of the estimation methods using the median of 100 ensemble simulations with the 100 perturbed input time series. The values for the applied calibration uncertainty and precision are taken from the technical manual of the IMKO TRIME©-PICO32 soil moisture sensor (http://www.imko.de/en/products/soilmoisture/ soil-moisture-sensors/trimepico32).

A common procedure with environmental measurements for dealing with precision errors is smoothing of the measured time series (Li et al., 2002; Peters et al., 2013), which we also reproduced by additionally applying a moving average filter on the disturbed soil moisture time series.

2.5 Evaluation criteria

A successful model should be able to reproduce the first and second moment of the distribution of the observed values (Gupta et al., 2009), and we used a similar approach to assess the quality of the methods for estimating the total evapotranspiration and the sink term profiles. The first and the second moment refer to the mean and the standard deviation. Additionally, the correlation coefficient evaluates whether the model is able to reproduce the timing and the shape of observed time series. To compare the applicability and the quality of the four methods, we use three performance criteria suggested by Gupta et al. (2009): (i) the correlation coefficient (R), (ii) the relative variability measure (RV) and (iii) the bias (b), which are described in this section. The comparison is based on daily values.

First, we use R to estimate the strength of the linear correlation between estimated ($\tilde{\ }$) and synthetic values:

$$R = \frac{\mathrm{Cov}(\tilde{x}, x)}{s_x \cdot s_{\tilde{x}}}, \qquad (14)$$

where "Cov" is the covariance of estimated and observed (synthetic) values, and s_x and $s_{\tilde{x}}$ are the standard deviations of synthetic and estimated values, respectively. The variable x stands for any of the variables of interest, such as total evapotranspiration or $z_{25\%}$. R ranges between -1 and $+1$. The closer R is to 1, the better the estimate.

Second, we use the relative variability in estimated and synthetic data (RV) to determine the ability of the particular method to reproduce the observed variance (Gupta et al., 2009):

$$\mathrm{RV} = \frac{s_{\tilde{x}}}{s_x}. \qquad (15)$$

RV values around 1 indicate a good estimation procedure.

Third, we use the relative bias (b) to describe the mean systematic deviation between estimated ($\tilde{\ }$) and observed (synthetic) values, which is never captured by R:

$$b = \frac{\bar{\tilde{x}} - \bar{x}}{\bar{x}} \cdot 100(\%), \qquad (16)$$

where $\bar{\tilde{x}}$ and $\bar{x}$ are the means of the estimated and synthetic data, respectively. The best model performance is reached if the bias is close to zero.

3 Results

In total, we compared synthetic evapotranspiration rates from 33 consecutive days in July/August 2009. Evapotranspiration could not be estimated for days with rainfall using either the sssl or ssml method, nor with the msml regression. Therefore, we excluded all days with rainfall from the analysis for all considered methods. In Sect. 3.1 and 3.2 we first consider the performance of the estimation methods on undisturbed synthetic time series, i.e., we ignore measurement errors or assume they do not exist. The influence of measurement errors is investigated in Sect. 3.3.

3.1 Evapotranspiration derived by soil water content measurements

The performance of the data-driven methods depends strongly on the complexity of the respective method, which substantially increases with a higher degree of complexity. However, the influence of the measurement frequency differs considerably among the four methods.

The im predicted the daily evapotranspiration for a measurement frequency of 12 h with a very small relative bias of 0.89 %, which is the best value of all investigated methods. Additionally, the im reaches the best R value ($R = 0.99$) for all measurement frequencies (Table 2), and closely follows the 1 : 1 line between synthetic and estimated evapotranspiration (Fig. 3a, b). However, the RV and the relative bias indicate better prediction with decreasing measurement frequency.

The second-best method is the msml regression, in particular when applied for high temporal resolution measurements (1 and 3 h). There, the bias is comparatively small ($\pm 20\,\%$) and the correlation between synthetic (observed) and estimated values is relatively high ($R = 0.58$ and $R = 0.71$ for 1 and 3 h resolution, respectively). Also, the msml results match the 1 : 1 line well between synthetic and estimated evapotranspiration (Fig. 3a, b).

The sssl and the ssml methods show a weaker performance compared to the more complex im and msml methods. Neither of them follows the 1 : 1 line well between synthetic and estimated evapotranspiration (Fig. 3a, b). Regardless, they could reproduce the synthetic evapotranspiration with a relatively high linear correlation (Table 2), and comparable bias to the regression method, in particular for the range of intermediate measurement frequencies. However, values for the RV are comparatively large, in particular for the ssml method. Interestingly, the model performance criteria of the simpler sssl method show only minor differences between the particular temporal resolutions, and overall the sssl method performs better than the ssml method. Note that both water balance methods (sssl and ssml) overestimate the evapotranspiration at the beginning of the study period (Fig. 3c, d), which was marked by greater vertical flow between top soil and deeper soil due to preceding rainfall events.

Our results also show that less complex data-driven methods also perform better at higher temporal resolution (1 and 3 h), except for the ssml method. In contrast, the im is better at predicting evapotranspiration when a coarse measurement

Table 2. Comparison of the model performance of the four data-driven methods for reproducing daily evapotranspiration for the particular time resolution of soil moisture measurements. The model performance is expressed as correlation coefficient R, relative variability in simulated and reference values (RV), and relative bias (b) for the period 25 July–26 August 2009. Days on which rainfall occurs were excluded for the data analysis.

	Single-step, single-layer water balance			Single-step multi-layer water balance			Multi-step, multi-layer regression			Inverse model		
Δt (h)	R	RV	b (%)	R	RV	b (%)	R	RV	b (%)	R	RV	b (%)
1	0.77	1.51	−38.6	0.64	3.32	54.2	0.58	1.54	−22.9	0.99	0.78	−41.5
3	0.75	1.54	−38.6	0.66	3.37	46.8	0.71	1.03	20.3	0.99	0.97	−18.2
6	0.75	1.69	−35.9	0.67	3.52	36.4	0.78	1.87	86.5	0.99	1.03	−7.6
12	0.75	1.44	−38.6	0.70	3.49	37.1	0.85	4.22	202.4	0.99	1.04	0.89
24	0.58	1.76	−37.3	0.53	3.72	26.4	–	–	–	0.99	1.11	3.5

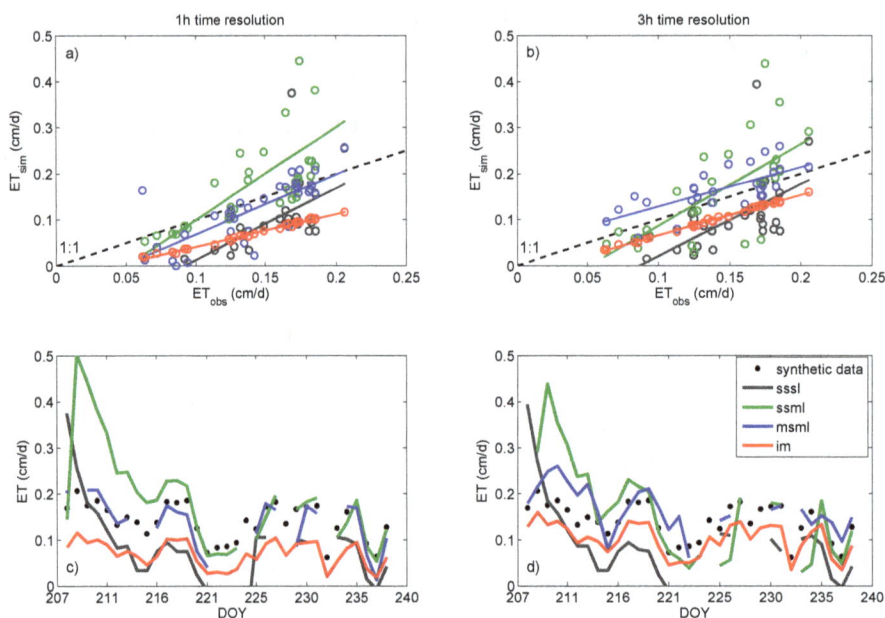

Figure 3. Top: comparison of synthetic (ET_{obs}) and estimated (ET_{sim}) values of daily evapotranspiration for hourly (**a**) and 3-hourly (**b**) observation intervals of soil water content measurements. Bottom: comparison of synthetic and estimated time series of daily evapotranspiration (ET) for hourly (**c**) and 3-hourly (**d**) observation intervals of soil water content measurements (25 July to 26 August 2009). Missing values are times when rainfall and percolation appeared. An estimation of evapotranspiration was not possible with the single-step, single-layer (sssl) water balance; the single-step, multi-layer (ssml) water balance; or the multi-step, multi-layer (msml) regression for these days.

frequency is used. Further, the results indicate that the estimated actual evapotranspiration becomes more accurate with increasing model intricacy, and with vertical flow accounted for.

3.2 Root water uptake profiles estimated with three different data-driven methods

The ssml, msml and im method appropriate for determining root water uptake profiles by inclusion of all available measurements over depth. Table 3 summarizes the model applicability to estimate the depths at which 25, 50 and 90 % of water extraction occurs (later stated as $z_{25\%}$, $z_{50\%}$ and $z_{90\%}$). Here, we used the standard deviation $s_{\bar{x}}$ instead of the relative

variability to evaluate the observed variance. This criterion was chosen because the standard deviation of the synthetic reference values is approximately zero and thus the RV is increasing, which is not practical for the method evaluation. The criteria are shown for the respective best achieved model performance (1 h – ssml and msml; 24 h – im).

Again, the quality of predicting the sink term distribution depends on the method complexity and increases with increasing complexity. The most complex im delivers the best prediction of sink term distribution for a temporal resolution of 24 h. The depth above which 50 % of water extraction occurs ($z_{50\%}$) could be predicted with a bias of less than 2 % (Table 3) and for $z_{90\%}$; the relative bias increased only

Table 3. Comparison of model performance for reproducing the sink term profile (single-step, multi-layer water balance; multi-step, multi-layer regression; and inverse model). Depths where 25, 50 and 90 % water extraction occurs were regarded. Mean synthetic (syn.) depth and mean estimated (est.) depth describe the mean depth over 33 days where water extraction occurs. b is the relative bias and $\widetilde{s}$ is the standard deviation of the estimated values. Larger width of the black arrow denotes higher accuracy of the model results.

Time resolution of measurements	Single-step, multi-layer water balance 1 h			Multi-step, multi-layer regression 1 h			Inverse model 24 h		
Criterion	$z_{25\%}$	$z_{50\%}$	$z_{90\%}$	$z_{25\%}$	$z_{50\%}$	$z_{90\%}$	$z_{25\%}$	$z_{50\%}$	$z_{90\%}$
Mean syn. depth (cm)	8.1	17.1	55.6	8.1	17.1	55.6	8.1	17.1	55.6
Mean est. depth (cm)	10.8	28.5	101.9	9.7	13.9	63.8	8.2	17.3	57.3
b (%)	33	74	83	−14	−21	15	0.75	1.05	2.97
$\widetilde{s}$	4.07	12.31	57.89	1.69	4.01	25.83	1.81	4.08	68.26

slightly to approximately 3 %. Indeed, these comparatively accurate results are to be expected due to the two intrinsic assumptions: (1) the required soil hydraulic parameters for the implemented soil water flow model are exactly known, and (2) the measurement uncertainty of the soil sensors is zero.

The regression method (msml) also delivers good estimations of sink term profiles over the entire soil column (Table 3 and Fig. 4), although it manages without any intrinsic assumptions. Figure 4 shows that the msml regression overestimates the sink term at the intermediate depths. The maximum relative bias is about −21 % at $z_{50\%}$. Overall, the msml regression is applicable for determining the mean sink term distribution with an acceptable accuracy.

The ssml-estimated sink terms correspond only weakly to the synthetic ones, and the relative bias is lowest for $z_{25\%}$ with 33 % but increases strongly for $z_{50\%}$ and $z_{90\%}$ (Table 3). Moreover, the standard deviations of the predictions are substantial at most measurement depths (Table 3, Fig. 4). Because of these large variations in sink term distribution, the prediction of sink term profiles becomes imprecise. Thus for the chosen simulation experiment, the ssml method is not applicable for deriving the sink term from soil water content measurements.

3.3 Influence of soil moisture sensor uncertainty on root water uptake estimation

We only evaluated the influence of measurement errors for two methods (msml and im). The single-layer approach was omitted since it does not allow for estimation of the sink term profile, and ssml was omitted since the estimation of the sink term profile was already inappropriate when ignoring measurement errors (see Sect. 3.2).

The influences of soil moisture sensor uncertainties differ considerably among the investigated methods. The msml method predicted the median daily evapotranspiration with precision uncertainty, calibration uncertainty and a combination of both reasonably well (Fig. 5). For all three types of uncertainty, the correlation between synthetic (observed)

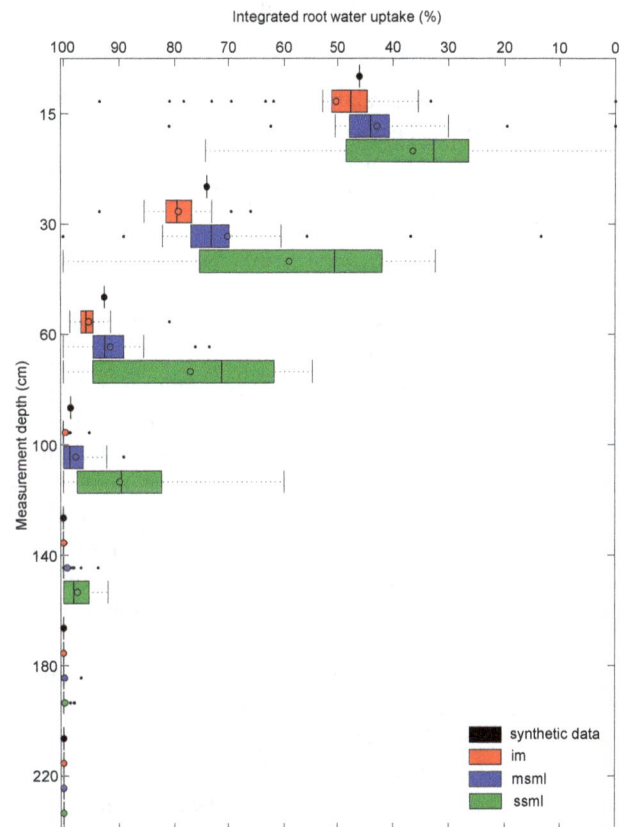

Figure 4. Box plots of the estimated daily percentage of integrated sink term. Colors are assigned as follows: synthetic values are black, the im is red, the msml regression is blue and the ssml water balance is green. The percentage of integrated sink term is shown for all measurement locations over the soil column. The circles show the mean values; the vertical line depicts the median and the 25 and 75 % percentile. Values are given for the respective underlying time resolution which achieved the best results according to Table 3 (ssml – 1 h; msml – 1 h; im – 24 h).

Table 4. Comparison of the model performance with considering soil moisture measurement uncertainties for the msml regression and the im for reproducing daily evapotranspiration and the mean depths where 25, 50 and 90 % water extraction occurs. The model performance is expressed as correlation coefficient (R), relative variability in simulated and reference values (RV) and relative bias (b) for the period 25 July to 26 August 2009. The precision uncertainty is abbreviated as prec err, the calibration uncertainty as cali err, and the combined uncertainty as com err. The relative bias for reproducing evapotranspiration is abbreviated as b_{ET}, and is abbreviated as $b_{25\%}$, $b_{50\%}$ and $b_{90\%}$ for reproducing mean depths where 25, 50 and 90 % water extraction occurs, respectively.

Time resolution of measurements	Multi-step, multi-layer regression 1 h			Inverse model 24 h		
Criterion	prec err	cali err	com err	prec err	cali err	com err
R	0.90	0.89	0.91	−0.027	0.847	−0.054
RV	1.35	1.50	1.35	1.51	1.25	1.85
Median bias b_{ET} (%)	−6.2	−4.9	−6.1	−10.3	498.1	483.3
Median bias $b_{25\%}$ (%)	19.6	3.6	19.5	25.2	531.1	405.1
Median bias $b_{50\%}$ (%)	28.0	5.4	27.7	42.0	622.4	659.1
Median bias $b_{90\%}$ (%)	80.8	27.7	84.7	128.5	757.6	569.0

and estimated values is relatively high (around $R = 0.9$, Table 4). Also, with respect to the median relative bias (%), the three cases differ only marginally ($|b| = 7\%$, Table 4). Interestingly, the calibration uncertainty showed the lowest impact on the predicted evapotranspiration, with a median bias of about −5 % for the respective 100 ensemble calculations (Fig. 5).

Additionally, the bias is also used to compare the predicted relative water extraction depths ($z_{25\%}$, $z_{50\%}$ and $z_{90\%}$) (Fig. 6). The uncertainty caused by the calibration of the sensor shows the least differences to the observed values below 10 %. These results are similar to those from simulations with soil moisture without any introduced measurement uncertainty. Further, the uncertainties caused by the precision of the sensors have the highest impact on predicted root water uptake patterns. It turns out that the relative uncertainty increases with increasing depth (decreasing sink term or rather water extraction, Fig. 6a).

Interestingly, the im shows worse model performances than the msml regression for all three types of uncertainty. Although, the predicted evapotranspiration from soil moisture with precision uncertainty is close to the observed values (Fig. 5), it differs around days when rainfall occurs (DOY 225, 230 and 234). This results in underestimation of evapotranspiration during these times and a weak correlation (Table 4), but an acceptable relative bias of about −10 %. In contrast, for the calibration uncertainty it is the other way around. Here, the correlation is relatively high ($R = 0.85$) but evapotranspiration is greatly overestimated ($b = 498\%$). A combination of both uncertainty sources does not further increase the overall error, but does combine both weaknesses to an overall poor estimation (Table 4).

The sensitivity to the type of uncertainty concerning prediction of sink term patterns is shown in Fig. 6b and Table 4. Similar to the msml regression, the im is able to handle uncertainties in sensor precision to predict root water uptake

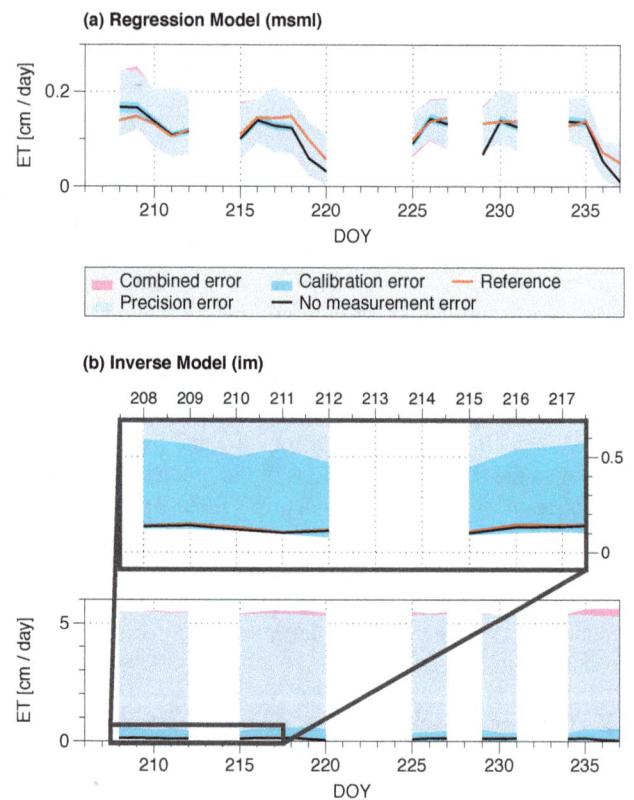

Figure 5. Influence of soil moisture uncertainty on evapotranspiration estimated with the msml regression model (**a**) and the im (**b**). The red line is the evapotranspiration from the synthetic data (Reference). The colored bands indicate the 95 % confidence intervals.

depths, whereas uncalibrated sensors lead to considerable increases in relative bias. Overall, the simpler msml regression method shows a higher robustness against measurement uncertainties than the more complex im.

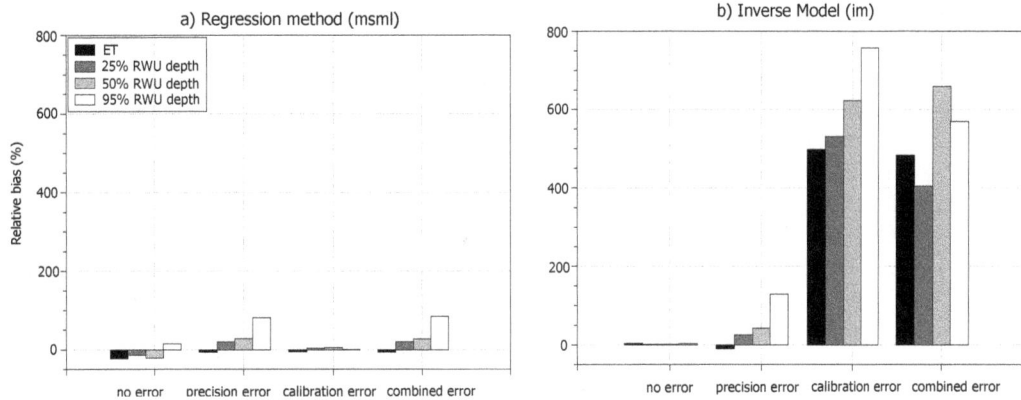

Figure 6. Comparison of the mean relative bias between synthetic and predicted values of evapotranspiration and the mean depths where 25, 50, 90 % of water extraction occurs for soil moisture time series: without uncertainty (no error), precision uncertainty (precision error), calibration uncertainty (calibration error) and precision and calibration uncertainty (combined error) for the msml regression (**a**) and the im (**b**).

4   Discussion

We tested the application of several methods deriving, based on the soil water balance, how much water was extracted from the soil by evapotranspiration and how the extraction profile (sink term profile) changed with soil depth. The bases for all methods are time series of volumetric soil water content derived from measurements, although some methods require more information on soil properties, in particular the inverse model (im). None of the methods relies on a priori information on the shape of the sink term profile, nor do any of them make any assumptions on it being constant with time. This is the great advantage of these methods over others (Dardanelli et al., 2004; McIntyre et al., 1995; Hopmans and Bristow, 2002; Zuo et al., 2004). Since only changes in soil water content are considered, none of the investigated methods distinguish between soil evaporation and root water uptake. For the same reason, none of the water balance methods can be applied during times of fast soil water flow, for example during or after a rainfall event.

We used synthetic soil water content "observations" to validate the model results. This procedure has the great advantage that the "true" water flow and sink term profiles are perfectly known, including the nature of data uncertainty with regard to calibration error and sensor precision. However, our model only accounts for vertical matrix flow, notably neglecting horizontal heterogeneity, which may be an additional challenge for deriving evapotranspiration in real-world situations. Thus, additional tests of the methods in controlled field conditions, like with large lysimeters, and comparison with additional data, like isotope profiles, are necessary to confirm our results.

In the first part of the paper, we investigated how well all methods reproduced the sink term profile and total evapotranspiration when assuming that the measurements of soil water content were free of measurement errors, i.e., they were well calibrated and measured precisely. Even in this idealistic setting, the investigated methods performed very differently, most prominently depending on whether or not vertical flow could be accounted for by the method. The methods showing the greatest deviation between the "observed" (synthetic) evapotranspiration and sink term profiles were those not accounting for vertical flow within the soil (sssl and ssml methods). In those simpler soil water balance methods, any change in soil moisture is assigned only to root water uptake (Rasiah et al., 1992; Musters et al., 2000; Hupet et al., 2002). However, even several days after a rainfall event, the vertical matrix flow within the soil can be similar in magnitude to the root water uptake (Schwärzel et al., 2009), and this leads to considerable overestimation of the sink term when soil water flow is not accounted for. This error adds up when the sink term is integrated over depth and leads to a bias in the evapotranspiration estimate, which is the case for the ssml method.

This distinction between vertical soil water flow and water extraction is the major challenge when applying water balance methods, because these fluxes occur concurrently during daytime (Gardner, 1983; Feddes and Raats, 2004). The regression method (msml) avoids this problem by considering vertical soil water fluxes, estimated from change in soil water content during nighttime. Li et al. (2002) used a similar approach to derive transpiration and root water uptake patterns from soil moisture changes between different times of the day. This direct attribution of nighttime change in soil water content to soil water flow inherently assumes that both nighttime evapotranspiration and hydraulic redistribution are negligible. Li et al. (2002) measured nocturnal sap flow in order to ascertain that nighttime transpiration was insignificant. Also, in lysimeters, the weight changes can be used to validate the assumption. This assumption is the main drawback of this method, in contrast to the large advantage that it requires very limited input data, especially no a priori information on soil properties. In contrast, the im approach inferred evapotranspiration and sink term patterns with greater quality when soil water content measurements were free of error. However, because our analysis uses model-generated time series of soil water content in order to mimic measurements, the soil properties of the original "experiment" are completely known, which is not usually the case in natural conditions. Usually, soil hydraulic parameters have to be estimated by means of a calibration procedure. This process is non-trivial and limited by the non-uniqueness of the calibrated parameters (Hupet et al., 2003), which results in uncertainties in simulated soil water fluxes and root water uptake rates (Duan et al., 1992; Musters and Bouten, 2000; Musters et al., 2000; Hupet et al., 2002, 2003). This reliance of the im approach on precise knowledge of the soil environment is the main drawback of that approach.

Several studies on estimation of root water uptake profiles focused on uncertainties related to calibrated parameters of soil and the root water uptake models (Musters and Bouten, 2000; Musters et al., 2000; Hupet et al., 2002, 2003). When data and models are used, uncertainties arise not from soil parameter uncertainty but in fact already evolve during the measurement process of the environmental data (Spank et al., 2013). Thus, in the second part of this paper, we investigated how measurement noise (precision), wrong sensor calibration (accuracy) and their combination reflect on the derivation of evapotranspiration and sink term patterns from soil water content measurements. We only performed this analysis for the two methods which performed satisfactorily without sensor errors: the msml regression method and the im. In this more realistic setting, the simpler regression method (msml) performed much better than the im. The latter was strongly affected by inaccurate or lack of site-specific calibration. This "calibration error" renders the evolution of the vertical potential gradients and soil moisture profile inconsistent with the evolution of the vertical sink term distribution, and thus introduces grave overestimation of root water uptake and evapotranspiration for the considered time steps (Fig. S2). Generally, the prediction of the im improves when longer evaluation periods are considered (cf. Zuo and Zhang, 2002), and therefore the calibration error may become less prominent when considering time steps of several days as done in Zuo and Zhang (2002). Compared to the effect of calibration, the sensor precision had a much smaller effect. Thus, the im may be applicable and should be tested in situations where all sensors in the profile are well calibrated. A further improvement of the im could be achieved by smoothing the measured soil water content profiles via a polynomial function to get an accurate and continuous distribution of soil water contents as done in Li et al. (2002) and Zuo and Zhang (2002).

The msml regression model was overall more robust towards the investigated measurement errors. It was barely affected by calibration error but was somewhat affected by sensor precision. This is expected, since the sensor calibration only improves the absolute values of the measurements, and does not affect the course of the soil moisture desiccation. The case is different for uncertainty due to sensor precision, which results in higher deviations between observed and predicted sink term uptake patterns (Fig. 6). As this method uses linear regression on the temporal evolution of soil water contents, the quantity of root water uptake depends on the gradient of the slopes. Those slopes are strongly influenced by the random scatter of data points, which is characteristic for sensor noise. Using the smallest time step of 1 h, we could estimate the relative depth where 50 % of water extraction occurs with a bias less than 30 %. Using higher time resolution with several measurements per hour or several minutes and noise-reducing filters (Li et al., 2002; Peters et al., 2013) would likely further improve this result. This method should be further evaluated with lysimeters in order to test its application in controlled but more realistic environments.

Furthermore, our study demonstrates that measured soil moisture time series already include information on evapotranspiration and root water uptake patterns. This has already been stated by Musters and Bouten (2002) as well as Zuo and Zhang (2002). Contrary to these studies, where only temporal resolutions of 1 day or more are investigated, we additionally looked at measurement time intervals in the range of hours. Our results confirm that different methods require measurements with different temporal resolutions. The more simple msml regression model showed better applicability for measurements taken with an interval less than 6 h. These results are similar to Breña Naranjo et al. (2011) for a water balance method. The higher time resolution better reflects the temporal change in evapotranspiration, which may be considerable over the course of a day (Jackson et al., 1973). Conversely, the im works better for coarser temporal resolution for the case that soil water content measurements are error-free. If a possible measurement error is considered, coarser temporal resolutions are also better suitable to estimate evapotranspiration and root water uptake. With a higher temporal resolution (here 1 day instead of several hours) the total evapotranspiration and sink term also increases (integrated over the entire time). Therefore, the iteration of the im procedure could determine the sink term with a higher accuracy.

Another important prerequisite besides temporal resolution of the soil moisture time series is the adequate number of soil moisture measurements over the entire soil column to well capture the very nonlinear depth profile of water removal from the soil. This becomes most obvious when comparing the results from the simple single-layer water balance method (sssl) with the multi-layer (ssml) one. The prediction of the single-layer model is dominated by the specific depth at which the single sensor is located, and how much it is affected by root water uptake. In the presented case it strongly underestimated overall evapotranspiration because it observes only one part of the sink term profile and omits both the much more elevated uptake in the top soil and the deep uptake below the measurement depth. In contrast to that, the multi-layer method reproduces better the time series of evapotranspiration, because it samples the uptake profiles more holistically. Similarly, Schwärzel et al. (2009) and Clausnitzer et al. (2011) also found that high spatial resolution of water content sensors allow for a more reliable determination of evapotranspiration. Important consideration should be given to the very shallow soil depths, representative of the pure soil evaporation process ($z < 5$ cm), which are notoriously undersampled due to technical limitations. This may lead to underestimation of evaporation and therefore evapotranspiration in all investigated water balance applications.

Our results show that water balance methods have potential to be applied for derivation of water extraction profiles, but they also suggest that their application may be challenging in realistic conditions. In particular, im has great potential, in theory, but obtaining information of the soil environment with sufficient accuracy may be unrealistic. The msml regression method is particularly promising, as it requires little input and is comparably robust towards measurement errors. Further tests in controlled environments and ideally in concert with isotope studies should be conducted to further test the application of these methods in real-world conditions.

The great advantage of all considered methods is that they do not require a priori information about total evapotranspiration or the shape of the root water uptake profiles. Root water uptake moves up or down depending on soil water status (Lai and Katul, 1998; Li et al., 2002, Doussan et al., 2006; Garrigues et al., 2006), and many existing approaches are unable to account for this dynamic of root water uptake. Root water extraction profiles are central topics in ecological and ecohydrological research on resource partitioning (e.g., Ogle et al., 2004; Leimer et al., 2014; Schwendenmann et al., 2014) and drivers for ecosystem structure (Arnold et al., 2010). Water balance methods are potential tools for comparing those extraction profiles between sites and thus contributing to ecohydrological process understanding.

5   Conclusions

The aim of this study was to evaluate four water balance methods of differing complexity to estimate sink term profiles and evapotranspiration from volumetric soil water content measurements. These methods do not require any a priori information of root distribution parameters, which is the advantage compared to common root water uptake models. We used artificial data of soil moisture and sink term profiles to compare the quality of the estimates of those four methods. Our overall comparison involved the examination of the impact of measurement frequency and model intricacy, as well as the uncertainties of soil moisture sensors on predicting sink term profiles. For the selected dry period of 33 days and under consideration of possible measurement un-

certainties the multi-step, multi-layer (msml) regression obtained the best estimation of sink term patterns. In general, the predictions with the four data-driven methods show that these methods have different requirements on the measurement frequency of soil moisture time series and on additional input data like precipitation and soil hydraulic parameters. Further, we were able to show that the more complex methods like the msml regression and the inverse model (im) predict evapotranspiration and the sink term distribution more accurately than the simpler single-step, single-layer (sssl) water balance and the single-step, multi-layer (ssml) water balance.

Unfortunately, the estimations of the im are strongly influenced by the uncertainty of measurements. Moreover, numerical soil water flow models like the im require a large amount of prior information (e.g., boundary conditions, soil hydraulic parameters) which is usually not available in sufficient quality. For example, the soil hydraulic parameters have to be calibrated before use, which introduces additional uncertainties in the parameter sets. It is important to keep this in mind while comparing the im with the msml regression model, especially in light of the influence of measurement uncertainties.

Our results show that highly resolved (temporal and spatial) soil water content measurements contain a great deal of information which can be used to estimate the sink term when the appropriate approach is used. However, we acknowledge that this study using numerical simulations is only a first step towards the application on real field measurements. The msml regression model has to be tested with real field data, especially with lysimeter experiments. Lysimeters allow for closing of the water balance and validation with measured evapotranspiration, while soil water content measurements can be conducted in a similar way to field experiments. With such experiments, the proposed method can be evaluated in an enhanced manner.

Appendix A

Table A1. Nomenclature.

b	relative bias (%)
d_t	length of active transpiration period over a day (h)
$d_{z,i}$	thickness of soil layer i (m)
DOY	day of year
e	difference in observed and estimated soil water content in the inverse model
E	evapotranspiration ($\mathrm{mm\,h^{-1}}$ or $\mathrm{cm\,d^{-1}}$)
E_s	bare soil evaporation ($\mathrm{mm\,h^{-1}}$)
E_t	transpiration ($\mathrm{mm\,h^{-1}}$)
$\widetilde{E}$	estimated evapotranspiration ($\mathrm{mm\,h^{-1}}$)
h	soil matric potential (m)
i	soil layer index
j	time step index
$K(h)$	hydraulic conductivity ($\mathrm{m\,s^{-1}}$)
K_{sat}	saturated hydraulic conductivity ($\mathrm{m\,s^{-1}}$)
m_{tot}	slope of fitted linear function on $\theta(t)$
m_{extr}	slope of fitted linear function on $\theta(t)$ due to sink term
m_{flow}	slope of fitted linear function on $\theta(t)$ due to vertical soil water flow
n_{vG}	van Genuchten parameter (−)
NSE	Nash–Sutcliffe efficiency criterion
P	precipitation ($\mathrm{mm\,h^{-1}}$)
q	percolation ($\mathrm{mm\,h^{-1}}$)
RV	relative variability
S	sink term in Richards equation ($\mathrm{s^{-1}}$)
S_i	discretized sink term in the soil layer i ($\mathrm{m\,s^{-1}}$)
$\widetilde{S}$	estimated sink term ($\mathrm{m\,s^{-1}}$)
s	standard deviation
t	time (s)
Δt	time step (h)
v	iteration step number (–)
$\bar{x}$	mean value
x	observed (synthetic) value
$\widetilde{x}$	estimated values
z	vertical coordinate (m)
z_r	active rooting depth (cm)
$z_{25\%}$	depth up to which 25 % of root water uptake occurs (cm)
$z_{50\%}$	depth up to which 50 % of root water uptake occurs (cm)
$z_{90\%}$	depth up to which 90 % of root water uptake occurs (cm)
α	van Genuchten parameter ($\mathrm{m^{-1}}$)
θ	volumetric soil water content ($\mathrm{m^3\,m^{-3}}$)
θ_r	residual volumetric soil water content ($\mathrm{m^3\,m^{-3}}$)
θ_s	saturated volumetric soil water content ($\mathrm{m^3\,m^{-3}}$)
$\widetilde{\theta}$	estimated volumetric soil water content ($\mathrm{m^3\,m^{-3}}$)
$\Delta\theta$	deviation in volumetric soil water content over time ($\mathrm{m^3\,m^{-3}}$)
ε_{ZZ}	decision criterion for termination of the iteration process (inverse model from Zuo and Zhang, 2002)
$\varepsilon_{GH,i}$	decision criterion for termination of the iteration process in the inverse model proposed here

Acknowledgements. Financial support through the "ProExzellenz" Initiative from the German federal state of Thuringia to the Friedrich Schiller University Jena within the research project AquaDiva@Jena for conducting the research is gratefully acknowledged. This work was also financially supported by the Deutsche Forschungsgemeinschaft (DFG) within the project "The Jena Experiment". M. Guderle was also supported by the International Max Planck Research School for Global Biogeochemical Cycles (IMPRS-gBGC). We thank the editor, Nadia Ursino, for handling the manuscript and the two anonymous referees for their helpful comments. We also thank Maik Renner, Kristin Bohn, and Marcel Bechmann for fruitful discussions on an earlier version of this manuscript.

Edited by: N. Ursino

References

Allen, R. G., Pereira, L. S., Raes, D., and Smith, M.: Crop evapotranspiration: Guidelines for computing crop requirements, FAO Irrigation and Drainage Paper No. 56, FAO, Rome, Italy, 1998.

Arnold, S., Attinger, S., Frank, K., and Hildebrandt, A.: Uncertainty in parameterisation and model structure affect simulation results in coupled ecohydrological models, Hydrol. Earth Syst. Sci., 13, 1789–1807, doi:10.5194/hess-13-1789-2009, 2009.

Asbjornsen, H., Goldsmith, G. R., Alvarado-Barrientos, M. S., Rebel, K., Van Osch, F. P., Rietkerk, M., Chen, J., Gotsch, S., Tobón, C., Geissert, D. R., Gómez-Tagle, A., Vache, K., and Dawson, T. E.: Ecohydrological advances and applications in plant-water relations research: a review, J. Plant Ecol., 4, 3–22, doi:10.1093/jpe/rtr005, 2011.

Bechmann, M., Schneider, C., Carminati, A., Vetterlein, D., Attinger, S., and Hildebrandt, A.: Effect of parameter choice in root water uptake models – the arrangement of root hydraulic properties within the root architecture affects dynamics and efficiency of root water uptake, Hydrol. Earth Syst. Sci., 18, 4189–4206, doi:10.5194/hess-18-4189-2014, 2014.

Breña Naranjo, J. A., Weiler, M., and Stahl, K.: Sensitivity of a data-driven soil water balance model to estimate summer evapotranspiration along a forest chronosequence, Hydrol. Earth Syst. Sci., 15, 3461–3473, doi:10.5194/hess-15-3461-2011, 2011.

Chanzy, A., Gaudu, J. C., and Marloie, O.: Correcting the temperature influence on soil capacitance sensors using diurnal temperature and water content cycles (Basel, Switzerland), Sensors, 12, 9773–9790, doi:10.3390/s120709773, 2012.

Chapin, F. S., Matson, P. A., and Mooney H. A.: Principles of Terrestrial Ecosystem Ecology, Springer-Verlag, New York, ISBN 0-387-95439-2, 2002.

Clausnitzer, F., Köstner, B., Schwärzel, K., and Bernhofer, C.: Relationships between canopy transpiration, atmospheric conditions and soil water availability – Analyses of long-term sapflow measurements in an old Norway spruce forest at the Ore Mountains/Germany, Agr. Forest Meteorol.,151, 1023–1034, doi:10.1016/j.agrformet.2011.04.007, 2011.

Clothier, B. E. and Green, S. R.: Rootzone processes and the efficient use of irrigation water, Agr. Water Manage., 25, 1–12, doi:10.1016/0378-3774(94)90048-5, 1994.

Coelho, F. and Or, D.: A parametric model for two-dimensional water uptake intensity by corn roots under drip irrigation, Soil Sci. Soc. Am. J., 60, 1039–1049, 1996.

Dardanelli, J. L., Ritchie, J. T., Calmon, M., Andriani, J. M., and Collino, D. J.: An empirical model for root water uptake, Field Crops Res., 87, 59–71, doi:10.1016/j.fcr.2003.09.008, 2004.

Davis, S. D. and Mooney, H. A.: Water use patterns of four co-occurring chaparral shrubs, Oecologia, 70, 172–177, doi:10.1007/BF00379236, 1986.

Doussan, C., Pierret, A., Garrigues, E., and Pagès, L.: Water uptake by plant roots: II – Modelling of water transfer in the soil root-system with explicit account of flow within the root system - Comparison with experiments, Plant Soil, 283, 99–117, doi:10.1007/s11104-004-7904-z, 2006.

Duan, Q., Sorooshian, S., and Gupta, V.: Effective and Efficient Global Optimization for Conceptual Rainfall-Runoff Models, Water Resour. Res., 28, 1015–1031, 1992.

Feddes, R. A. and Raats, P. A. C.: Parameterizing the soil-water-plant root system, in: Unsaturated-zone Modeling: Progress, Challenges and Applications, edited by: Feddes, R. A., de Rooij, G. H., and van Dam, J. C., Kluwer Academic Publishers, Dordrecht, the Netherlands, 95–141, 2004.

Feddes, R. A., Hoff, H., Bruen, M., Dawson, T., De Rosnay, P., Dirmeyer, P., Jackson, R. B., Kabat, P., Kleidon, A., Lilly, A., and Pitman, A. J.: Modeling root water uptake in hydrological and climate models, B. Am. Meteorol. Soc., 82, 2797–2809, 2001.

Gale, M. R. and Grigal, D. K.: Vertical root distributions of northern tree species in relation to successional status, Can. J. For. Res., 17, 829–834, 1987.

Gardner, W. R.: Soil properties and efficient water use: An overview, in: Limitations to efficient water use in crop production, edited by: Taylor, H. M., Jordan, W. R., and Sinclair, T. S., ASA-CSSA-SSSA, Madison, USA, 45–64, 1983.

Garrigues, E., Doussan, C., and Pierret, A.: Water Uptake by Plant Roots: I – Formation and Propagation of a Water Extraction Front in Mature Root Systems as Evidenced by 2D Light Transmission Imaging, Plant Soil, 283, 83–98, doi:10.1007/s11104-004-7903-0, 2006.

Green, S. R. and Clothier, B. E.: Root water uptake by kiwifruit vines following partial wetting of the root zone, Plant Soil, 173, 317–328, 1995.

Green, S. R. and Clothier, B. E.: The root zone dynamics of water uptake by a mature apple tree, Plant Soil, 206, 61–77, 1999.

Gupta, H. V., Kling, H., Yilmaz, K. K., and Martinez, G. F.: Decomposition of the mean squared error and NSE performance criteria: Implications for improving hydrological modelling, J. Hydrol., 377, 80–91, doi:10.1016/j.jhydrol.2009.08.003, 2009.

Hamblin, A. and Tennant, D.: Root length density and water uptake in cereals and grain legumes: how well are they correlated?, Aust. J. Agr. Res., 38, 513–527, doi:10.1071/AR9870513, 1987.

Hildebrandt, A. and Eltahir, E. A. B.: Ecohydrology of a seasonal cloud forest in Dhofar: 2. Role of clouds, soil type, and rooting depth in tree-grass competition, Water Resour. Res., 43, 1–13, doi:10.1029/2006WR005262, 2007.

Hopmans, J. W. and Bristow, K. L.: Current capabilities and future needs of root water and nutrient uptake modeling, Adv. Agron., 77, 104–175, 2002.

Hupet, F., Lambot, S., Javaux, M., and Vanclooster, M.: On the identification of macroscopic root water uptake parameters from soil water content observations, Water Resour. Res., 38, 1–14, doi:10.1029/2002WR001556, 2002.

Jackson, R. B., Candell, J., Ehleringer, J. R., Mooney, H. A., Sala, O. E., and Schulze, E. D.: A global analysis of root distributions for terrestrial biomes, Oecologia, 108, 389–411, 1996.

Jackson, R. B., Sperry, J. S., and Dawson, T. E.: Root water uptake and transport: using physiological processes in global predictions, Trends Plant Sci., 5, 482–488, 2000.

Jackson, R. D., Kimball, B. A., Reginato, R. J., and Nakayama, F. S.: Diurnal soil-water evaporation: time-depth-flux patterns, Soil Sci. Soc. Am. Pro., 37, 505–509, doi:10.2136/sssaj1973.03615995003700040014x, 1973.

Khalil, M., Sakai, M., Mizoguchi, M., and Miyazaki, T.: Current and prospective applications of Zero Flux Plane (ZFP) method, J. Jpn. Soc. Soil Phys., 95, 75–90, 2003.

Kollet, S. J.: Influence of soil heterogeneity on evapotranspiration under shallow water table conditions: transient, stochastic simulations, Environ. Res. Lett., 4, 035007, doi:10.1088/1748-9326/4/3/035007, 2009.

Kosugi, Y. and Katsuyama, M.: Evapotranspiration over a Japanese cypress forest, II. Comparison of the eddy covariance and water budget methods, J. Hydrol., 334, 305–311, 2007.

Kuhlmann, A., Neuweiler, I., van der Zee, S. E. A. T. M., and Helmig, R.: Influence of soil structure and root water uptake strategy on unsaturated flow in heterogeneous media, Water Resour. Res., 48, W02534, doi:10.1029/2011WR010651, 2012.

Lai, C. T. and Katul, G.: The dynamic role of root-water uptake in coupling potential to actual transpiration, Adv. Water Resour., 23, 427–439, doi:10.1016/S0309-1708(99)00023-8, 2000.

Lee, A.: Movement of water through plants, Pract. Hydropon. Greenhous., 50, GRODAN, http://www.grodan.com/files/Grodan/PG/Articles/2009/Movement_of_water_through_plants.pdf (last access: September 2014), 2009.

Leimer, S., Kreutziger, Y., Rosenkranz, S., Beßler, H., Engels, C., Hildebrandt, A., Oelmann, Y., Weisser, W. W., Wirth, C., Wilcke, W.: Plant diversity effects on the water balance of an experimental grassland, Ecohydrol., 7, 1378–1391, doi:10.1002/eco.1464, 2014.

Le Roux, X., Bariac, T., and Mariotti, A.: Spatial partitioning of the soil water resource between grass and shrub components in a West African humid savanna, Oecologia, 104, 147–155, 1995.

Li, K., Dejong, R., and Boisvert, J.: An exponential root-water-uptake model with water stress compensation, J. Hydrol., 252, 189–204, doi:10.1016/S0022-1694(01)00456-5, 2001.

Li, Y., Fuchs, M., Cohen, S., Cohen, Y., and Wallach, R.: Water uptake profile response of corn to soil, Plant Cell Environ., 25, 491–500, 2002.

Loheide, S. P.: A method for estimating subdaily evapotranspiration of shallow groundwater using diurnal water table fluctuations, Ecohydrology, 66, 59–66, doi:10.1002/eco.7, 2008.

Maruyama, A. and Kuwagata, T.: Diurnal and seasonal variation in bulk stomatal conductance of the rice canopy and its dependence on developmental stage, Agr. Forest Meteorol., 148, 1161–1173, doi:10.1016/j.agrformet.2008.03.001, 2008.

McIntyre, B. D., Riha, S. J., and Flower, D. J.: Water uptake by pearl millet in a semiarid environment, Field Crop. Res., 43, 67–76, 1995.

Musters, P. A. D. and Bouten, W.: Assessing rooting depths of an austrian pine stand by inverse modeling soil water content maps, Water Resour. Res., 35, 3041, doi:10.1029/1999WR900173, 1999.

Musters, P. A. D. and Bouten, W.: A method for identifying optimum strategies of measuring soil water contents for calibrating a root water uptake model, J. Hydrol., 227, 273–286, doi:10.1016/S0022-1694(99)00187-0, 2000.

Musters, P. A. D., Bouten, W., and Verstraten, J. M.: Potentials and limitations of modelling vertical distributions of root water uptake of an Austrian pine forest on a sandy soil, Hydrol. Process., 14, 103–115, 2000.

Ogle, K., Wolpert, R. L., and Reynolds, J. F.: Reconstructing plant root area and water uptake profiles, Ecology, 85, 1967–1978, 2004.

Peters, A., Nehls, T., Schonsky, H., and Wessolek, G.: Separating precipitation and evapotranspiration from noise – a new filter routine for high-resolution lysimeter data, Hydrol. Earth Syst. Sci., 18, 1189–1198, doi:10.5194/hess-18-1189-2014, 2014.

Plamboeck, A. H., Grip, H., and Nygren, U.: A hydrological tracer study of water uptake depth in a Scots pine forest under two different water regimes, Oecologia, 119, 452–460, 1999.

Rasiah, V., Carlson, G. C., and Kohl, R. A.: Assessment of functions and parameter estimation methods in root water uptake simulation, Soil Sci. Soc. Am., 56, 1267–1271, 1992.

Ravenek, J. M., Bessler, H., Engels, C., Scherer-Lorenzen, M., Gessler, A., Gockele, A., De Luca, E., Temperton, V. M., Ebeling, A., Roscher, C., Schmid, B., Weisser, W. W., Wirth, C., de Kroon, H., Weigelt, A., and Mommer, L.: Long-term study of root biomass in a biodiversity experiment reveals shifts in diversity effects over time, Oikos, 000, 1–9, doi:10.1111/oik.01502, 2014.

Roscher, C., Scherer-Lorenzen, M., Schumacher, J., Temperton, V. M., Buchmann, N., and Schulze, E. D.: Plant resource-use characteristics as predictors for species contribution to community biomass in experimental grasslands, Perspect. Plant Ecol., 13, 1–13, doi:10.1016/j.ppees.2010.11.001, 2011.

Ross, P. J.: Modeling soil water and solute transport – fast, simplified numerical solutions, Am. Soc. Agron., 95, 1352–1361, 2003.

Ross, P. J.: Fast Solution of Richards' Equation for Flexible Soil Hydraulic Property Descriptions, Land and Water Technical Report, CSIRO, 39/06, 2006.

Sánchez, C., Fischer, G., and Sanjuanelo, D. W.: Stomatal behavior in fruits and leaves of the purple passion fruit (*Passiflora edulis* Sims) and fruits and cladodes of the yellow pitaya [*Hylocereus megalanthus* (K. Schum ex Vaupel) Ralf Bauer], Agronomía Colombiana, 31, 38–47, 2013.

Schaap, M. G., Leij, F. J., and van Genuchten, M. T.: Rosetta: a computer program for estimating soil hydraulic parameters with hierarchical pedotransfer functions, J. Hydrol., 251, 163–176, doi:10.1016/S0022-1694(01)00466-8, 2001.

Schenk, H. J.: The shallowest possible water extraction profile: a null model for global root distributions, Vadose Zone J., 7, 1119–1124, doi:10.2136/vzj2007.0119, 2008.

Schneider, C. L., Attinger, S., Delfs, J.-O., and Hildebrandt, A.: Implementing small scale processes at the soil-plant interface – the role of root architectures for calculating root water uptake profiles, Hydrol. Earth Syst. Sci., 14, 279–289, doi:10.5194/hess-14-279-2010, 2010.

Schume, H., Hager, H., and Jost, G.: Water and energy exchange above a mixed European Beech – Norway Spruce forest canopy: a comparison of eddy covariance against soil water depletion measurement, Theor. Appl. Climatol., 81, 87–100, 2005.

Schwärzel, K., Menzer, A., Clausnitzer, F., Spank, U., Häntzschel, J., Grünwald, T., Köstner, B., Bernhofer, C., and Feger, K. H.: Soil water content measurements deliver reliable estimates of water fluxes: a comparative study in a beech and a spruce stand in the Tharandt forest (Saxony, Germany), Agr. Forest Meteorol., 149, 1994–2006, doi:10.1016/j.agrformet.2009.07.006, 2009.

Schwendenmann, L., Pendall, E., Sanchez-Bragado, R., Kunert, N., and Hölscher, D.: Tree water uptake in a tropical plantation varying in tree diversity: interspecific differences, seasonal shifts and complementarity, Ecohydrology, doi:10.1002/eco.1479, online first, 2014.

Seneviratne, S. I., Corti, T., Davin, E. L., Hirschi, M., Jaeger, E. B., Lehner, I., Orlowsky, B., and Teuling, A. J.: Investigating soil moisture-climate interactions in a changing Climate: a review, Earth-Sci. Rev., 99, 125–161, doi:10.1016/j.earscirev.2010.02.004, 2010.

Spank, U., Schwärzel, K., Renner, M., Moderow, U., and Bernhofer, C.: Effects of measurement uncertainties of meterological data on estimates of site water balance components, J. Hydrol., 492, 176–189, 2013.

Teuling, A. J., Uijlenhoet, R., Hupet, F., and Torch, P. A.: Impact of water uptake strategy on soil moisture and evapotranspiration dynamics during drydown, Geophys. Res. Lett., 33, L03401, doi:10.1029/2005GL025019, 2006a.

Teuling, A. J., Seneviratne, S. I., Williams, C., and Torch, P. A.: Observed timescales of evapotranspiration response to soil moisture, Geophys. Res. Lett., 33, L23403, doi:10.1029/2006GL028178, 2006b.

van Genuchten, M. T.: A closed-form equation for predicting the hydraulic conductivity of unsaturated soils, Soil Sci. Soc. Am. J., 44, 892–898, 1980.

Verhoef, A., Fernández-Gálvez, J., Diaz-Espejo, A., Main, B. E., and El-Bishti, M.: The diurnal course of soil moisture as measured by various dielectric sensors: effects of soil temperature and the implications for evaporation estimates, J. Hydrol., 321, 147–162, doi:10.1016/j.jhydrol.2005.07.039, 2006.

Vrugt, J. A., van Wijk, M. T., Hopmans, J. W., and Šimunek, J.: One-, two-, and three-dimensional root water uptake functions for transient modeling, Water Resour. Res., 37, 2457, doi:10.1029/2000WR000027, 2001.

Wilson, K. B., Hanson, P. J., Mulholland, P. J., Baldocchi, D. D., and Wullschleger, S. D.: A comparison of methods for determining forest evapotranspiration and its components: sap-flow, soil water budget, eddy covariance and catchment water balance, Agr. Forest Meteorol., 106, 153–168, 2001.

Zuo, Q. and Zhang, R.: Estimating root-water-uptake using an inverse method, Soil Sci., 167, 561–571, 2002.

Zuo, Q., Meng, L., and Zhang, R.: Simulating soil water flow with root-water-uptake applying an inverse method, Soil Sci., 169, 13–24, doi:10.1097/01.ss.0000112018.97541.85, 2004.

Zwieniecki, M. A., Thompson, M. V., and Holbrook, N. M.: Understanding the Hydraulics of Porous Pipes: Tradeoffs Between Water Uptake and Root Length Utilization, J. Plant Growth Regul., 21, 315–323, 2003.

9

Reimagining the past – use of counterfactual trajectories in socio-hydrological modelling: the case of Chennai, India

V. Srinivasan

Ashoka Trust for Research in Ecology and the Environment, Royal Enclave Sriramapura, Jakkur Post, Bangalore, Karnataka, India

Correspondence to: V. Srinivasan (veena.srinivasan@atree.org)

Abstract. The developing world is rapidly urbanizing. One of the challenges associated with this growth will be to supply water to growing cities of the developing world. Traditional planning tools fare poorly over 30–50 year time horizons because these systems are changing so rapidly. Models that hold land use, economic patterns, governance systems or technology static over a long planning horizon could result in inaccurate predictions leading to sub-optimal or paradoxical outcomes. Most models fail to account for adaptive responses by humans that in turn influence water resource availability, resulting in coevolution of the human–water system. Is a particular trajectory inevitable given a city's natural resource endowment, is the trajectory purely driven by policy or are there tipping points in the evolution of a city's growth that shift it from one trajectory onto another?

Socio-hydrology has been defined as a new science of water and people that will explicitly account for such bidirectional feedbacks. However, a particular challenge in incorporating such feedbacks is imagining technological, social and political futures that could fundamentally alter future water demand, allocation and use. This paper offers an alternative approach – the use of counterfactual trajectories – that allows policy insights to be gleaned without having to predict social futures. The approach allows us to "reimagine the past"; to observe how outcomes would differ if different decisions had been made.

The paper presents a "socio-hydrological" model that simulates the feedbacks between the human, engineered and hydrological systems in Chennai, India over a 40-year period. The model offers several interesting insights. First, the study demonstrates that urban household water security goes beyond piped water supply. When piped supply fails, users turn

to their own wells. If the wells dry up, consumers purchase expensive tanker water or curtail water use and thus become water insecure. Second, unsurprisingly, different initial conditions result in different trajectories. But initial advantages in piped infrastructure are eroded if the utility is unable to expand the piped system to keep up with growth. Both infrastructure and sound management decisions are necessary to ensure household water security although the impacts of mismanagement may not manifest until much later when the population has grown and a multi-year drought strikes. Third, natural resource endowments can limit the benefits of good policy and infrastructure. Cities can boost recharge through artificial recharge schemes. However, cities underlain by productive aquifers can better rely on groundwater as a buffer against drought, compared to cities with unproductive aquifers.

1 Introduction

The world's population is rapidly urbanizing. One of the challenges associated with this growth will be to supply water to rapidly growing cities of the developing world. With growing population size and density, more water must be sourced from outside the boundaries of the cities and wastewater collected, treated and released safely into the environment (Lundqvist et al., 2003; McDonald et al., 2011). However, many developing cities are not equipped to meet even current demands let alone future growth. Inadequate and unreliable piped supply in developing world cities has measurable impacts on human well-being (Baisa et al., 2010; Srinivasan et al., 2010b). Although many developing world cities have not achieved reliable water supply, this is not an inevitable trajectory, i.e. not all developing urban systems end up becoming unreliable. For instance, some water Asian utilities (McIntosh, 2014) which have experienced high rates of population growth have managed the transition to "24/7" piped supply.

This paper addresses questions on how urban water systems evolve. Given a set of initial conditions, is a particular trajectory inevitable, or are there tipping points in the city's growth that shift it from one trajectory onto another? If so, are these tipping points influenced by government policy? Are there path dependencies such that early decisions constrain possibilities later?

1.1 Review of methodological approaches

Urban water systems are not pristine, natural systems; they are shaped both by societal decisions on infrastructure, governance, pricing and so forth, as well as the natural resource endowments of the region. Reflecting this, there is a long history of interdisciplinary research in urban water resource management. Traditionally, the focus of this type of research has been on policy prescription and/or infrastructure planning (Gober and Kirkwood, 2010; Brown et al., 2012; Ward et al., 2006). Researchers use economic analyses and water resource system models to make the case for new infrastructure projects, demand-side management programmes or alternative pricing policies. Such studies can broadly be categorized under Integrated Water Resources Management (IWRM) or Integrated Assessment (IA). They identify stakeholder priorities, and then integrate multiple scales of system and agent behaviour by drawing on the relevant disciplines within and across the human and natural sciences to explore alternative management options (Jakeman and Letcher, 2003; Gober et al., 2011). The purpose of such modelling efforts is usually to influence management decisions and understand trade-offs over a range of ecological, social and economic considerations. The role of the scientist in this endeavour is to enable decision-makers to decide how to manage the system better (Liu et al., 2008).

However, in the developing world, traditional planning tools fare poorly over 30–50 year time horizons. Here, systems are changing so quickly that holding land use, irrigation, agricultural technology, economic activity or technology static over the model period results in paradoxical outcomes (Sivapalan et al., 2014). As new technologies develop, users adapt to unreliable water supply. Adaptive responses by humans (acting individually and collectively) in turn may alter the watershed hydrology and consequently water availability. These bi-directional feedbacks often result in unexpected emergent behaviour. Many water managers fail to account for these complexities.

To address this challenge, socio-hydrology (Sivapalan et al., 2012) has been proposed as a "new science of humans and water systems". Socio-hydrology involves understanding the dynamics of coupled human–water systems over large spatial and temporal scales. In addition to studying specific sites, a central goal of socio-hydrology is to build a general theory of coupled human–water systems. This necessitates the inclusion of feedbacks between climate, land use, technology and social systems (Thompson et al., 2013) across multiple scales, sectors and agents in order to explain, in the most meaningful but parsimonious way, trajectories exhibited by coupled human–water systems. Such an improved understanding of the interactions between water and society can be used to improve decision making in the medium to long term (Clark and Clarke, 2011).

Recent discussions on socio-hydrologic methods within the scholarly community suggest that a diverse set of ideas exist on what socio-hydrologic modelling entails. Socio-hydrologic modelling includes a wide range of tools from "toy" models that do not aim to simulate a *specific* human–water system (Di Baldassarre et al., 2013) to coupled models that link agent-based and hydrologic models and validate them against detailed empirical observations. Each of these approaches has advantages and disadvantages. Toy models are relatively inexpensive to develop and are by design abstract and generalizable. However, they run the danger of predicting dynamics that are not in fact observed anywhere in the real world. This is particularly true of models of human behaviour, which are difficult to characterize in general terms. In contrast, "real world" models coupling agent behaviour to hydrologic models that are carefully calibrated and tested against empirical observations may yield reliable results for a particular site, but often lack abstraction and comparability beyond that study site. A third category, "stylized models" (Chakravorty and Umetsu, 2003; Kilgour and Dinar, 2001) offers a compromise between detail and generalizability. Such models have been used by economists in both natural resources and other contexts. A stylized model is a simplified representation of the real world that aims to replicate the *essential dynamics* observed in one or more study sites, but does not attempt to calibrate and validate every variable.

Methodologically, this paper illustrates how a stylized, socio-hydrologic model that explores bi-directional feedbacks between the societal, engineered and hydrologic components of water systems might be applied to achieve insights into household water security in developing urban regions. It presents a model of the coupled human–water system in a single case study site, Chennai, India, over a time span of 30–40 years and explores what factors drive the changes over time. The paper begins with an exploration of possible socio-hydrologic modelling approaches. Next it describes the case study and the stylized model of the water system. The model results then explore the actual trajectory as well as some alternative trajectories and the implications for household water security.

1.2 Socio-hydrologic model conceptualization

There are several challenges in attempting a stylized socio-hydrologic model of the type proposed herein. First, deciding which outcomes are worth explaining, i.e. the act of "framing", involves making choices on which problems are worth focusing on and which linkages to include or exclude (Lane, 2014). With coupled human–water systems, the decision of what to study does not emerge from the theoretical frameworks of a single discipline. Instead, it involves making a judgement about *which and whose* problems matter (Lane, 2014) and how to model them. This is critical because if socio-hydrologic models are intended to feed into the policy process, researchers cannot truly remain "external" observers of the system. As Schlueter et al. (2012) point out, human societies are reflexive and respond in unpredictable ways to new information. As a result, the very process of deciding what to model, which variables are static and which ones may be changed in the model, can influence which policy options get communicated and debated – a self-fulfilling prophecy. As a result, the researcher is not an impartial observer, but (albeit unintentionally) a social engineer too (Lane, 2014). Second, once the "system of interest" is extended beyond the biophysical or engineered sub-systems, every aspect of human society: culture, politics, economic trends, technology, social movements and so forth, and every related subsector such as energy, food, public health and biodiversity, is a candidate for inclusion. How can the socio-hydrology model avoid spiralling into a general system dynamics model of the whole world? Third, preventing the future from looking mostly like the past is a non-trivial challenge. How can a researcher "imagine" feedbacks and thresholds that go beyond what has occurred in the past and ensure that the model can accommodate the widest possible range of possible trajectories? In summary, the socio-hydrologic modeller must make a choice about how to frame the problem, decide which human well-being/biophysical outcomes are worth studying, and allow the system to evolve beyond trajectories that have occurred in the past.

With regard to the first challenge of framing the research questions, one approach that has been suggested is to embed the research process within a stakeholder dialogue and let the definitions and questions of interest emerge from these consultations (Tidwell and Van Den Brink, 2008; van den Belt et al., 2010). However, it is not always feasible to embed every research project within a stakeholder process. Instead, to ensure that the research is usable (Dilling and Lemos, 2011), in the present study, the variables, feedbacks and outcomes were chosen by referencing contemporary debates over hot to manage water and through one-on-one interviews with stakeholders and experts. Additionally, the original framework was validated in an expert consultation meeting held at Chennai in 2006. Moreover, a key contribution of this work is that it simulates *past counterfactual trajectories*; i.e. asking if the current water situation would be different if dif-

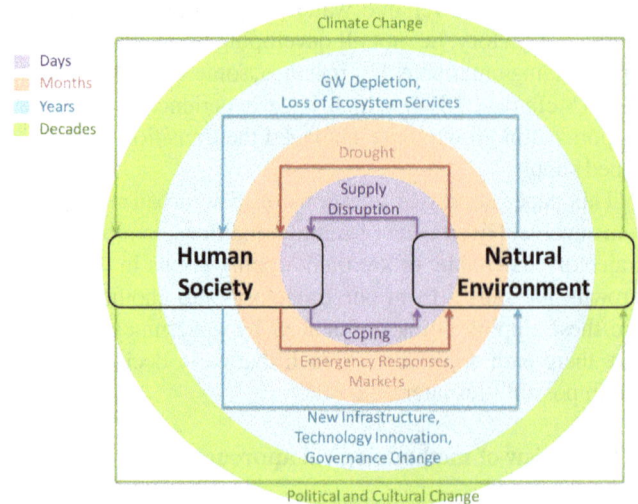

Figure 1. Feedbacks in coupled human–water systems.

ferent decisions had been made in the past. Since the focus is on past trajectories, the study sidesteps the "researcher as social engineer" problem to some extent. The second challenge involves deciding which feedbacks to include. Lane (2014) argues that predictive socio-hydrological models are challenging because social futures are not well defined. The position taken here is that the feedbacks and sub-systems simulated depend on the time-span of the model, which in turn depend on the scale of system behaviour that needs to be understood (see Fig. 1). For shorter time periods of about a year (e.g. a specific drought event), infrastructure, economic activity, and political structures can be held constant, though water availability and markets may change. Over a decade or two (e.g. the planning horizon for a water resources agency), infrastructure and politics would change and some incremental improvements in technology and market adjustments would occur, but it would be reasonable to assume that the structure of an economy or cultural beliefs are likely to be the same. Over a hundred years (e.g. in making decisions over major infrastructure projects), all these factors along with hydro-climatic patterns are likely to change. In this study, the temporal choice of 30–40 years dictates which feedbacks are appropriate.

The third challenge involves designing models that can accommodate a broader range of feedbacks than have been observed in the past. Focusing on past counterfactual trajectories mitigates this concern somewhat. Counterfactual trajectories use actual rainfall data, political and technology changes that occurred over the period of the model. Only policy variables are allowed to change. In the model presented here, the choice of counterfactual trajectories was based on contemporary debates on how urban water supply should be managed. In recent years, many Indian scholars and practitioners have begun questioning the wisdom that all urban water needs must be met through 24/7, potable piped supply imported from outside the city. They point out that inadequate piped supply does not automatically mean that users do not get enough water to meet their needs. Meeting a portion of urban water needs from local supply or self-supply may be an acceptable or at least realistic alternative (Shah, 2013). Already, many users rely on their own private wells for at least the non-potable component of their needs (Shaban and Sharma, 2007). Taking this into account, the model allows for multiple source dependence.

As the scenarios considered involve a range of water provision options going beyond piped supply, a metric that goes beyond engineering measures of piped supply reliability that could allow comparison over time and alternative trajectories was needed. In recent years "water security" has emerged as a new organizing idea in the water sector, encompassing both human and ecological concerns over multiple spatial and temporal scales. However, in practice, "water security" has been extremely difficult to operationalize (Cook and Bakker, 2012). Based on a broad review of studies on water security, Cook and Bakker (2012) suggest that the concept is best used to guide the selection of narrower, case-specific indices that may be used in policy, modelling or empirical research. In this paper, the term "household water security" is applied at the household scale to refer to the "quantity of water used by the household when all available sources of water are pooled". The evolution of household water security is traced over a 40-year period from 1965 to 2005 using a stylized model of Chennai India

2 Methods

2.1 Case study: Chennai, India

Chennai, formerly Madras, is India's fourth largest city, located in the southern state of Tamil Nadu. As per the 2011 Census, about 8.9 million people resided in the urban agglomeration, which includes peri-urban areas, towns and villages. Chennai lies in the rain-shadow region of the Western Ghats and is dependent on the northeast monsoon – a series of tropical depressions between October and December which deliver large quantities of rain over a few rainy days. Although the city receives almost 1250 mm of rain annually, the rainfall is irregular and episodic.

Unlike other Indian cities, Chennai does not have much of a pre-colonial history. The city of Madras developed around the port and the military establishment of Fort St. George (Gopakumar, 2011). Until about 1870, the population was dependent on privately dug wells or public wells and tanks. Organized water supply to the British colonies was commenced in 1872. As the city grew, three reservoirs were acquired or constructed between 1944 and 1972, to bring the combined storage capacity of Chennai's reservoirs to about 175 million cubic metres (MCM). Following hydrogeological investigations by UNDP between 1966 and 1969,

well fields in the Araniar-Kosathalaiyar Basin (A.K. Basin) located north of Chennai were developed for abstracting groundwater. The Chembarambakkam Lake, another small peri-urban reservoir, was acquired for city water supply after its irrigation command area disappeared due to urbanization (Metrowater, 2011) by 2000.

The biggest augmentation to Chennai's water supply occurred via the so-called "Telugu Ganga" project. An agreement was signed jointly by riparian governments of Maharashtra, Karnataka and Andhra Pradesh in 1976 to allocate about 420 MCM annually of Krishna River water to Chennai. Initial works for supplying water under the Telugu Ganga scheme were completed in 1996. Water began to be delivered into Chennai's reservoir system through a 152 km long open canal. This design allowed for significant losses along the way, both through direct lifting and seepage, as the water flows through the drought-prone regions of Andhra Pradesh before reaching the Tamil Nadu border. So only a fraction of the water actually reaches Chennai. Another project, the intra-state Veeranam Water Supply Project, was implemented in 2004 as additional source of water to Chennai. The project supplies 180 million litres per day (MLD) of water to Chennai by drawing water from Veeranam Lake in the Cauvery Basin (Metrowater, 2011) in Tamil Nadu.

The major challenge of Chennai's water supply system and consequently its vulnerability has thus been its inability to store monsoon waters for supply throughout the year. Even today, Chennai's reservoir storage capacity remains very low by Western standards. Even as Chennai's population has almost tripled since 1965, very little new reservoir storage was added. The two most recent projects, Veeranam and Telugu Ganga, did involve "new" reservoir capacity for Chennai, but the reservoirs associated with these projects are controlled by other agencies and Chennai must negotiate with farmers (in the case of Veeranam) and Andhra Pradesh (in the case of Telugu Ganga) to secure releases. This means that the reservoirs are not necessarily managed to optimize Chennai's needs.

Throughout its history, Chennai has been water scarce. Despite this, urban piped water supply has remained unmetred. Rationing rather than pricing has remained the dominant mode of controlling water use. Even today only a fraction of households are metered, but water is supplied for only a few hours each day. Because water supply is unreliable, more than two-thirds of Chennai's households have private wells as a supplementary water source (Srinivasan et al., 2010a). Peri-urban towns and villages are served by several different agencies. Some of these receive bulk supply from Chennai's water utility, while others rely entirely on bore wells. Overall, as the city continues to grow outward rapidly, peri-urban areas are increasingly groundwater dependent. It is expected that peri-urban villages and towns will eventually be supplied with water and sewerage services via the city municipal supply agency, but this is likely to further strain the limited reservoir capacity of the city.

2.2   Model conceptualization and parameterization

The model described here is a "stylized" version of a detailed, spatially explicit coupled human–hydrologic model developed and published previously (Srinivasan et al., 2010a). The previous coupled model was run and calibrated using a variety of hydrologic and socioeconomic data. The model was calibrated for the period from 2002 to 2006, which included both one of the worst droughts and one of the wettest years in historical record. As a result, the model was able to capture both hydrologic and social responses to drought. Longer-term parameters such as reservoir storage, the poor state of the piped supply system and household dependence on private wells were taken as given and constant over the 5-year period.

This present study uses the previous model as a starting point. The parameters for shorter-term processes are imported from the earlier model (Srinivasan et al., 2010a). These include the user demand function, the response of the aquifer system to recharge and pumping, the rainfall-inflow model into the reservoir system, reservoir operations, user behaviour and the functioning of the tanker market. However, the present study explores coevolutionary, temporal dynamics over a much longer period. As the model is run over a longer period, it incorporates additional feedbacks representing slower decadal-scale processes, which were held constant in the previous model. The new feedbacks added include growing income, penetration of indoor plumbing, private wells, the impact of pricing policies on the water utility's finances and thus its ability to expand and maintain infrastructure. The socio-hydrologic model first replicates the actual trajectory of Chennai's water supply system during 1965–2006. Next, "counterfactual" trajectories that might have occurred if different decisions had been taken are explored. The spatial component has been eliminated entirely and the model only focuses on the core urban area of 176 km^2.

The urban water system is conceptualized through a set of equations and feedbacks (see Fig. 2). A key element of this model is the integration of the hydrologic system, the engineered water delivery system and household-level decision making. The reservoir and aquifer system, population and user investments in bore wells were all stock variables. Water supply, pipeline leakages and water extraction and use by users are flow variables. In setting up the model, rainfall, demography, economic growth, prices and user preferences were assumed to be exogenous or external to the model. It was assumed that the presence or absence of water was not a significant determinant in Chennai's growth, which instead was driven by larger macro-economic factors. Investments in infrastructure both by the water utility and private users were, however, determined within the model. The initial conditions – reservoir capacity, reservoir storage, the coverage and efficiency of the piped system in 1965 – were based on actual data. Actual water tariffs fixed by the government were used.

In terms of users, a distinction was made between households with in-house plumbing ("Tap" households) versus those who access water manually from standpipes and public wells ("NonTap" households) for a variety of reasons (Strand and Walker, 2005). This was necessary because both population growth and changing lifestyles are contributing to the city's demand for water. This categorization also allows different demand and supply functions to be used for the two types of users. Poorer households lacking indoor plumbing use water very differently and have a much lower willingness to pay. They also face different resource constraints; they must store water in pots and use it manually rather than from a tap. Consequently they face a higher cost of water (Zérah, 2000; Pattanayak et al., 2005). Wealthier households can invest in underground sumps and pumps and thus face a lower marginal cost of water.

The model incorporates feedbacks between multiple spatial scales (utility versus household) and temporal scales (decadal versus daily). To achieve this, both city-scale long-term decisions (infrastructure investments in reservoirs and piped infrastructure) and short-term decisions (reservoir releases) as well as household-scale long-term investments (private wells) and short-term decisions (cutting back on consumption and procuring water from alternate sources) were considered.

In the long-term, as the city grows, the urban water utility makes decisions about expanding and maintaining the piped infrastructure and reservoirs depending on the financial resources available to it. The model simplistically assumes that improved finances will actually result in better infrastructure (i.e. the money will not all be lost to corruption). In the short term, the water utility makes decisions on how to allocate the resources available to it given the infrastructure available. The state of the infrastructure thus determines how much water is available to households in each period.

Households make independent decisions on how to cope with the available supply. If piped supply infrastructure is insufficient or degraded, households may invest in private wells and underground sump storage. These coping investments allow households to diversify the sources available to them when water shortages occur and enhance their own water security. Moreover, these investments are "sticky" – once made they permanently alter their choice set. In each time period, households optimize their daily water use based on the available quantity and cost of water from different sources. However, the options available to households in the short term are contingent on their long-term coping investments. Households may self-supply from private wells or purchase water from tankers. Thus each feedback —slow or fast — is associated with biophysical changes and socioeconomic changes (Table). These interact to generate emergent behaviour.

The socio-hydrologic model consists of a number of linked sub-models each consisting of one or more equations (Fig. 2). A complete description of the sub-models with the equations is presented in Appendix A.

Table 1. Details of feedbacks between human and natural sub-systems.

Feedback	Sub-models	Description of feedback	Slow/Fast
Nature − > Nature	Climate − > Reservoir	Increase in rainfall => increased inflows into the reservoir system	Fast
	Climate − > Aquifer	Increase in rainfall => increased aquifer recharge	Fast
Nature − > Human	Reservoir − > City Water Supply	Decrease in reservoir storage => cutbacks in piped supply	Fast
	Aquifer − > User Agent	Drop in Groundwater table => more wells going dry	Fast
Human − > Nature	User Agent − > Aquifer	Decrease in piped water availability => more groundwater extraction	Fast
	User Agent − > Aquifer	Increase in private wells => more groundwater extraction	Slow
	Infrastructure − > Aquifer	Improved piped infrastructure => less pipeline leakage into groundwater	Slow
Human − > Human	City Water Supply − > User Agent	Decreased piped supply => switch to private sources like wells and tankers	Fast
	Infrastructure − > City Water Supply	Increase in utility revenues => improved pipeline infrastructure, reservoir capacity	Slow
	Infrastructure − > User Agent	Decreased piped supply => increased drilling of private wells	Slow

- The *Climate Sub-model* specifies the rainfall in Chennai. For the purposes of this study, historical rainfall data were used.

- The *Population Sub-model* specifies the number of Tap and NonTap HH in Chennai. Population growth and rate of increase in Tap HH was based on actual historical data and were assumed to be the same for all trajectories, i.e. it was assumed that water availability does not significantly influence either population growth or the number of households investing in indoor plumbing.

- The *Reservoir Sub-model* estimates storage in the reservoir system at the end of each month. In the historical trajectory, data on reservoir storage, inflows, rainfall and diversions were available and were used to derive the rainfall-inflow and storage-diversion relationships.

- The *City Water Supply Sub-model* distributes the amount of water available in the reservoir system between Tap and NonTap HH. Based on interviews with city water utility engineers, it was assumed that the amount released from the reservoir system is a fixed fraction of reservoir storage at the beginning of the month.

- The *Infrastructure Sub-model* determined the rate of deterioration or improvement in pipelines and thus the pipeline leakage over time as well as the amount of new reservoir storage added based on the how much the tariff exceeds or falls short of the long-run marginal cost of supply.

- The *Aquifer Sub-model* simulates water levels in the aquifer as a bathtub. The depth of the water table in the aquifer is thus a linear function of the total aquifer storage. Given the distribution of well-depths in Chennai, the fraction of wells that go dry is calculated. The amount of groundwater extracted in each period is obtained from the User Agent and Tanker Sub-models.

- The *User Agent Sub-model* was the representation of user (households). It was assumed they make two types of decisions. In the short term, they decide what sources of water to use in a given time period given their income and past investments in wells, sumps etc. In the long term, households must decide whether to get a connection and drill a private well. It was assumed that when piped water supply drops below quantity, a fraction of piped households will drill wells.

- The *Tanker Sub-model* estimates the size of the tanker market by multiplying the number of households purchasing tanker water with the quantity of water each household purchases.

- The *Cost of Water Sub-model* is estimated as the total amount for water divided by the total water use by all households.

3 Model results

The model explored three coevolutionary trajectories that Chennai's water system could have followed. In all three, Chennai's population and economic growth were assumed to be exogenous, i.e. independent of the water situation. The number of households almost tripled from 400 000 to about 1.1 million and fraction of households with indoor plumbing also increased from half of all households in 1965 to al-

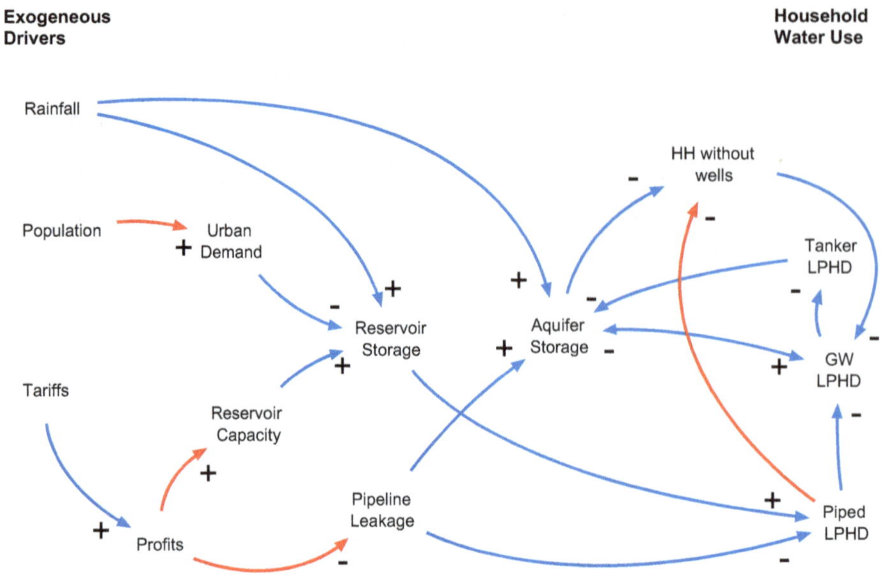

Figure 2. Feedbacks in Chennai's coupled human–water system. The red arrows represent slow feedbacks, while the blue arrows represent fast feedbacks. LPHD is litres per household per day.

most 70 % in 2006. All three scenarios use the same actual historical rainfall scenario. It may be observed that Chennai experienced several prolonged multi-year droughts. The first multi-year drought occurred between 1985 and 1990 and the second one, between 1999 and 2004 (Fig. 3). In all three scenarios, incomes were assumed to grow at about the same rate, so that more and more households were able to afford sumps, bore wells and indoor plumbing and thus the fraction of "Tap" households increased over time. All prices represent real prices in 2005. i.e. inflation is not explicitly modelled.

The first, the current trajectory, is called "Low initial reservoir storage, no metering, flat price". In this trajectory, Chennai starts in 1965 with a relatively low level of surface storage and a flat-rate tariff which does not allow cost recovery. Over time the piped system cannot be maintained; pipeline leakages become worse and less and less of the water reaches users. Very little new storage is added. This scenario essentially replicates historical reservoir storage, tariff and population.

The second, called the "High initial reservoir storage, volumetric tariffs" is a counterfactual trajectory. The initial reservoir capacity in 1965 is about 2.5 times the actual 1965 reservoir capacity. The tariff is high enough to cover both the short- and long-run cost of piped supply and so the infrastructure keeps pace with the population. Reservoir storage gradually increases and pipeline leakage decreases and stabilizes at 5 % over time.

The third, called the "High initial reservoir storage, no metering, flat price" is a another counterfactual trajectory, in which the city starts with 2.5 times the actual 1965 reservoir capacity, but a flat-rate tariff policy does not allow cost

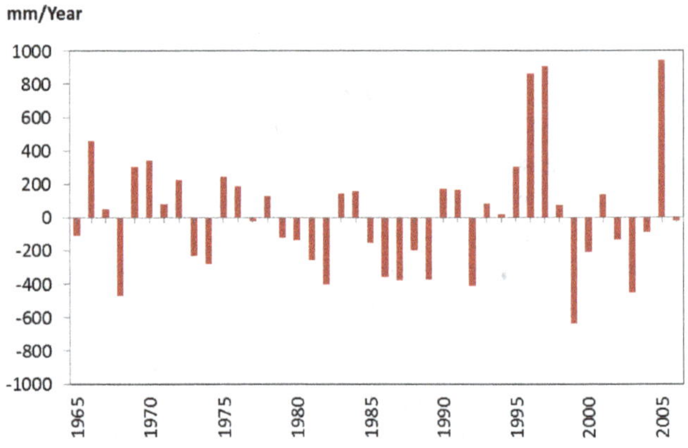

Figure 3. Deviation from average annual rainfall of 1261 mm.

recovery. In this scenario, reservoir capacity stays frozen at 1965 levels and pipeline leakage worsens gradually.

For each trajectory, three types of results are presented: (a) long-term infrastructure changes over time because of investments by the water utility and households; (b) short-term changes in water availability in the reservoir system and aquifer which depend on the infrastructure available as well as rainfall in a given year and (c) short-term changes in water consumed by and costs to households.

3.1   Current trajectory: low initial storage, no metering, flat price

The driving assumptions in this scenario are that the utility starts out with very little reservoir storage but also has no system of metering. The coevolution of the system is presented in a series of graphs (Fig. 4a–i). The flat-rate tariff system does not allow the city to invest in expensive infrastructure projects or maintain the piped network. As a result, reservoir capacity increased by just 20 % even as population almost tripled (Fig. 4a). Pipeline leakage also worsened, increasing from 20 % in 1965 to almost 40 % in 2005 (Fig. 4b). Moreover, in order to serve the growing population with the same level of storage, the utility became more and more aggressive in its management of the reservoir system. The inability to increase reservoir storage along with increased pipeline leakage results in piped supply becoming very unreliable over time (Fig. 4f).

When households do not receive reliable piped supply, it is economically rational for them to invest in private wells. At first, only the wealthiest few households could afford wells, but well ownership gradually increased over time as incomes rose (economic growth was assumed exogenous to the model), (Fig. 4c). Households who had wells were able to use wells whenever piped supply fell short. When only a few households had wells, the aquifer was able to buffer them over a multi-year drought. However, as more and more wells were drilled, the groundwater level dropped faster and faster in drought periods and more households became tanker dependent (Fig. 4g, h). As tanker water is much more expensive than all other sources of water, the cost of water rose sharply during droughts and households became more water insecure (Fig. 4i).

Overall, the historical trajectory is the story of a shift from public investment in reservoirs and piped infrastructure to private investment in wells. However, the common pool groundwater resource which was able to support a few well owning households, got depleted with increases in population.

3.2 Counterfactual trajectory 1: high initial reservoir storage, volumetric tariffs

The second trajectory is based on different initial conditions. The initial reservoir storage is assumed to be 2.5 times Chennai's actual storage in 1965 and comparable to many cities in developed countries. In this case, piped supply is assumed to be fully metered and priced at Rs 13/kL, above the long-run marginal cost of supply. The additional revenue is assumed to be used to maintain the system, expand reservoir storage and inter-basin transfer projects. In this trajectory, the reservoir is assumed to be managed carefully and releases are matched to meet urban demand. However, because water is metered and priced, consumers have incentives to invest in water use efficiency measures. The utility is able to successfully control demand. The reservoir does not dry up as often.

Except for the severe multi-year drought during the 1980s, when piped supply becomes slightly unreliable, Chennai by and large enjoys secure piped supply (Fig. 5). Although the rate of ownership of private wells increases during the drought, well-drilling stops once piped supply is re-stored; very few new wells are dug. As very few consumers depend on private wells, the aquifer storage does not fluctuate much. No tanker market develops. Consumers are able to satisfy their needs with piped supply in almost all periods. This is the trajectory that most developed world cities have been able to follow. Although consumers incur much higher costs each month over the 40-year period, there is very little variability in the cost of water.

3.3 Counterfactual trajectory 2: high initial reservoir storage, no metering, flat price

The third trajectory for Chennai begins with robust infrastructure, but in this case water is charged at a flat rate and is not metered. As a result, the utility is unable to expand in response to demand or maintain the piped network, which gradually deteriorates. In each consecutive drought, the city is unable to control demand and the reservoir dries up.

For the first 35 years, from about 1965 to 2000, the city does not feel the effects of the weak tariff policy (Fig. 6). When the multi-year drought strikes in the early 2000s, the aquifer no longer has the buffering ability – 70 % of the Tap households have wells by this time. The higher initial reservoir storage helps the city. Only 10 % of the wells go dry and a small tanker market develops – not as severe as the historical trajectory. The cost of water rises because some users must depend on tankers.

3.4 Sensitivity analysis

The model was found to be sensitive to aquifer parameters. Overall, parameters affecting the aquifer were much more sensitive than the reservoir system. Model results were sensitive to "natural" parameters such as the specific yield and "policy-relevant" parameters such as the recharge rate.

In particular, increasing specific yield improves the buffering capacity of the aquifer because groundwater levels do not drop as much for a given level of extraction. Fewer wells go dry during droughts and fewer users are forced to buy expensive tanker water. This suggests that cities which are underlain by less productive aquifers (e.g. the hard rock aquifers found in peninsular India) are likely to be worse off compared to cities underlain by productive aquifers, as private wells are less able to provide a supplementary source of water. Similarly, increasing the proportion of rainfall that recharges the aquifer boosts the buffering capacity of the aquifer. The benefits of improving recharge through artificial recharge and household rainwater harvesting have been discussed elsewhere (Srinivasan et al., 2010b). The sensitivity analysis is consistent with the previous finding that boosting recharge reduces the tanker market size during droughts.

The model is also sensitive to the user demand function. For instance, if the demand function is changed so that households consume 33 % less water, the water table does not reduce as much, and the tanker market virtually disap-

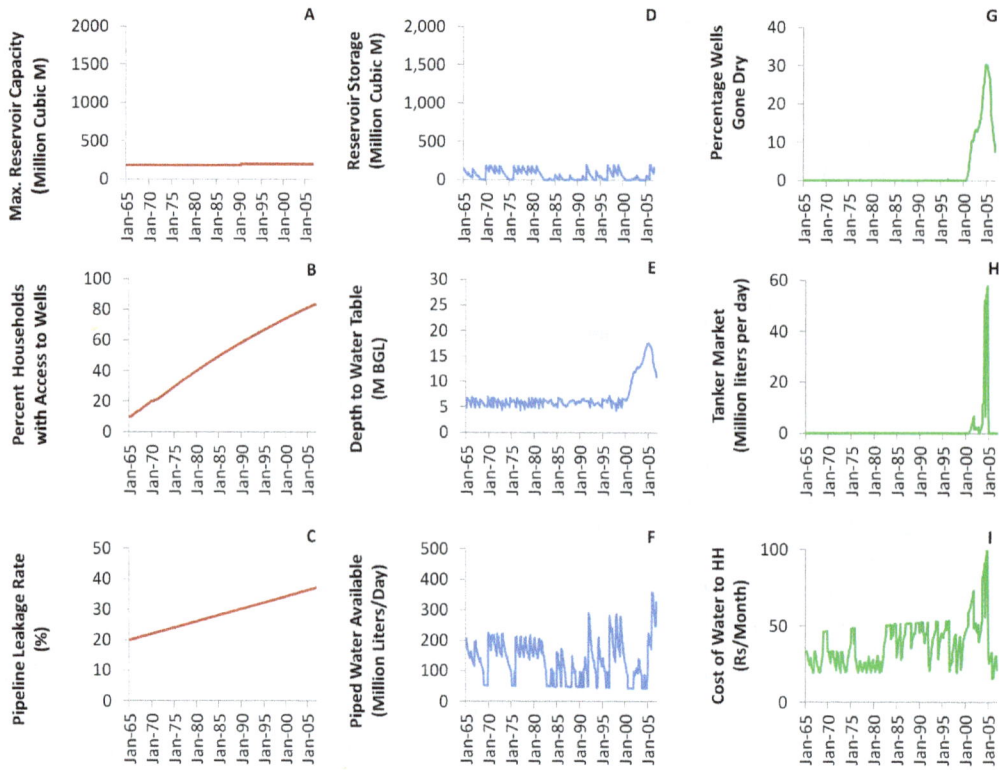

Figure 4. Current trajectory: low initial reservoir storage, no metering, flat price.

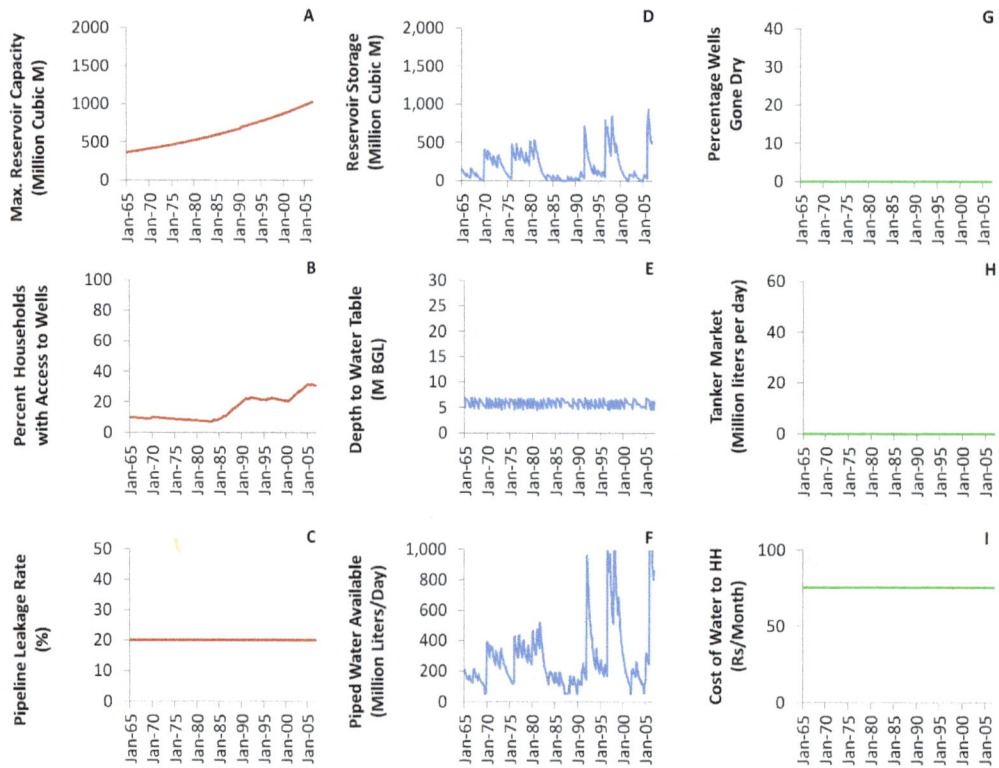

Figure 5. Counterfactual trajectory 1: high initial reservoir storage, volumetric tariffs.

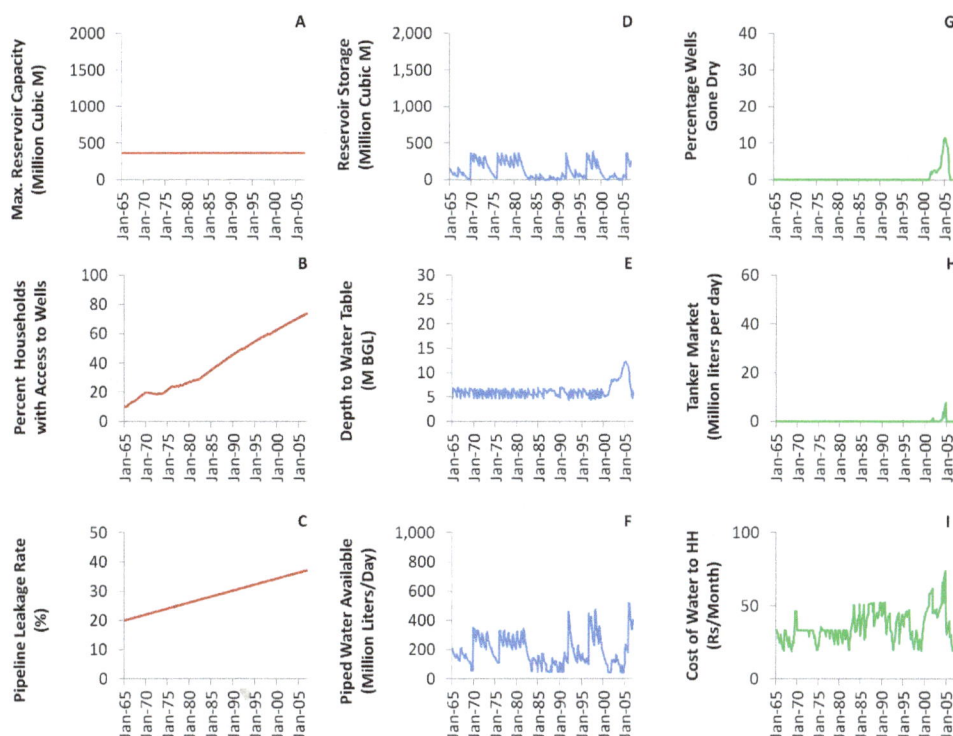

Figure 6. High initial reservoir storage, no metering, flat price.

pears. The demand function used was a simplistic model based on the Chennai household survey (Srinivasan et al., 2010a). However, to our knowledge, no study has successfully modelled water demand under supply constrained conditions, with multiple source dependence in the developing world. This suggests that additional research on user demand is much needed. From a policy perspective, the result shows that any reductions in groundwater extractions such as demand-side management and wastewater recycling would yield significant benefits.

Interestingly, the model is relatively insensitive to increasing reservoir storage. A 50 % increase in initial reservoir storage in 1965 yields only marginal benefits during the prolonged drought of the 1980s. Significant additional reservoir storage is needed to completely prevent piped supply shutdowns during multi-year droughts.

3.5 Model limitations

Any model of a complex real-world system, including the one presented in this paper is likely to suffer from limitations and it is worth reflecting on what effects this may have on the conclusions.

First, the model presented herein is weak on politics. A key assumption is that human responses to water scarcity are primarily techno-economic. While most households in Chennai do indeed respond by making coping investments in sumps and wells or purchasing water from tankers, water users are also citizens who engage in the political process. Indeed, in my own field investigations I encountered several examples of communities, particularly slums, using a range of strategies to lobby the local government to improve water supply. However, it was also clear that there were no universal factors that could predict why some slums were better at securing access to water than others. This suggests that there are inherent limits to quantitative approaches to modelling water security at the household scale.

Second, the narrow definition of household water security (in terms of the average cost of water) overlooks the nuances of reliability, inequity and uncertainty of the amount and timing of supply. Yet, studies show that certain sections of society are disproportionately affected by uncertain timings of supply because of lost wages from waiting for water. Moreover, average costs obscure distributional differences. As socio-hydrology evolves as a field, greater attention to normative lenses and how the choice of outcomes influences the conclusions drawn is needed.

Third, the model presented in this paper makes an assumption that demographic and economic growth are not limited by water scarcity; these were assumed to be exogenous to the model. While there is insufficient evidence on how unreliable water supply might limit long-term economic growth, additional research exploring these feedbacks is warranted. It is difficult to imagine that a city with no water could grow as quickly as one with abundant water supply.

Finally, ideally, socio-hydrologic models should be developed in consultation with stakeholders to frame the research questions, determine which dynamics are essential to replicate, which thresholds are important and which tradeoffs are acceptable. Although the present study is grounded in extensive interactions with domain experts, water managers and users at the study site, the intuition derived from these interactions was not formally codified. It is therefore always possible for the model to be biased by the values and training of the researcher.

4   Conclusions

In the developing world, traditional planning tools fare poorly over a 30–50 year time horizon. In most developing regions, water systems are changing so quickly that holding land use, economic patterns, governance systems or technology static over a 50-year period results in inaccurate predictions leading to sub-optimal paradoxical outcomes. Many water managers fail to account for impacts of the adaptive responses by humans that could result in unexpected outcomes.

Socio-hydrology has been defined as a new science of water and people to precisely address this problem. The goal is to explicitly account for such bi-directional feedbacks and improve predictive insight. While there are several challenges socio-hydrologic modellers face in framing the problem and choosing which outcomes are worth studying, perhaps the biggest challenge is imagining technological, social and political futures. If technology, social preferences, the structure of the economy or governance systems change, these could fundamentally alter future water demand, allocation and use. So while it is necessary to ask decision makers to examine alternative futures and figure out policies that might get us there, it can be an abstract, perilous process.

This paper offers an alternative approach – the use of counterfactual trajectories – that allows policy insights to be gleaned without having to predict the social futures. The approach allows us to instead "reimagine the past"; to observe how outcomes would differ if different decisions had been made in the past. Because the focus is on the recent past, the results could be applicable in other regions facing similar decisions.

A stylized, socio-hydrologic model that explores bi-directional feedbacks between the societal, engineered and hydrologic components of water systems is applied to achieve insights into household water security in developing urban regions, using the case study of Chennai, India. The model includes both "fast" processes such as short-term reservoir management and source switching by consumers; as well as "slow" processes such as long-term investments in infrastructure by the water utility (pipes and reservoirs) as well as users (wells, piped connections). On the one hand, the water utility's investments in pipes and reservoir storage constrains the water available to households in a given pe-

riod. On the other hand, lack of water availability in a given period prompts a fraction of the users to drill wells. Additionally, the dynamics observed in the model are influenced by the biophysical constraints of the aquifer and watershed hydrology.

This paper presents an example of a socio-hydrologic modelling study, which can model coevolutionary, emergent behaviour. In contrast to traditional water resources management studies, the goal is not to prescribe policy. The model allows the water utility to develop reservoir storage based on the utility's finances. It also allows households to make private coping investments. Household water security evolves based on infrastructure and pricing by the water utility and corresponding coping investments by consumers. Thus, for instance, whereas "optimal" reservoir storage is usually prescribed by a water resources management model, reservoir storage is an emergent property of the system. Instead, the objective is to explore alternative trajectories that a water supply system might have followed.

Two counterfactual trajectories are explored in addition to the actual historical trajectory. The model results offer interesting insights into urban household water security in developing water systems. First, household water consumption in Chennai goes beyond piped water supply; instead, the aquifer acts as a backstop source. When piped supply fails users first turn to their own wells. When their wells dry up, a tanker market develops. When consumers are forced to purchase expensive tanker water, they become water insecure. Second, not unexpectedly, different initial conditions result in different trajectories However, initial advantages in infrastructure are eroded if the utility's management is weak and it is unable to expand or maintain the piped system to keep up with growth. Both infrastructure and management decisions are necessary to ensure household water security. Indeed, if storage capacity has to keep up with demand, Chennai's reservoir storage would need to be ten times the actual storage today and comparable to cities like Boston, MA. This raises the issue of path dependence and the extent to which such increases in reservoir storage are feasible in the current socio-political climate. Even if full metering and a rational tariff policy were followed, emerging social movements in the 1980s over resettlement and environmental concerns of dam-building may have limited Chennai's options as some studies have shown (Feldman, 2009). Third, the effects of weak management and inability to expand reservoir capacity do not manifest right away. Instead, the situation deteriorates over time and the impacts of bad policy may not manifest until much later when the population has grown and a major multi-year drought strikes.

Appendix A

The equations used to specify the model are described below in detail. In describing the model, the subscript t (referring to the model time period of 1 month) is skipped to improve readability. The convention used in describing the variables is as follows: variables prefixed with "Total" refer to city-wide quantities measured in MLD – e.g. "TotalCityDmd" is the total water demand for the city. Variables referring to household level supply, demand and consumption in litres per day are prefixed with "T" or "NT" for Tap and NonTap households respectively and suffixed with "Dmd", "Sup" and "Use" depending on whether they refer to quantity demanded (if supply were unconstrained), quantity available and quantity actually used. For instance, TPipSup, TPipDmd and TPipUse refer to supply, demand and use from piped supply by Tap households. Reservoir and aquifer models stocks and flows are in m^3 and m^3 per month respectively.

Population Sub-model: population growth in Chennai was based on actual historical projections. The average household size of 4.5 persons per household based on the 2001 Census of India data for Chennai, was assumed to hold good for all households. It was assumed that households gradually converted from NonTapHH to TapHH as they became wealthier; i.e. indoor plumbing gradually increased. The total number of households in Chennai is the sum of the number of Tap and NonTap households:

$$TotalHH = TapHH + NonTapHH. \tag{A1}$$

The fraction of NonTap households (households lacking indoor plumbing) dropped over time from half of all households in 1965 to 33 % in 2005. The increase in indoor plumbing was linked to economic growth rather than water availability and was therefore treated as being exogenous to the model.

Reservoir Sub-model: the reservoirs receive inflows from the local watershed and water from the Telugu Ganga scheme is also delivered into the reservoirs. Local inflows were modelled as an exponential function of monthly rainfall R:

$$\delta S = k e^{\lambda R} + TG - W - Ev - O, \tag{A2}$$

where k (1.22) and λ (0.017) are empirically derived constants from the historical rainfall–runoff relationship if rainfall R is in mm month^{-1} and inflows are in Mm^3 month^{-1}. S is the total reservoir storage in cubic metres at the beginning of the period. TG is actual inflow received from the Telugu Ganga project at the state border in m^3 month^{-1}, W is the water supply released from the reservoir for urban supply and Ev is the average lake evaporation calculated from Lake Evaporation data in m^3 month^{-1}. O is the spills from the reservoir downstream; any inflow in excess of maximum reservoir storage ResMax is assumed to be released downstream.

City Water Supply and Distribution Sub-model: the City Water Supply Model distributes the amount of water avail-

able in the reservoir system between Tap and NonTap HH. Based on interviews with city water utility engineers, it was assumed that the amount released from the reservoir system (W) is a fixed fraction (p %) of storage S. It is interesting to note that the fraction p did not turn out to be constant over time. In order to match observed storage, p had to be increased over time. Throughout the particularly wet decade of the 1970s, p was approximately 4 %. After 1980, p had gradually increased to 7 % by 2005 suggesting that the reservoir management became more aggressive to meet the increased demand while reservoir storage remained the same:

$$TotalCitySup = (p.S + TotalImports) \times Cf. \tag{A3}$$

The water released from the reservoir system is shared between Chennai and the surrounding towns – only a fraction Cf is supplied within the city. Based on historical data this was averaged to be about 66 % The rest goes to industrial and commercial bulk supply and nearby towns. The amount of water available for piped supply includes diversions from the reservoir system plus imports from an intra-state scheme and well-fields:

$$TotalPipSup = TotalCitySup \times (1 - LeakRate). \tag{A4}$$

A percentage of the water, LeakRate, is lost via pipeline leakage and in turn recharges the shallow aquifer. The rest, TotalPipSup, is distributed via the piped distribution system:

$$TotalCityDmd = Ind+ \tag{A5}$$
$$\frac{(TapHH \times TPipDmd + NonTapHH \times NTHPDmd) * 1.2}{10^6}.$$

The TotalCityDmd is based on how much water would be demanded by households and commercial establishments if supply were unconstrained (i.e. everyone can get as much water as they want). This was estimated based on the demand function explained later (Eq. A21). The factor of 1.2 includes Commercial and Industrial demand, assumed to be 20 % of domestic demand:

$$Shutdown = \begin{cases} 1, & \text{if TotalPipSup} < \text{TotalCityDmd} \times 0.1 \\ 0, & \text{otherwise.} \end{cases} \tag{A6}$$

It is assumed that if the amount available from all sources drops below 10 % of the city's demand for water, the piped supply system shuts down and the scarce supply is distributed via tankers without any leakage loss.

The water is delivered to NonTap households via standpipes and Tap households via private piped connections. However, because the two types of consumers access water very differently the water available to them must be modelled separately. Water in standpipes is manually collected during the few hours when the pipes have water in them. In contrast, for private connections, the water is pumped to overhead tanks and flows by gravity to the taps in the house whenever they are turned on. Owing to the lack of storage and effort involved in hauling water around, users who depend on standpipes generally end up accessing much less water.

NonTap households must collect the water manually during the hours of supply (even in the wettest periods, Chennai does not receive 24-hour water supply). During droughts, the city cuts back on hours of supply further. Therefore, in the model, the amount of water accessed by NonTap households depends on how many hours of piped supply is provided, which in turn depends on the availability of water in the piped supply system (TotalPipSup). It is assumed that the rest of the available piped water is equally distributed among Tap households. To simulate this we assumed that the total water supply could be translated to hours of supply based on an empirically derived equation:

$$HrsSup = \begin{cases} \frac{TotalPipSup}{TotalCityDmd} \times 4, & \text{if } Shutdown = 0 \\ 0, & \text{if } ShutDown = 1. \end{cases} \quad (A7)$$

HrsSup represents the number of hours of piped supply. Non-Tap users accessing water via standpipes can only receive water during hours of supply. The amount of water theoretically available to NonTap users in litres per day is given as

$$NTHPSup = \frac{HrsSupply \times 60}{10} \times 15. \quad (A8)$$

NonTap users dependent on standpipes are hit hardest by cutbacks. It takes roughly 10 min to fill a pot of water including time wasted and in transitions between people. Each pot holds 15 L:

$$NTHPUse = min(NTHPSup, NTHPDmd). \quad (A9)$$

A key assumption is that NonTap households are both supply and demand constrained. They will queue up to collect water but only until their demand is satisfied. Beyond this even if hours of supply are expanded, they will not use more water:

$$TotalHPUse = \frac{NTHPUse \times NonTapHH}{10^6} \quad (A10)$$

$$TPipSup = \frac{\frac{1}{1.2}(TotalPipSup - TotalHPUse).10^6}{TapHH}. \quad (A11)$$

Assuming that NonTap users wait in line to get water and fill every available pot during the hours water is available, the rest of the water is delivered into the sumps of all piped users. After accounting for the 20 % supplied to commercial and industrial users, the amount of piped water supply available to each TapHH can be calculated.

Infrastructure Sub-model: it was assumed that the deterioration in pipelines and thus increase in leakage over time is proportional to the difference between the tariff and the short-run marginal cost of supply. If the tariff exceeds the operation and maintenance cost (OMCost is Rs 12/kL), then the pipeline leakage gradually improves at a rate proportional to the difference or surplus revenue (RevSurplus) earned on

each unit of water delivered:

$$RevSurplus = \frac{PipedPrice - OMCost}{OMCost} \quad (A12)$$

$$\delta LeakRate = C1 \times RevSurplus. \quad (A13)$$

The constant $C1$ was chosen so that the piped system leakage rate deteriorates from 20 % in 1965 to 37 % by 2007 under the current flat-rate tariff structure.

Similarly, it is assumed that if there are surplus revenues, new reservoir capacity can be added to keep pace with population growth:

$$\delta ResMax = C2 \times RevSurplus \times \kappa(TotalHH). \quad (A14)$$

The constant $C2$ was chosen such that reservoir capacity increases at a rate to maintain the initial per capita reservoir capacity, when water is priced to be Rs 15/kL (i.e. Rs 3/kL more than O & M costs). Reservoir inflows are also proportionately increased – i.e. the reservoirs are assumed to have their own catchments which generated inflows using the same equations as the existing reservoir system.

Aquifer Sub-model: the aquifer is simulated as a simple bathtub, so the water level in the aquifer is a simple linear function of the total aquifer storage with specific yield (SYield) of 10 %. When the aquifer is fully saturated, saturated thickness is assumed to be 20 m. The area (ChennaiArea) is $176 \times 10^6 \text{ m}^2$. Thus, the volume of water stored in the aquifer in m^3 when fully saturated is

$$MaxGW = 20 \times SYield \times ChennaiArea. \quad (A15)$$

The groundwater balance equation is

$$\delta GW = PR + RR - SS - TankerUse - GWUse. \quad (A16)$$

PR in m^3 per month is the pipeline recharge depends on leakage from pipelines. RR is rainfall recharge in m^3 per month, which is defined as 10 % of rainfall. Rainfall is in mm month^{-1}. SS is defined as sub-surface flow to the ocean or baseflow to the river which occurs whenever GW is completely saturated. TankerUse and GWUse represent groundwater extractions by tankers and households respectively. These are discussed in the User Agent section:

$$PR = \begin{cases} \frac{TotalCitySup \times 30 \times LeakRate}{1000}, & \text{if } Shutdown = 0 \\ 0, & \text{otherwise} \end{cases} \quad (A17)$$

$$RR = Rainfall/1000 * ChennaiArea * 0.1. \quad (A18)$$

The depth to water is based on the volume dewatered over the period of the model plus an assumed initial depth to water in 1965 (5 m b.g.l.):

$$DepthToGW = \frac{(MaxGW - GW)}{ChennaiArea \times SYield} + 5. \quad (A19)$$

User Agent Sub-model: a key feature of the model was the representation of user behaviour. Users (households) make two types of decisions. In the short-term, they decide what sources of water to use in a given time period given their income and past investments in wells, sumps etc. In the long-term, households decide what types of investments to make in the water system. In every period when piped supply is less than consumer demand, some fraction of the households (5 % per year or 0.41 % per month is assumed) drill new wells. Long-term investments such as wells are "sticky" and once made remain in place even if they are not used. They permanently alter the options and incentive structure to households:

$$\delta\text{Wells} = \begin{cases} \text{TapHH} * 0.041, & \text{if TPipSup} < \text{TPipDmd} \\ 0, & \text{otherwise.} \end{cases}$$
$$(A20)$$

The short-term consumption model recognizes that households are often supply constrained and must cope with water shortages. In the short term, households have options – i.e. they can switch sources or buy water. Consumption is constrained both by supply (amount of water available) as well as demand (amount of water they are willing to consume). In the short term, users need a small amount for their potable needs (about 10 L per capita per day). After allocating the best quality water for their potable needs, they allocate the cheapest available source for their non-potable needs:

$$\log(7Q) = \alpha \log\left(\frac{\text{Price}}{1000}\right) + \gamma \log(\text{HHSize}) + \delta I + \kappa, \quad (A21)$$

where α (the price elasticity of water) is -0.49, γ (the coefficient of household size) is 0.48 and δ (the income elasticity) is 0.19. Because α, γ and δ are exponents, they are unitless. A weekly demand function estimated from the household data set was divided by 7 to obtain the daily household water demand Q in litres per day (Srinivasan et al., 2010a).

In understanding short-term user decisions, users were assumed to be rational fully informed agents. The primary principle of the user agent model is that users will use up the lowest cost source accessible to them before moving to the next cheapest source. In other words, water consumption is driven by price and supply constraints. In developing the user agent model, only sources reported in the household survey were included. For instance, no households reported purchasing water from neighbours or using public surface water sources like ponds or temple tanks; so these were not included.

The price of water varies by source (piped, well, and tanker) as described in later sections. HHSize is the average household size in Chennai, which is 4.5 people. I is a binary income variable simply coded as high or low, and N is the number of members in the household. As the presence of indoor plumbing is linked to household wealth, Tap and NonTap categories also serve as way to categorize rich and poor users. Thus, $I = 1$ for Tap households and 0 for NonTap households. Thus, in the model, the demand for each source is different for Tap and NonTap households.

Tap and NonTap households were each assumed to access three sources of water: the piped supply, groundwater (own or shared wells) and purchased water from tankers. For any source, the amount used is the lesser of what is available and what is demanded at the marginal price. The model uses a simple allocation rule to decide which sources the users will use; consumers will use as much of the cheapest source available, then move to the next cheapest source. Thus the inputs to the user agent model are a price, a quantity demanded at that price and quantity available for each of the three sources.

Quantity Demanded: users rank sources in terms of cost from least to most expensive. Purchased tanker water (at Rs 60/kL) is always the most expensive and is the last possible resort. Between groundwater and piped supply, users pick whichever is cheaper. If users have wells, they will compare the marginal cost of groundwater (Rs 7/kL) to the cost of piped water. If piped water is charged at a flat rate, users only take into account the cost of pumping the water to their overhead tank (Rs 2/kL); but if piped supply is metered they must pay the volumetric tariff which may be anything between Rs 5/kL and Rs 15/kL. For each source, the maximum quantity the user would demand at that price was estimated using the demand function in Eq. (A21). For instance, for flat-rate piped supply the quantity demanded was 615 L day^{-1} (TPipDmd), for wells 356 L day^{-1} (TWelDmd) and for tankers 91 L day^{-1} (TTanDmd). The quantity demanded by NonTap households for standpipes and shallow bore wells use was 180 L day^{-1} (NTHPDmd and NTWelDmd).

Quantity Availability: the water available from different sources is obtained from the reservoir and aquifer models. The model assumes a maximum demand for each source of water based on its cost. Then the model calculates the potentially available supply from that source – NonTap use from handpumps (HPUse) and Tap piped supply (PipedSup) as calculated earlier. To estimate water available from wells, Tap households were classified into households without wells or whose wells went dry and those whose with functioning wells. It is assumed that the wells which have not gone dry will yield enough water to meet domestic water needs. So water availability from wells was simply assumed to be 0 for households lacking wells, but households with functioning wells were assumed to be able to satisfy all their residual water demand from wells. Similarly, water available from tankers is assumed to be infinite for all practical purposes.

Tap households – Case 1: if piped supply is the cheapest source and the household has a functioning well, then the household will use all available piped water before turning on their well:

$$1: \begin{cases} \text{TPipUse} = \min(\text{TPipSup}, \text{TPipDmd}) \\ \text{TWelUse} = \max(\text{TWelDmd} - \text{TPipUse}, 0) \\ \text{TTanUse} = 0. \end{cases} \quad (A22)$$

Tap households – Case 2: if piped supply is the cheapest source and the household has no functioning well, then the household will use all available piped water before purchasing tanker water. However, they will only purchase tanker water if there is any "residual demand" after the available piped water supply is used up:

$$2 : \begin{cases} \mathrm{TPipUse} = \min(\mathrm{TPipSup}, \mathrm{TPipDmd}) \\ \mathrm{TWelUse} = 0 \\ \mathrm{TTanUse} = \max(\mathrm{TTanDmd} - \mathrm{TPipUse}, 0). \end{cases} \quad (A23)$$

The percentage of households with dry wells is obtained from the distribution below using the empirically derived equation.

Tap households – Case 3: if well supply is the cheapest source and the household has a functioning well, it is assumed the household will continue to use some piped water (assumed 75 L per household per day) for drinking, cooking and other kitchen uses:

$$3 : \begin{cases} \mathrm{TWelUse} = \mathrm{TWelDmd} - 75 \\ \mathrm{TPipUse} = 75 \\ \mathrm{TTanUse} = 0. \end{cases} \quad (A24)$$

NonTap households – Case 4: NonTap households follow a similar strategy preferring public standpipes if available. However, if supply is restricted, they will use shallow bore well handpumps (locally called "India Mark Pumps"). These shallow bore wells are typically easily accessible on every street, but the quality is not as good as piped water. Bore well handpumps usually function if the water table is shallow (defined in the model as < 15 m). Any residual demand will be met by tankers.

Water use from public handpumps (standpipes) has already been defined in Eq. (A9). Water available and used from shallow bore wells is given as follows:

$$\mathrm{NTWelSup} = \begin{cases} 100\,000, & \text{if DepthtoGW} < 15 \\ 0, & \text{Otherwise} \end{cases} \quad (A25)$$

$$4 : \begin{cases} \mathrm{NTapWelUse} = \\ \min(\mathrm{NTWelDmd} - \mathrm{NTHPUse}, \mathrm{NTWelSup}) \\ \\ \mathrm{NTTanUse} = \\ \max(\mathrm{NTTanDmd} - \mathrm{NTHPUse} - \mathrm{NTWelUse}, 0), \end{cases}$$
$$(A26)$$

where NTHPUse is as defined in Eq. (A9).

The total quantity of the groundwater extractions by users and tanker operators is estimated by summing of HH tanker demand and HH well use across all household types (TapHH with wells, TapHH without wells and NonTapHH). These extractions (TotalTankerUse and TotalGWUse respectively) feed back into the Aquifer model.

Cost of Water Sub-model: because quantity of water consumed is a function of marginal price, monthly cost of water was assumed to be a reasonable indicator of water security. The cost of water is simply the total amount paid divided by the total water use by all HH in Chennai.

Acknowledgements. I thank Sally Thompson, who provided extensive comments on an earlier draft of this paper. I am grateful to the two anonymous reviewers, reviewer Pat Gober and editor Murugesu Sivapalan for their insightful comments. Their comments were crucial in a substantial revision of the paper.

The original Chennai study was funded by an Environmental Ventures Project Grant from the Stanford Woods Institute for the Environment and The Teresa Heinz Environmental Scholars Program. Subsequent research on urbanization and water security is being funded by IDRC, Canada grant number 107086-001 titled "Adapting to Climate Change in Urbanizing Watersheds (ACCUWa) in India".

All errors and weaknesses in the model and arguments remain mine.

Edited by: M. Sivapalan

References

Baisa, B., Davis, L. W., Salant, S. W., and Wilcox, W.: The welfare costs of unreliable water service, J. Developm. Econom., 92, 1–12, doi:10.1016/j.jdeveco.2008.09.010, 2010.

Brown, C., Ghile, Y., Laverty, M., and Li, K.: Decision scaling: Linking bottom-up vulnerability analysis with climate projections in the water sector, Water Resour. Res., 48, W09537, doi:10.1029/2011WR011212, 2012.

Chakravorty, U. and Umetsu, C.: Basinwide water management: a spatial model, J. Environ. Econom. Manage., 45, 1–23, 2003.

Clark, J. R. A. and Clarke, R.: Local sustainability initiatives in English national parks: what role for adaptive governance?, Land Use Pol., 28, 314–324, 2011.

Cook, C. and Bakker, K.: Water security: debating an emerging paradigm, Global Environ. Change, 22, 94–102, 2012.

Di Baldassarre, G., Viglione, A., Carr, G., Kuil, L., Salinas, J. L., and Blöschl, G.: Socio-hydrology: conceptualising human-flood interactions, Hydrol. Earth Syst. Sci., 17, 3295–3303, doi:10.5194/hess-17-3295-2013, 2013.

Dilling, L. and Lemos, M. C.: Creating usable science: Opportunities and constraints for climate knowledge use and their implications for science policy, Global Environ. Change, 21, 680–689, doi:10.1016/j.gloenvcha.2010.11.006, 2011.

Feldman, D. L.: Preventing the repetition: Or, what Los Angeles' experience in water management can teach Atlanta about urban water disputes, Water Resour. Res., 45, W04422, doi:10.1029/2008WR007605, 2009.

Gober, P. and Kirkwood, C. W.: Vulnerability assessment of climate-induced water shortage in Phoenix, Proc. Natl. Acad. Sci., 107, 21295–21299, doi:10.1073/pnas.0911113107, 2010.

Gober, P., Wentz, E. A., Lant, T., Tschudi, M. K., and Kirkwood, C. W.: WaterSim: a simulation model for urban water planning in Phoenix, Arizona, USA, Environ. Plann. Part B, 38, 197–215, doi:10.1068/b36075, 2011.

Gopakumar, G.: Transforming urban water supplies in India: the role of reform and partnerships in globalization, Routledge, 2011.

Jakeman, A. J. and Letcher, R. A.: Integrated assessment and modelling: features, principles and examples for catchment management, Environ. Modell. Softw., 18, 491–501, 2003.

Kilgour, D. M. and Dinar, A.: Flexible water sharing within an international river basin, Environ. Resour. Econom., 18, 43–60, doi:10.1023/A:1011100130736, 2001.

Lane, S. N.: Acting, predicting and intervening in a socio-hydrological world, Hydrol. Earth Syst. Sci., 18, 927–952, doi:10.5194/hess-18-927-2014, 2014.

Liu, Y., Gupta, H., Springer, E., and Wagener, T.: Linking science with environmental decision making: Experiences from an integrated modeling approach to supporting sustainable water resources management, Environ. Modell. Softw., 23, 846–858, doi:10.1016/j.envsoft.2007.10.007, 2008.

Lundqvist, J., Appasamy, P., and Nelliyat, P.: Dimensions and approaches for Third World city water security, Philos. Trans. Roy. Soc. London B, 358, 1985–1996, 2003.

McDonald, R. I., Douglas, I., Revenga, C., Hale, R., Grimm, N., Grönwall, J., and Fekete, B.: Global urban growth and the geography of water availability, quality, and delivery, Ambio, 40, 437–446, 2011.

McIntosh, A. C.: Urban Water Supply and Sanitation in Southeast Asia, Asian Development Bank, Philippines, available at: http://admin.indiaenvironmentportal.org.in/files/file/urban-water-supply-sanitation-southeast-asia.pdf, last access: 5 January 2014.

Metrowater: Chennai Metropolitan Water Supply and Sewerage Board: Water Supply System, available at: http://www.chennaimetrowater.tn.nic.in/departments/operation/%develowss.htm, 2011.

Pattanayak, S. K., Yang, J.-C., Whittington, D., and Bal Kumar, K.: Coping with unreliable public water supplies: Averting expenditures by households in Kathmandu, Nepal, Water Resour. Res., 41, W02012, doi:10.1029/2003WR002443, 2005.

Schlueter, M., McAllister, R. R. J., Arlinghaus, R., Bunnefeld, N., Eisenack, K., Hoelker, F., Milner-Gulland, E. J., Müller, B., Nicholson, E., Quaas, M., and Stöven, M.: New Horizons for Managing The Environment: A Review of Coupled Social-Ecological Systems Modeling, Nat. Resour. Model., 25, 219–272, doi:10.1111/j.1939-7445.2011.00108.x, 2012.

Shaban, A. and Sharma, R. N.: Water consumption patterns in domestic households in major cities, Econ. Polit. Weekly, 42, 2190–2197, 2007.

Shah, M.: Water: Towards a paradigm shift in the twelfth plan, Economic & Political Weekly, 48, 40–52, 2013.

Sivapalan, M., Savenije, H. H. G., and Blöschl, G.: Socio-hydrology: A new science of people and water, Hydrol. Process., 26, 1270–1276, doi:10.1002/hyp.8426, 2012.

Sivapalan, M., Konar, M., Srinivasan, V., Chhatre, A., Wutich, A., Scott, C. A., Wescoat, J. L., and Rodríguez-Iturbe, I.: Socio-hydrology: Use-inspired water sustainability science for the Anthropocene, Earth's Future, 2, 225–230, 2014.

Srinivasan, V., Gorelick, S. M., and Goulder, L.: A hydrologic-economic modeling approach for analysis of urban water supply dynamics in Chennai, India, Water Resour. Res., 46, W07540, doi:10.1029/2009WR008693, 2010a.

Srinivasan, V., Gorelick, S. M., and Goulder, L.: Sustainable urban water supply in south India: Desalination, efficiency improvement, or rainwater harvesting?, Water Resour. Res., 46, W10504, doi:10.1029/2009WR008698, 2010b.

Strand, J. and Walker, I.: Water markets and demand in Central American cities, Environ. Develop. Econom., 10, 313–335, 2005.

Thompson, S., Sivapalan, M., Harman, C., Srinivasan, V., Hipsey, M., Reed, P., Montanari, A., and Blöchl, G.: Developing predictive insight into changing water systems: use-inspired hydrologic science for the Anthropocene, Hydrol. Earth Syst. Sci., 17, 5013–5039, doi:10.5194/hess-17-5013-2013, 2013.

Tidwell, V. C. and Van Den Brink, C.: Cooperative modeling: linking science, communication, and ground water planning, Groundwater, 46, 174–182, doi:10.1111/j.1745-6584.2007.00394.x, 2008.

van den Belt, M., Kenyan, J. R., Krueger, E., Maynard, A., Roy, M. G., and Raphael, I.: Public sector administration of ecological economics systems using mediated modeling, Ann. New York Acad. Sci., 1185, 196–210, doi:10.1111/j.1749-6632.2009.05164.x, 2010.

Ward, F. A., Booker, J. F., and Michelsen, A. M.: Integrated economic, hydrologic, and institutional analysis of policy responses to mitigate drought impacts in Rio Grande Basin, J. Water Resour. Plann. Manage., 132, 488–502, doi:10.1061/(ASCE)0733-9496(2006)132:6(488), 2006.

Zérah, M.-H.: Household strategies for coping with unreliable water supplies: the case of Delhi, Habitat Int., 24, 295–307, doi:10.1016/S0197-3975(99)00045-4, 2000.

10

Variations in quantity, composition and grain size of Changjiang sediment discharging into the sea in response to human activities

J. H. Gao[1]**, J. Jia**[2]**, Y. P. Wang**[1]**, Y. Yang**[1]**, J. Li**[3]**, F. Bai**[3]**, X. Zou**[1]**, and S. Gao**[1]

[1]Ministry of Education Key Laboratory for Coast and Island Development, Nanjing University, Nanjing 210093, China
[2]State Research Centre for Island Exploitation and Management, Second Institute of Oceanography,
State Oceanic Administration, Hangzhou 310012, China
[3]Qingdao Institute of Marine Geology, Qingdao 266071, China

Correspondence to: J. H. Gao (jhgao@nju.edu.cn) and Y. P. Wang (ypwang@nju.edu.cn)

Abstract. In order to evaluate the impact of human activities (mainly dam building) on the Changjiang River sediment discharging into the sea, the spatial–temporal variations in the sediment load of different tributaries of the river were analyzed to reveal the quantity, grain size and composition patterns of the sediment entering the sea. The results show that the timing of reduction in the sediment load of the main stream of the Changjiang was different from those associated with downstream and upstream sections, indicating the influences of the sub-catchments. Four stepwise reduction periods were identified, i.e., 1956–1969, 1970–1985, 1986–2002, and 2003–2010. The proportion of the sediment load originating from the Jinsha River continuously increased before 2003; after 2003, channel erosion in the main stream provided a major source of the sediment discharging into the sea. In addition, in response to dam construction, although mean grain size of the suspended sediment entering the sea did not change greatly with these different periods, the inter-annual variability for sediment composition or the relative contributions from the various tributaries changed considerably. Before 2003, the clay, silt and sand fractions of the river load were supplied directly by the upstream parts of the Changjiang; after 2003, although the clay component may still be originating mainly from the upstream areas, the source of the silt and sand components have been shifted to a large extent to the river bed erosion of the middle reach of the river. These observations imply that the load, grain size and sediment composition deposited over the coastal and shelf water adjacent to the river mouth may have changed rapidly recently, in response to the catchment changes.

1 Introduction

Recently, the global sediment flux into the sea has drastically decreased under the influence of human activities (Vörösmarty et al., 2003; Walling, 2006), resulting in considerable changes in the geomorphology and eco-environment of estuarine, coastal and continental shelf regions (Syvitski et al., 2005; Gao and Wang, 2008; Gao et al., 2011). Thus, the source–sink processes and products of the catchment–coast system, including those associated with sediment transport pathways from catchment to continental margins under the impact of climate change and human activities, have received increasing attention (Driscoll and Nittrouer, 2002; Gao, 2006).

Because marine deposits consist of the materials from different sub-catchments, variations in the sediment characteristics at the deposition site should result from both sediment load reduction and alterations in sediment grain size, as well as the proportion of the sedimentary materials from different tributaries (which will be referred to as sediment composition in the present study). With regard to the sediment load reduction, there have been studies about the impact of human activities (particularly large hydrologic projects), analyzing long-term variation trends for a number of representative rivers (e.g., Milliman, 1997; Syvitski, 2003; Syvitski and Saito, 2007; Milliman and Farnsworth, 2011; Yang et al., 2011). However, less attention has been paid to the variations in the grain size and sediment composition in response to human activities, as well as to their sedimentological and environmental consequences. The importance of these two

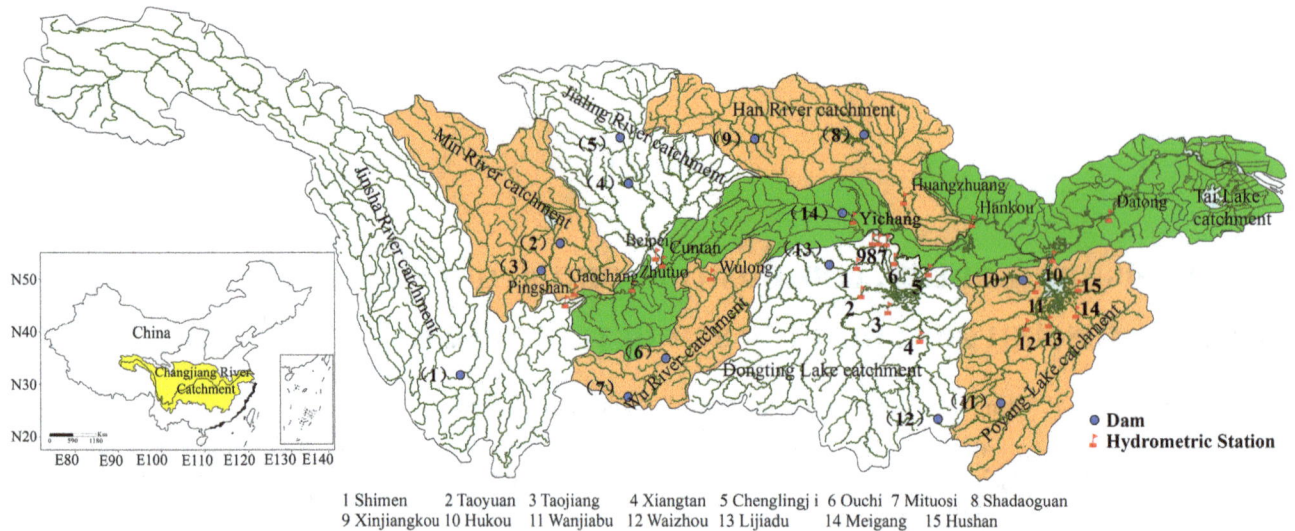

Figure 1. The Changjiang catchment and location of the hydrologic stations for the Changjiang catchment. The numeric symbols in the figure denote some important reservoir sites, including (1) Er'tan, (2) Heilongtan, (3) Tongjiezi, (4) Shengzhong, (5) Baozhushi, (6) Wujiangdu, (7) Puding, (8) Danjiangkou, (9) Ankang, (10) Zhelin, (11) Wan'an, (12) Dongjiang, (13) Jiangya and (14) Three Gorges Dam.

aspects lies in that they reflect the sediment contribution of different sub-catchments to the marine deposits and determine the geochemical and sediment dynamic characteristics (Gao, 2007). Therefore, knowledge about the catchment sediment characteristics for different periods is critical for an accurate analysis of the sediment source/distribution pattern over the estuary and coast–continental shelf regions, and for an improved prediction of the response of the marine sedimentary system to climate change, sea level change and human activities.

The Changjiang is one of the largest rivers in the world. A part of the sediment from the Changjiang catchment has formed a large sub-aqueous delta system of around $10\,000\,km^2$ (Milliman et al., 1985); the remainder escapes from the delta, being transported to the Yellow Sea, the East China Sea and the Okinawa Trough, thereby exerting a considerable impact on the sedimentological and biochemical conditions of these areas (Liu et al., 2007; Dou et al., 2010). Recently, the sediment load of the Changjiang into the sea was reduced considerably in response to dam emplacement and soil water conservation projects (Yang et al., 2002). Dai et al. (2008) estimated that the contribution of dam construction and the water and soil conservative measures accounted for $\sim 88\%$ and $15 \pm 5\%$ of the decline in sediment influx, respectively. The Changjiang catchment consists of numerous tributaries, which are characterized by different rock properties and climate conditions. On the other hand, the intensity and duration of human activities of these tributaries are also varied, which leads to different spatial–temporal patterns of the sediment yield of the sub-catchments (Lu et al., 2003). Thus, the sediment supply of each tributary to the main stream of the Changjiang also changed with time. Furthermore, dam construction and land cover changes also have an important impact on changes in sediment grain size for the tributaries and main stream of the Changjiang (Zhang and Wen, 2004). Therefore, the sediment contribution made by the different tributaries to the sea-going sediment load, the grain size and sediment composition may vary at the same time for the decrease in the total sediment load of the Changjiang River.

In order to evaluate the impact of human activities (mainly dam construction) on the quantity, composition and grain size of the Changjiang sediment discharging into the sea, we attempt to (1) analyze the effect of dam emplacement on the sediment load of different tributaries; (2) identify the spatial–temporal variation patterns of sediment load within the main stream of the Changjiang associated with dam emplacement; (3) reveal the quantity, grain size and composition features of the sea-going sediment during different periods; and (4) delineate the variations in sediment load originating from the tributaries of the Changjiang for different historical times.

2 Regional setting

The Changjiang, with a drainage basin area of approximately $1.80 \times 10^6\,km^2$, originates in the Qinghai–Tibet Plateau and flows 6300 km eastward toward the East China Sea. The upper reach of the river, from the upstream source to the Yichang gauging station (Fig. 1), is the major sediment-yielding area of the entire catchment (Shi, 2008). The main upstream river has four major tributaries, i.e., the Jinsha, Min, Jialing and Wu rivers. The upper reach region is typically mountainous, with an elevation exceeding 1000 m a.s.l. (above sea level) (Chen et al., 2001). The mid-lower reach extends from Yichang to the Datong gauging station, with three large inputs joining the main stream in this section: the Dongting Lake drainage basin, the Hanjiang River, and the Poyang Lake drainage basin. The catchment area of this section mainly comprises alluvial plains and low hills with elevations of less than 200 m (Yin et al., 2007). Dongting Lake is the second largest freshwater lake in China, and part of the main river flow enters Dongting Lake via five different entrances. Four tributaries enter Dongting Lake from the south and southwest, and water from Dongting Lake flows into the Changjiang main river channel at the Chenglingji gauging station (Dai et al., 2008). Dongting Lake was a major sink of the upstream sediment of the Changjiang and, due to sediment decrease from the upstream Changjiang, has become a weak sediment source to its downstream sections (Dai and Liu, 2013). Poyang Lake is the largest freshwater lake in China, and it directly exchanges and interacts with the river. Poyang Lake receives runoff from five smaller tributaries (the Gan, Fu, Xin, Rao, and Xiu rivers) and discharges freshwater into the Changjiang at Hukou (Shankman et al., 2006). The estuarine reach of the Changjiang extends from Datong (tidal limit) to the river mouth. The Datong gauging station is the last station along the Changjiang before going to the sea, and its hydrological records are often used to derive a representative sediment flux of the Changjiang into the adjacent East China Sea.

Due to intensified human activities, the catchment forest vegetation degenerated continuously, with a large-scale reduction of forest cover in the Changjiang catchment (Xu, 2005), resulting in serious deterioration of the ecological environment (Lu and Higgitt, 2000). Starting in the late 1980s, a major soil conservation campaign was implemented in high sediment yielding regions of the upper Changjiang catchment. However, due to the highly variable natural conditions of the tributaries, the effect of this campaign was different in every upstream tributary. For example, most of the Jialing River catchment is characterized by hilly areas, with the potential for severe slope erosion (Zhang and Wen, 2004), yet its vegetation restoration rate is quite high due to the humid climate; hence, the effect of vegetation recovery on the reduction of slope erosion is prominent (Lei et al., 2006). As such, the sediment yield of these parts of the Jialing River catchment has rapidly decreased since the soil conservation campaign began in the 1980s (BSWC, 2007). However, downstream Jinsha River has a different situation. This section, 782 km in length, is the main sediment yield area; although its area only accounts for 7.8 % of the upstream Changjiang, the average annual sediment load reaches 35.50 % of the quantity at the Yichang station (Zhang and Wen, 2004). The high and steep mountains here are characterized by landslides and debris flow, reducing the effect of vegetation restoration (Lei and Huang, 1991; Yang, 2004). Therefore, the water and soil erosion prevention scheme works neither in the Jinsha River nor in the Jialing River

(BSWC, 2007); in the former case, reservoir interception is still the dominant factor of the sediment load reduction.

3 Material and method

3.1 Data sources

3.1.1 Water discharge and sediment load data

A long-term discharge and sediment monitoring program for the entire catchment has been implemented since the 1950s by the Changjiang Water Resource Commission (CWRC) under the supervision of the Ministry of Water Resources, China (MWRC). The monitoring data of each station include field survey and measurement of water discharge, suspended sediment concentration, suspended sediment load and suspended sediment grain size, in accordance with China's national data standards (Ministry of Water Conservancy and Electric Power, 1962, 1975); 10–30 vertical profiles within the water column were established for the measurements of each river cross-section, with the number of profiles varying with the width of the river. For each profile, the flow velocity are measured (using a direct reading current meter) at different depths (normally at surface, 0.2, 0.6, 0.8 H and the bottom, where H is the water depth). Meanwhile, the water mass of these layers is sampled for the measurements of the suspended sediment concentration (using filtration) and grain size (using the suspension settling method). Such measurements are repeated daily at each station. The homogeneity and reliability of the hydrological data, with an estimated daily error of 16 % (Wang et al., 2007), has been strictly examined by the CWRC before release. The data for the period of 1956–2001 were either published in the Yangtze River Hydrological Annals or provided directly by the CWRC. After 2002, these hydrological data were reported in the Bulletin of China River Sediment published by the Ministry of Water Resources, China (BCRS, 2002–2010; available at: http://www.mwr.gov.cn/zwzc/hygb/zghlnsgb/).

We acquired the annual sediment load data from 26 hydrological stations distributed in the main reach and seven of the tributaries (for the location of these stations, see Fig. 1). The data set for these gauging stations covers a 55 year period (i.e., 1956–2010).

3.1.2 Dam data

In the present study, the reservoirs with a storage capacity of greater than 0.01 km³ (i.e., large and medium sized reservoirs according to the MWRC) are considered. Data on reservoir emplacement during 1949–2001 were obtained from the MWRC (2001), and those built during 2002–2007 were obtained from the annual reports published by the MWRC (http://www.mwr.gov.cn/zwzc/hygb/slbgb/). In total, we counted 1132 large and medium sized reservoirs located within the Changjiang catchment, of which 1037 reservoirs are situated upstream of the Datong station (Fig. 1b). The database includes information on reservoir storage capacity, construction and impoundment time.

In the present study, the reservoir storage capacity index (RSCI) is defined as the ratio of the reservoir storage capacity to the annual average water discharge of the contributed catchment; thus, the total RSCI of a catchment is the ratio of total capacity of reservoir to the annual average water discharge.

3.2 Analytical methods

The Mann–Kendall (M–K) test is a nonparametric method, which has been used to analyze long-term hydrometeorological temporal series (Mann, 1945; Kendall, 1955). This test does not assume any distribution form for the data and is as powerful as its parametric competitors (Serrano et al., 1999). Trend analysis of the sediment load changes was conducted based on this method. Before using the M–K test, the autocorrelation and partial autocorrelation functions were used to examine the autocorrelation of all the hydrological data. The results indicated that there was no significant autocorrelation in the data. The modified M–K method was used to analyze variations in the sediment load data: $X_t = (x_1, x_2, x_3 \ldots x_n)$, where the accumulative number m_i for samples for which $x_i > x_j$ ($1 \leq j \leq i$) was calculated, and the normally distributed statistical variable d_k was expressed as (Hamed and Rao, 1998)

$$d_k = \sum_{i=1}^{k} m_i. \tag{1}$$

The mean and variance of the normally distributed statistic d_k were defined as

$$E[d_k] = \frac{k(k-1)}{4}, \tag{2}$$

$$\mathrm{Var}[d_k] = \frac{k(k-1)(2k+5)}{72}. \tag{3}$$

Then, the normalized variable statistical parameter UF_k was calculated as

$$\mathrm{UF}_k = \frac{d_k - E[d_k]}{\sqrt{\mathrm{var}[d_k]}}, \tag{4}$$

where UF_k is the forward sequence. The backward sequence UB_k was obtained using the same equation but with a retrograde sample. The C values calculated with progressive and retrograde series were named as C_1 and C_2. The intersection point of the two lines, C_1 and C_2 ($k = 1, 2 \ldots n$) was located within the confidence interval, providing the beginning of the step change point within the time series. Assuming a normal distribution at the significant level of $P = 0.05$, a positive Mann–Kendall statistics C larger than 1.96 indicates a significant increasing trend, while a negative C value with an absolute value lower than 1.96 indicates a significant decreasing trend.

4 Results

4.1 Stepwise variations in the reservoir storage capacity of the tributaries

The total RSCI values of the seven tributaries and the main stream of the Changjiang reveal stepwise increasing trends (Fig. 2). The variations in reservoir storage capacity of the four upstream tributaries indicated that the total RSCI of the Min River catchment is low (1.72 % in 2010) and those of the Jialing and Wu rivers rapidly increased in 1985. In response to the construction of the Er'tan reservoir, the total RSCI of the Jinsha River also rose considerably in 1998. As a result of the increase in the reservoir storage capacity of these four rivers, the total RSCI of the Changjiang catchment, upstream of the Yichang station with increases of 2.8 % in 1985 and 16.0 % in 2003, also showed the stepwise patterns. The middle reaches of the Changjiang catchment consisted of three major tributaries, namely, the Han River, Dongting Lake and Poyang Lake. The total RSCI of the Han River began to increase in 1966, and greatly rose in 1968. In addition, the rapid increment in the total RSCI of the Poyang and Dongting lakes were also present in 1972 and 1985, respectively. Generally, as a consequence of dam construction, the total RSCI of the Changjiang upstream of the Datong greatly increased in 1969 and 2003.

The changes of the total RSCI and sediment load of the tributaries and the whole Changjiang catchment indicate that the stepwise decrease of sediment load is apparently related to the significant increase of the total RSCI. In the case of the Yichang and Datong stations, over the last few decades, there has been a significant negative correlation between average sediment load and total RSCI at both the Yichang and Datong stations (Fig. 3), which reflected the impact dams have on the sediment load.

4.2 Spatial–temporal sediment load variations within the catchment

The trends, derived on the basis of the M–K method, of sediment load of the seven tributaries (Figs. 4 and 5) indicate that the downstream sediment load began to decrease earlier than the upstream sediment load. Due to the sediment load reduction of the Jialing and Wu rivers, the total sediment load of the four upstream rivers began to decrease in 1984. In the middle stream of the Changjiang, the sediment load for the Han River and the Dongting and Poyang lakes began to reduce in 1966, 1984 and 1985, respectively; the M–K trends of sediment load of the three sub-catchments exhibited significant trends of decrease (at the 95 % confidence level) in 1970, 1995 and 2000, respectively.

Due to the different patterns of sediment load variations of the seven sub-catchments, there were significant spatial–temporal differences in the sediment load variations of the mainstream Changjiang: the sediment load began to decrease

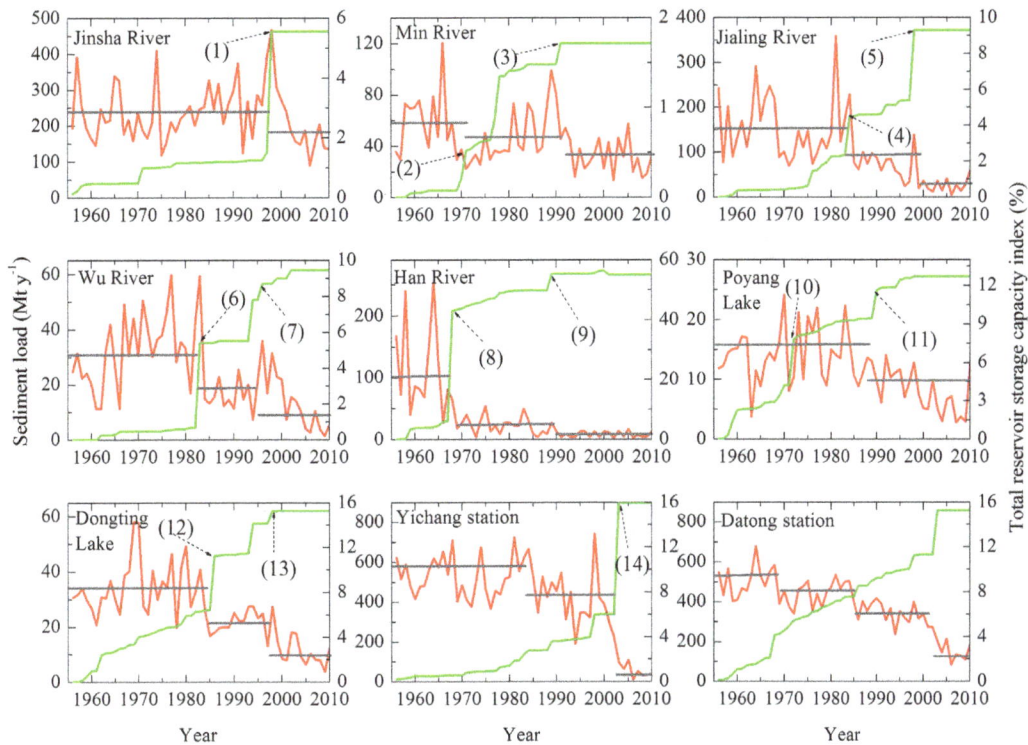

Figure 2. Relationship between the reduction in sediment load (red line) and the total reservoir storage capacity index (green line) in the tributaries and the main stream. Numeric symbols represent reservoirs listed in Fig. 1.

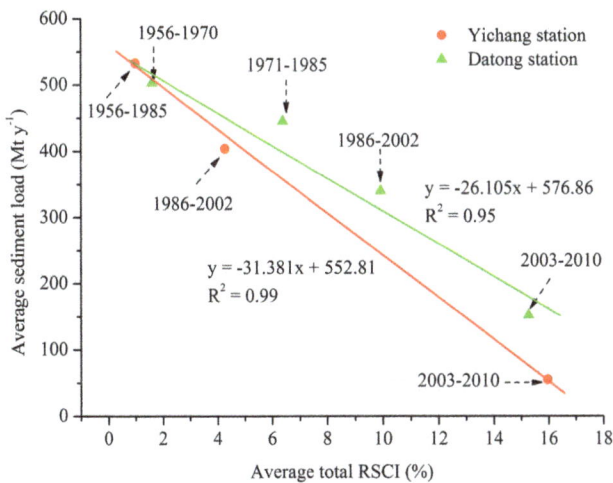

Figure 3. The relationship between average sediment load and total RSCI of different periods at the Yichang and Datong stations.

later at upstream locations than at downstream locations. The sediment load upstream of the Yichang station began to reduce in 1985, with a 95 % confidence level for the year of 1996. Impacted by sediment load decreasing of the Han River, beginning in 1966, the sediment load reduction trends of the middle reach (Hankou–Datong stations) were observed in 1969. Furthermore, as a result of sediment load

reduction of upstream and middle-reach tributaries in 1985, the sediment load of the middle reach of the Changjiang began to further decrease in 1985. The M–K trends of sediment load of the Datong, Hankou and Yichang stations are associated with a 95 % confidence level in 1989, 1997 and 1996, respectively.

4.3 Stepwise reduction of the sediment load entering the sea

The M–K trends of sediment load variation at the Datong station showed that 1969 and 1985 are two critical temporal nodes, reflecting the beginning time of sediment load decrease. While the M–K trends of the sediment load passes the 95 % confidence test for 1989, another important time node (2003) is not shown in the M–K trends of sediment load of the Datong station. Taking into account the great impact of the Three Gorges Dam (TGD) on the sediment load decrease of the main stream Changjiang (Hu et al., 2011), the variations of the sediment load entering the sea could be divided into four stepwise reduction stages, namely, 1956–1969, 1970–1985, 1986–2002 and 2003–2010.

The variations of sediment load discharging into the sea, measured at the Datong station, indicated that although the sediment load of the Datong station, 503 Mt yr^{-1} on average, exhibited fluctuations from 1956 to 1969, with the quantity generally remaining at a high level (Table 1). The Han River

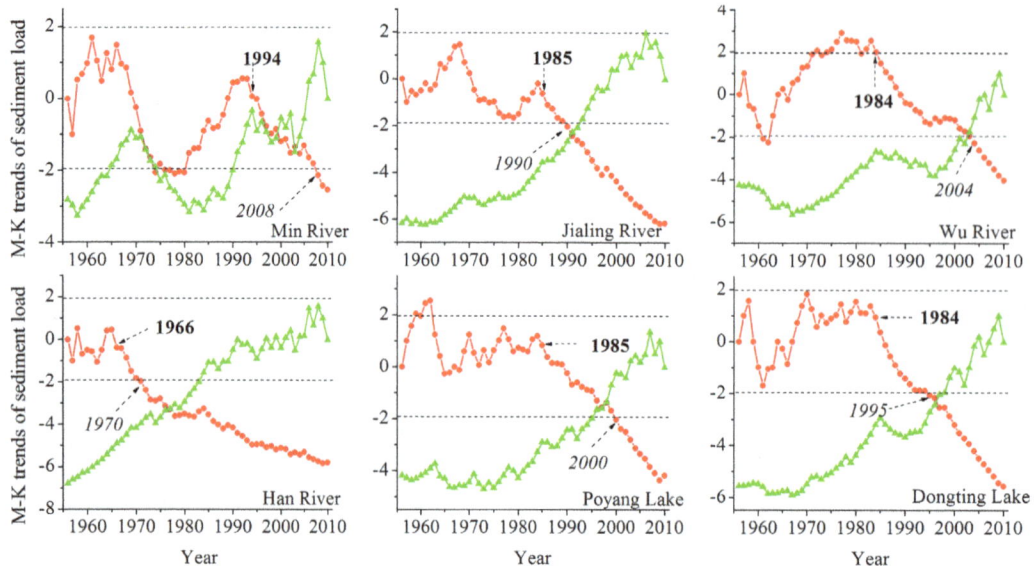

Figure 4. Mann–Kendall trends of the sediment load for the Jinsha, Min, Jialing, Wu, and Han rivers and the Poyang and Dongting lakes systems. The red bullets and green triangles denote C_1 and C_2, respectively. The bold represents the beginning time of sediment load decreasing, and the italics represent the time when the M–K trends of the sediment load pass the 95 % confidence test.

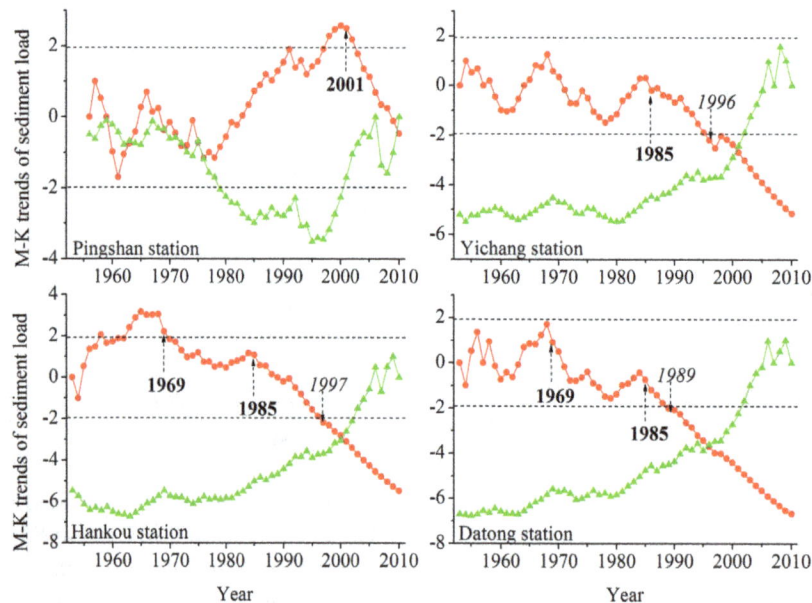

Figure 5. M–K trends of the sediment load for different gauging stations of the Changjiang main river. The red bullets and green triangles denote C_1 and C_2, respectively. The bold represents the beginning time of sediment load decreasing, and the years in italics denote the time when the M–K trends of the sediment load pass the 95 % confidence test.

was once the most important sediment source of the middle-reach Changjiang (Yin et al., 2007); however, since the annual sediment load supplied by the Han River decreased by 95 Mt, the sediment load of the Datong station was reduced to 445 Mt in the period of 1970–1985. Previous studies elsewhere have suggested that the sediment load from the Changjiang entering the sea began to decrease in the 1980s

(Yang et al., 2002); however, we would propose that such a decreasing trend already occurred in as early as 1970, and the impact of the reduced sediment load of the Han River on the overall sediment flux of the Changjiang was neglected in these previous studies. The sediment load for the upstream Changjiang had a decreasing trend starting in 1985; in terms of the quantity, there was a reduction from 533 Mt yr^{-1} dur-

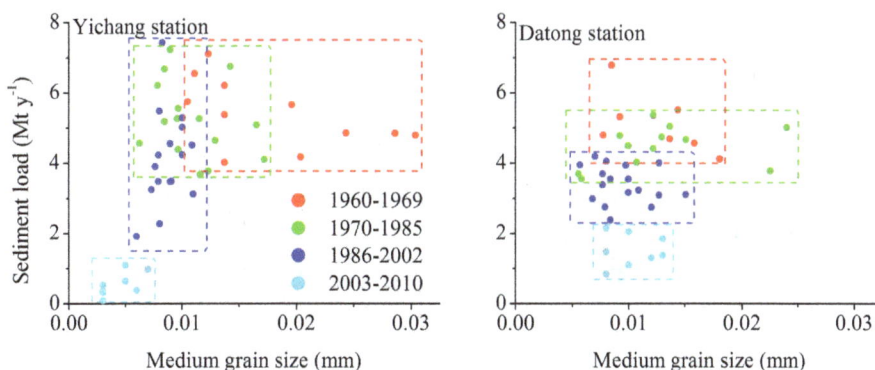

Figure 6. Relationship between the medium grain size of suspended sediments and the sediment load during different periods at the Yichang and Datong stations. Data are not available for the Datong station in 1968–1970, 1972–1973, and 1975.

ing 1956–1985 to 404 Mt yr^{-1} during 1986–2002. The seagoing sediment load of the Changjiang became less than 340 Mt yr^{-1} during this period. With the emplacement of the TGD in 2003, the sediment load upstream of the Changjiang decreased to 55 Mt yr^{-1} during 2003–2010, with the sediment discharge into the sea being around 152 Mt yr^{-1} (Table 1).

Overall, four stepwise reduction stage periods of the seagoing sediment load were observed, namely, 1956–1969, 1970–1985, 1986–2002 and 2003–2010. Further, the sediment load decrease may be related to sediment load decrease of different tributaries: the reduction during 1970–1985 was correlated with the Han River, while the upstream tributaries (mainly the Jialing and Wu rivers), together with the sub-catchment of the middle reach (mainly Poyang Lake), were responsible for the decrease during 1970–1985. The sediment load decrease during 2003–2010 resulted mainly from the emplacement of the TGD.

4.4 Variations in the grain size of the sediment entering the sea

Since most of the coarse-grained sediment is intercepted by reservoirs, the sediment grain size downstream of the reservoirs becomes significantly finer (Xu, 2005). The variation in the medium grain size (D_{50}) of suspended sediment at the Yichang station (Fig. 6) indicates that the average value of D_{50} was 0.017 mm in 1960–1969, 0.012 mm in 1970–1985, 0.009 mm in 1986–2002 and 0.004 mm in 2003–2010, suggesting that the sediment grain size from the upstream Changjiang exhibited a continuous decreasing trend. In contrast, the decreasing trend of D_{50} for the Datong station was not as significant as that of the Yichang station during these four stages: the average D_{50} in 1960–1969 (0.12 mm) was similar to that in 1970–1985 (0.13 mm), with a slight decrease for the year of 2002 (0.09 mm) and for the period of 2003–2010 (0.10 mm).

In addition, the degree of inter-annual variation in the upstream sediment grain size continuously decreased during the

Table 1. The mean value of sediment load of the Changjiang main river during different periods.

Time	Ping Shan station Mt yr^{-1}	Yichang station Mt yr^{-1}	Hankou station Mt yr^{-1}	Datong station Mt yr^{-1}
1956–1969	232	547	461	503
1970–1985	226	521	426	445
1986–2002	275	404	331	340
2003–2010	151	55	118	152

four periods at the Yichang station, i.e., the range of D_{50} variations is gradually narrowed (with a continuously reduced standard deviation), and the distribution range of the D_{50} data and sediment load moves towards the side of finer grain sizes; however, such a change is not so significant at the Datong station (Fig. 6). The sediment grain size variations of the two stations also indicated that the average value of D_{50} for the Yichang station was greater than that for the Datong station in 1960–1969, but the two stations had similar values in 1970–1985 and 1986–2002; after 2003, the average value of D_{50} of the Yichang station was smaller than that of the Datong station. Furthermore, D_{50} ranged from 0.003–0.007 mm for the Yichang station and 0.008–0.013 mm for the Datong station, in 2003–2010, suggesting that the D_{50} variation range of the two stations did not overlap after 2003.

Compared with previous periods, after 2003, the clay and silt contents at the Yichang station greatly increased, and the sand fraction significantly decreased (Fig. 7). At the Datong station, however, although the sand fraction had no apparent variation trends, the clay content increased, and the silt content reduced. Furthermore, before 2003, the silt and clay contents did not differ much between the Yichang and Datong stations, and the sand content at the Yichang station was slightly greater than that at the Datong station; however, after 2003, the sand content at the Datong station became significantly greater than that at the Yichang station, and the

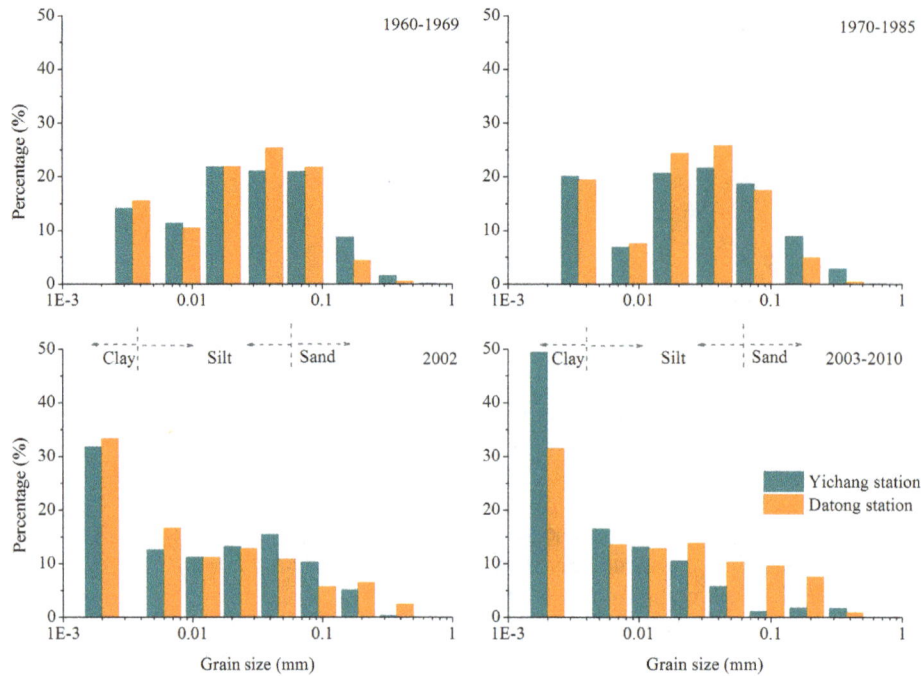

Figure 7. Distribution of the suspended sediment grain size of the Yichang and Datong stations in 1960–1969, 1970–1985, 2002 and 2003–2010.

clay content at the Datong station became lower than that at the Yichang station, implying that the sediment sources other than the seven tributaries supplied a sand fraction to the Yichang–Datong section of the Changjiang. This observation suggests that although the average value of the grain size of the sediment entering the sea during the different periods did not alter greatly, the inter-annual variation range, sediment components and the material sources changed considerably.

5   Discussion

As outlined above, the Changjiang sediment load is influenced by mixing of weathering products supplied by the different sub-catchments. The spatial–temporal differences among the sub-catchments, in terms of sediment load variations, caused the sediment load reduction and changes in the sediment composition. According to the concept of sediment budget (Houben, 2012), the following equation may be used to calculate the sediment balance of the main stream Changjiang:

$$\sum S_{input} = \Delta S + S_{output} = S_{Jinsha} + S_{Min} + S_{Jialing} + S_{Wu} + S_{Han} + S_{Poyang}, \tag{5}$$

where $\sum S_{input}$ is the contribution of the tributaries to the main stream sediment load, S_{output} is the sediment load entering the sea (measured at the Datong station), ΔS is the quantity of deposited (+)/eroded (−) sediment of the main

stream Changjiang and Dongting Lake. Thus, the contribution of the different tributaries to the overall sediment load can be expressed by

$$\frac{S_{Jinsha}}{S_{output}} + \frac{S_{Min}}{S_{output}} + \frac{S_{Jialing}}{S_{output}} + \frac{S_{Wu}}{S_{output}} + \frac{S_{Han}}{S_{output}}$$
$$+ \frac{S_{Poyang}}{S_{output}} - \frac{\Delta S}{S_{output}} = 1. \tag{6}$$

The calculated results indicated that (Table 2), in 1956–1969, the sediment load of the Datong station was mainly originated from the Jinsha, Jialing and Han rivers, with their contributions being 35.0, 24.3 and 19.0 %, respectively. As the sediment load of the Han River decreased, the Jinsha and Jialing rivers accounted for 46.7 and 27.6 %, respectively, in the sediment load at the Datong station during the 1970–1985 period, whereas the contribution from the Han River decreased to 5.8 %. During the 1986–2002 period, due to the reduced sediment yield in the Jialing River, the contribution of the Jinsha River to the sediment load at the Datong station further increased to 64.2 % and that of the Jialing River decreased to 15.0 %. The sediment composition changed considerably during the 2003–2010 period due to the TGD emplacement: the sediment proportion due to channel erosion of the main stream reached 48.3 % and the proportion of the Jinsha River decreased dramatically to 24.1 %. Furthermore, both the Jialing and Han rivers only contributed 5.3 % to the sediment load at the Datong station.

The above analysis indicates that as the sediment load at the Datong station decreased, although the average sediment

Table 2. The sediment contribution proportion (%) of different tributaries to the sediment load entering the sea of the Changjiang.

River/catchment	1956–1969	1970–1985	1986–2002	2003–2010
Jinsha River	35	46.7	64.2	24.1
Min River	8.8	8.6	10.1	6.1
Jialing River	24.3	27.6	15	5.3
Wu River	4.4	8.2	4.5	2.2
The total of the upstream four rivers	72.5	91.1	93.8	37.7
Han River	19	5.8	2.8	5.3
Channel erosion	6.1	0.9	1.1	48.3
Poyang Lake	2.4	2.2	2.3	8.7

Table 3. Annual quantities of clay, silt, and sand at the Yichang and Datong stations during different periods.

Time period	Clay ($Mt\,yr^{-1}$)		Silt ($Mt\,yr^{-1}$)		Sand ($Mt\,yr^{-1}$)	
	Yichang	Datong	Yichang	Datong	Yichang	Datong
1960–1969	78	78	297	291	172	134
1970–1985	105	86	257	257	159	102
1986–2002	128	113	212	174	63	50
2003–2010	27	48	25	77	3	27

grain size did not display clearly defined variations, the sediment composition changed considerably. Before 2003, the four rivers of the upstream Changjiang were the dominating sediment source to the sediment load entering the sea, and their total contribution was 72.5 % during 1956–1969, 91.1 % during 1970–1985 and 93.8 % during 1986–2002. In addition, during these periods, the variations in the sediment composition were mainly determined by the changes in the sediment contributions of the Jinsha, Jialing and Han rivers; i.e., with the sequential reduction in the sediment loads of the Han and Jialing rivers, the proportion of the sediment load originating from the Jinsha River continuously increased. However, after 2003, the sediment contribution of the upstream to the sediment load of the Datong station greatly decreased. The middle reach of the Changjiang became one of the major sinks of the upstream sediment (Yang et al., 2011); after 2003, channel erosion of the middle-reach main stream became the most important source of sediment load at the Datong station.

Apart from the dam interception effect, the soil conservation campaign starting in 1989 and implemented for the high sediment-yielding regions of the upper Changjiang basin (Hu et al., 2011) may be another factor accelerating the decreasing trend of the sediment grain size at the Yichang station. The different grain sizes of the suspended sediment at the Yichang and Datong stations indicate that the clay, silt and sand fluxes at the Yichang station were greater than those at the Datong station during the following periods: 1960–1969, 1970–1985 and 1986–2002 (Table 3); this implies that the sediment fractions of clay, silt and sand entering the sea were mainly originated from the upstream Changjiang,

with weak sediment exchange between the water column and the riverbed. After the emplacement of the TGD in 2003, the clay, silt and sand fractions originating from the upstream Changjiang decreased dramatically. With regard to the amount of sediment originating from Poyang Lake and the Han River to the main stream Changjiang, we may still use the sediment budget concept, to calculate the balance for the different sediment fractions for the Yichang–Datong reach:

$$S_{Yichang} + S_{Han} + S_{Poyang} = \Delta S + S_{datong}. \qquad (7)$$

The calculations show that the eroded sediment of the main river channel (between Yichang and Datong) and Dongting Lake contributed $13\,Mt\,yr^{-1}$ of clay, $43\,Mt\,yr^{-1}$ of silt and $20\,Mt\,yr^{-1}$ of sand to the sediment load at the Datong station in 2003–2010, which accounted for 27.1, 55.8 and 74.1 % of the corresponding sediment components of the Datong station. Taking into account the eroded sediment supply within the estuarine areas (Li, 2007), the percentages of the silt and sand fractions discharging into the sea, due to the material supply by the eroding main river channel, may exceed 55.8 and 74.1 %, respectively. These data imply that the clay fraction at the Datong station should be originated mainly from the upstream Changjiang, and the silt and sand fractions largely consisted of the eroded sediment of the middle-reach river channel.

6 Conclusions

1. The increment in reservoir storage capacity is significantly correlated with the decrease in the Changjiang sediment load, which reflected the impact of dams on the sediment load of the tributaries and the entire Changjiang catchment.

2. The patterns of sediment delivery from the sub-catchments of the Changjiang River have been changed, with significant spatial–temporal differences in the sediment load variations of the main stream Changjiang: four stepwise reduction stages were identified, i.e., 1956–1969, 1970–1985, 1986–2002 and 2003–2010. There was a lag of the decrease in the sediment load at upstream locations compared with those at downstream locations.

3. Before 2003, the variations in the sediment composition in the marine areas were mainly determined by the changes in the sediment contribution made by the Jinsha, Jialing and Han rivers. However, after 2003, channel erosion of the main stream Changjiang supplied around 48.3 % of the sediment load into the sea.

4. In response to dam construction, although mean grain size of the sediment entering the sea during the different periods did not show clearly defined variations, the inter-annual variations in terms of the size range, sediment components and source areas changed considerably.

5. Before 2003, the clay, silt and sand fractions entering the sea were mainly originated from the upstream regions of the river. In contrast, after 2003, the origin of the clay component of the sediment was dominated by the upstream areas, while the silt and sand component were mainly supplied by the eroding bed of the middle-reach main channel of the Changjiang River.

Acknowledgements. The study was supported by the National Basic Research Program of China (grant no. 2013CB956503) and the Natural Science Foundation of China (grant nos. 41376068 and 41476052).

Edited by: B. McGlynn

References

BSWC – Bulletin of soil and water conservation in China: Press of Ministry of Water Resources of the People's Republic of China, http://www.mwr.gov.cn/zwzc/hygb/zgstbcgb/, last access: 30 December 2007.

Chen, Z., Li, J., Shen, H., and Wang, Z. H.: Yangtze River of China: historical analysis of discharge variability and sediment flux, Geomorphology, 41, 77–91, 2001.

Dai, S. B., Lu, X. X., Yang, S. L., and Cai, A. M.: A preliminary estimate of human and natural contributions to the decline in sediment flux from the Yangtze River to the East China Sea, Quatern. Int., 186, 43–54, 2008.

Dai, Z. and Liu, J. T.: Impacts of large dams on downstream fluvial sedimentation: An example of the Three Gorges Dam (TGD) on the Changjiang (Yangtze River), J. Hydrol., 480, 10–18, 2013.

Dou, Y. G., Yang, S. Y., Liu, Z. X., Clift, P. D., Yu, H., Berne, S., and Shi, X. F.: Clay mineral evolution in the central Okinawa Trough since 28 ka: Implications for sediment provenance and paleoenvironmental change, Palaeogeogr. Palaeocl., 288, 108–117, 2010.

Driscoll, N. and Nittrouer, C.: Source to Sink Studies, Marg. Newslett., 5, 1–24, 2002.

Gao, S.: Catchment-coastal interaction in the Asia-Pacific region, in: Global change and integrated coastal management: the Asian-Pacific region, edited by: Harvey, N., Springer, Dordrecht, 67–92, 2006.

Gao, S.: Modeling the growth limit of the Changjiang Delta, Geomorphology, 85, 225–236, 2007.

Gao, S. and Wang, Y. P.: Changes in material fluxes from the Changjiang River and their implications on the adjoining continental shelf ecosystem, Cont. Shelf Res., 28, 1490–1500, 2008.

Gao, S., Wang, Y. P., and Gao, J. H.: Sediment retention at the Changjiang sub-aqueous delta over a 57 year period in response to catchment changes, Estuar. Coast. Shelf Sci., 95, 29–38, 2011.

Hamed, K. H. and Rao, A. R.: A modified Mann–Kendall trend test for autocorrelated data, J. Hydrol., 204, 182–196, 1998.

Houben, P.: Sediment budget for five millennia of tillage in the Rockenberg catchment (Wetterau loess basin, Germany), Quaternary Sci. Rev., 52, 12–23, 2012.

Hu, B. Q., Wang, H. J., Yang, Z. S., and Sun, X. X.: Temporal and spatial variations of sediment rating curves in the Changjiang (Yangtze River) basin and their implications, Quatern. Int., 230, 34–43, 2011.

Kendall, M. G.: Rank Correlation Methods, Griffin, London, 1955.

Lei, X. Z. and Huang, L. L.: Discussion of soil erosion mechanism in some areas of the upper Yangtze River, J. Sichuan Forest. Sci. Technol., 12, 9–16, 1991.

Lei, X. Z., Cao, S. Y., and Jiang, X. H.: Impacts of soil-water conservation in Jiangling River on sedimentation of the Three Gorges Reservoir. Journal of Wuhan University, Nat. Sci. Edit., 11, 922–928, 2006.

Li, L. Y.: The characteristics of water and sediment discharge and river channel evolution of Datong-Xuliujing section of the Changjiang, PhD Thesis, Hohai Univ. Nanjing, Nanjing, China, 2007.

Liu, J. P., Xu, K. H., Li, A. C., Milliman, J. D., Velozzi, D. M., Xiao, S. B., and Yang, Z. S.: Flux and fate of Yangtze River sediment delivered to the East China Sea, Geomorphology, 85, 208–224, 2007.

Lu, X. X. and Higgitt, D. L.: Estimating erosion rates on sloping agricultural land in the Yangtze Three Gorges, China, from caesium-137 measurements, Catena, 39, 33–51, 2000.

Lu, X. X., Ashmore, P., and Wang, J.: Sediment yield mapping in a large river basin: the Upper Yangtze, China, Environ. Modell. Softw., 18, 339–353, 2003.

Mann, H. B.: Nonparametric tests against trend, Econometrica, 13, 245–259, 1945.

Milliman, J. D.: Blessed dams or damned dams?, Nature, 388, 325–326, 1997.

Milliman, J. D. and Farnsworth, K. L.: River Discharge to the Coastal Ocean: A Global Synthesis, Cambridge Univ. Press, Cambridge, 2011.

Milliman, J. D., Shen, H. T., Yang, Z. S., and Meade, R. H.: Transport and deposition of river sediment in the Changjiang Estuary and adjacent continental-shelf, Cont. Shelf Res., 4, 37–45, 1985.

Ministry of Water Conservancy and Electric Power, P. R. C.: National Standards for Hydrological Survey (Vol. 1-3), China Industry Press, Beijing, 1962.

Ministry of Water Conservancy and Electric Power, P. R. C.: Handbook for Hydrological Survey (Vol. 1–3), Water Conservancy and Electric Power Press, Beijing, 1975.

MWRC – Ministry of Water Resources, China: The code for China Reservoir name, Chinese Water Conservancy and Hydroelectric Press, Beijing, 2001.

Serrano, V. L., Mateos, V. L., and García, J. A.: Trend analysis of monthly precipitation over the Iberian Peninsula for the period 1921–1995, Phys. Chem. Earth B, 24, 85–90, 1999.

Shankman, D., Heim, B. D., and Song, J.: Flood frequency in China's Poyang Lake region: trends and teleconnections, Int. J. Climatol., 26, 1255–1266, 2006.

Shi, C. X.: Scaling effects on sediment yield in the upper Yangtze River, Geogr. Res., 27, 800–811, 2008.

Syvitski, J. P. M.: Supply and flux of sediment along hydrological pathways: research for the 21st century, Global Planet. Change, 39, 1–11, 2003.

Syvitski, J. P. M. and Saito, Y.: Morphodynamics of deltas under the influence of humans, Global Planet. Change, 57, 261–282, 2007.

Syvitski, J. P. M., Vörömarty, C., Kettner, A. J., and Green, P.: Impact of humans on the flux of terrestrial sediment to the global coastal ocean, Science, 308, 376–380, 2005.

Vörösmarty, C. J., Meybeck, M., Fekete, B., Sharma, K., Green, P., and Syvitski, J. P. M.: Anthropogenic sediment retention: Major global impact from registered river impoundments, Global Planet. Change, 39, 169–190, 2003.

Walling, D. E.: Human impact on land–ocean sediment transfer by the world's rivers, Geomorphology, 79, 192–216, 2006.

Wang, Z. Y., Li, Y. T., and He, Y. P.: Sediment budget of the Yangtze River, Water Resour. Res., 43, W04401, doi:10.1029/2006WR005012, 2007.

Xu, J. X.: Variation in grain size of suspended load in upper Changjiang River and its tributaries by human activities, J. Sediment Res., 3, 8–16, 2005.

Yang, S. L., Zhao, Q. Y., and Belkin, I. M.: Temporal variation in the sediment load of the Yangtze River and the influences of human activities, J. Hydrol., 263, 56–71, 2002.

Yang, S. L., Milliman, J. D., Li, P., and Xu, K.: 50,000 dams later: Erosion of the Yangtze River and its delta, Global Planet. Change, 75, 14–20, 2011.

Yang, Z. S.: Soil erosion under different landuse types and zones of Jinsha River Basin in Yunnan Province, China, J. Mount. Sci., 1, 46–56, 2004.

Yin, H. F., Liu, G. R., Pi, J. G., Chen, G. J., and Li, C. G.: On the river-lake relationship of the middle Yangtze reaches, Geomorphology, 85, 197–207, 2007.

Zhang, X. and Wen, A.: Current changes of sediment yields in the upper Yangtze River and its two biggest tributaries, China, Global Planet. Change, 41, 221–227, 2004.

11

Linked hydrologic and social systems that support resilience of traditional irrigation communities

A. Fernald[1], S. Guldan[2], K. Boykin[1], A. Cibils[1], M. Gonzales[3], B. Hurd[1], S. Lopez[1], C. Ochoa[4], M. Ortiz[5], J. Rivera[6], S. Rodriguez[7], and C. Steele[1]

[1]College of Agricultural, Consumer and Environmental Sciences, New Mexico State University, Las Cruces, New Mexico, USA
[2]Sustainable Agriculture Science Center at Alcalde, New Mexico State University, Alcalde, New Mexico, USA
[3]Community & Regional Planning Program, University of New Mexico, Albuquerque, New Mexico, USA
[4]Department of Animal and Rangeland Sciences, Oregon State University, Corvallis, Oregon, USA
[5]New Mexico Acequia Association, Santa Fe, New Mexico, USA
[6]Center for Regional Studies, University of New Mexico, Albuquerque, New Mexico, USA
[7]Department of Anthropology, University of New Mexico, Albuquerque, New Mexico, USA

Correspondence to: A. Fernald (afernald@nmsu.edu)

Abstract. Southwestern US irrigated landscapes are facing upheaval due to water scarcity and land use conversion associated with climate change, population growth, and changing economics. In the traditionally irrigated valleys of northern New Mexico, these stresses, as well as instances of community longevity in the face of these stresses, are apparent. Human systems have interacted with hydrologic processes over the last 400 years in river-fed irrigated valleys to create linked systems. In this study, we ask if concurrent data from multiple disciplines could show that human-adapted hydrologic and socioeconomic systems have created conditions for resilience. Various types of resiliencies are evident in the communities. Traditional local knowledge about the hydrosocial cycle of community water management and ability to adopt new water management practices is a key response to disturbances such as low water supply from drought. Livestock producers have retained their irrigated land by adapting: changing from sheep to cattle and securing income from outside their livestock operations. Labor-intensive crops decreased as off-farm employment opportunities became available. Hydrologic resilience of the system can be affected by both human and natural elements. We find, for example, that there are multiple hydrologic benefits of traditional irrigation system water seepage: it recharges the groundwater that recharges rivers, supports threatened biodiversity by maintaining riparian vegetation, and ameliorates impacts of climate change by prolonging streamflow hydrographs. Human decisions to transfer water out of agriculture or change irrigation management, as well as natural changes such as long-term drought or climate change, can result in reduced seepage and the benefits it provides. We have worked with the communities to translate the multidisciplinary dimensions of these systems into a common language of causal loop diagrams, which form the basis for modeling future scenarios to identify thresholds and tipping points of sustainability. Early indications are that these systems, though not immune to upheaval, have astonishing resilience.

1 Introduction

In arid regions around the world, traditional irrigation systems have evolved to maintain community stability despite conditions of prolonged drought and climate variability. In the upper Rio Grande of the United States, *acequia* is the Spanish word for both the physical irrigation works and the water management institutions, governed by users who divert water from rivers and streams dependent on mountain snowpack in the uplands bioregion of north-central New Mexico and southern Colorado. The word acequia is originally derived from Arabic *as-sāqiya*, meaning a water conduit or irrigation canal. Transplanted to the New World from Iberia some 4 centuries ago, these community irrigation systems have developed complex self-maintaining interactions between culture and nature that enable drought survival while providing many other cultural, ecosystem, and economic benefits. Many of these benefits are tied to the connections between landscape and community. At the heart of the system, water from the river is diverted onto fields for crops, into ponds for animals, and across valleys supporting riparian vegetation. Grazing and wildlife on the upper watershed rely on forest and rangeland plant communities, the management of which determines runoff to the valley below. Community dynamics that use the valley and watershed for livelihoods determine water distribution that impacts the hydrologic cycle. Importantly, water that seeps from acequias and fields recharges groundwater that in turn provides river flow for downstream users (Fernald et al., 2010).

Starting in 1598, caravans of Hispano settlers and Mexican Indian allies traveled the Camino Real de Tierra Adentro heading north from Mexico City, traversed the Jornada del Muerto north of El Paso, Texas, and finally reached the Ohkay Owingeh Pueblo, north of Santa Fe, New Mexico at the confluence of the Rio Grande and the Rio Chama. Here, they set out to establish agricultural colonies in this northern frontier of New Spain in the search for perennial streams of water fed by snowpacks in the alpine sierras to the north. The colonists knew that in a desert environment there could be no irrigation without a snowpack, a natural system reservoir that accumulates and holds water during the winter months until the rising temperatures release the spring runoff needed to water crops and fields in the valley bottomlands.

As the first public works projects, the settlers diverted streams by constructing diversion dams and then hand dug acequia irrigation canals, on one or both banks, without the benefit of modern surveying equipment. These engineering works were of local physical design in order to operate as gravity flow systems, following the contours of the land with the goal of extending the boundaries of irrigation to the maximum extent before returning the canal back to the river. Due to variations in local topography, the design of each ditch system was unique, always tailored to local conditions and natural features in the landscape. This human engineering design, coupled with its associated social configuration, was specifically located and embedded in the hydrologic system. Acequias diverted and used surface water only when available and therefore made periodic adjustments dependent on water supply stored as snowpack in the upper watershed. Water management practices and operating procedures were adopted in response to natural system conditions during spring runoff, season to season, and over time these repeated adjustments produced a distinctive, community-based hydrosocial cycle.

Out of necessity, and as a potential contributor to the success of each system, the community of irrigators did not ad-

here to a prescribed set of regulations from a central authority; instead they negotiated institutional arrangements among the collective, operational rules that were specific to the water delivery requirements of the shared canal and its laterals. The taking of water by a collective enterprise carried forward into the local customs and traditions for water distribution and the operation and maintenance of the irrigation works, including repairs at the diversion structure in the river when needed and the annual rituals of ditch cleaning during early spring, just before the expected runoff season. This self-organized enterprise wedded the irrigators into a community system of water management that bonded them and formed a hydraulic society: a culture of water based on shared norms and mutualism. Rules for sharing evolved into a set of customary practices based on knowledge of the land, watershed, and water supply variability.

Today, these traditional communities are facing new socioeconomic and natural resource pressures that might impair their ability to function as originally designed. Like their counterparts elsewhere in the world, unpredictable and/or limited water supplies are a common challenge faced by traditional irrigation systems that depend on surface water and gravity flow technologies. Under conditions of warming temperatures and reduced snow storage during winter months, river flow is expected to decrease and snowmelt runoff has already begun to arrive earlier in the spring, trends that are expected to exacerbate water scarcity in the western United States (Barnett et al., 2008). Among other threats that acequias are facing today, most are directly or indirectly related to population growth and urban area expansion (Cox and Ross, 2011; Ortiz et al., 2007; Rivera, 1998).

In the face of these threats, we hypothesize that these acequia communities persevere by connecting variable hydrologic systems with adaptable human systems. These connections center upon the acequia itself, the physical structure to deliver water from river to field, and the human system to manage the water delivery and use (Fig. 1). We posit that the hydrology cannot be understood without the human connection and the human dynamic cannot be characterized without the hydrology, and the linkages are a source of continuity and longevity.

The goals of this study are (1) to connect human and natural systems using real data that cross over the intersecting disciplines, showing sustainability and adaptive capacity or nonsustainability and tipping points, and (2) identify resiliencies tied to hydrosocial changes that can be characterized only by including the multiple interacting disciplines of the hydrologic and human systems. For expected hydrosocial changes there are resident resiliencies that we measure and document. The system is complex and challenging to assess, highlighting the need for an interdisciplinary approach. Our intention is to show background history leading to the current situation and also examine the features that could impact resilience in the future. We try to identify the basic processes and the resulting impacts.

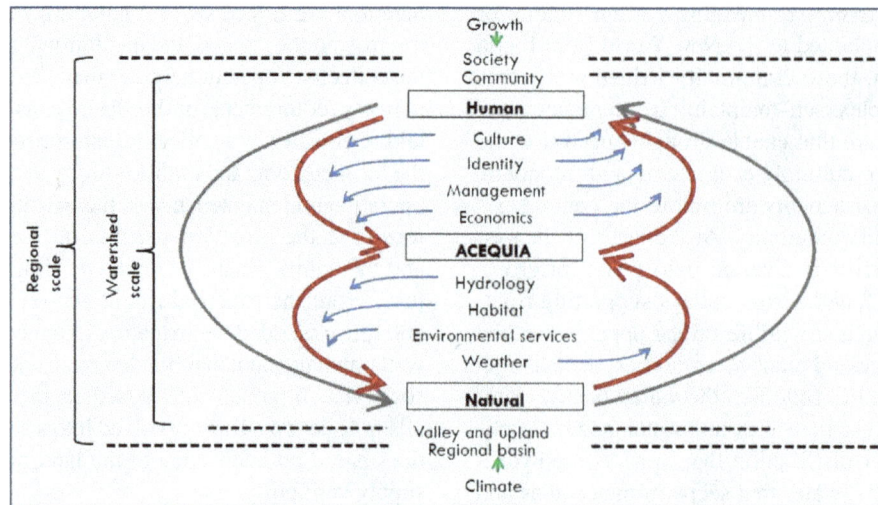

Figure 1. Acequia irrigation centered connections of human and natural systems (Fernald et al., 2012).

2   Resilience due to irrigation seepage

Acequia systems have multiple hydrologic benefits that accentuate resilience. Broadly speaking, the human system of acequia irrigation ditches impacts hydrologic processes by increasing water distribution across the irrigated landscape. Seepage from ditches and fields recharges groundwater, which in turn provides return flow to rivers. At the regional scale, return flow to the rivers delays the runoff hydrograph peak from the spring runoff period to later in the year. The hydrology of these rivers and valleys cannot be fully understood without including the human-created irrigation system. Recharge of aquifers for local groundwater use and downstream return flow benefits is an important aspect of the resiliency discussed below. Acequia seepage support of riparian areas that benefit habitat and biodiversity is another aspect of natural system resilience buttressed by human irrigation systems. These hydrologic and riparian habitat functions of acequia seepage are discussed in additional detail below.

2.1   Ditch and field seepage, groundwater recharge, and river return flow

In northern New Mexico, as in many areas of the southwestern United States and drylands worldwide, snowmelt runoff is the main source of streamflow during spring and summer, and agriculture is largely confined to narrow, irrigated floodplain valleys. In many of these agricultural valleys, river water is gravity-driven into irrigation canals (acequias) that run along the valley, where water is either diverted into smaller irrigation canals or applied directly to crop fields in the form of flood or furrow irrigation (Fernald et al., 2010). We studied water and communities of northern New Mexico (Fig. 2).

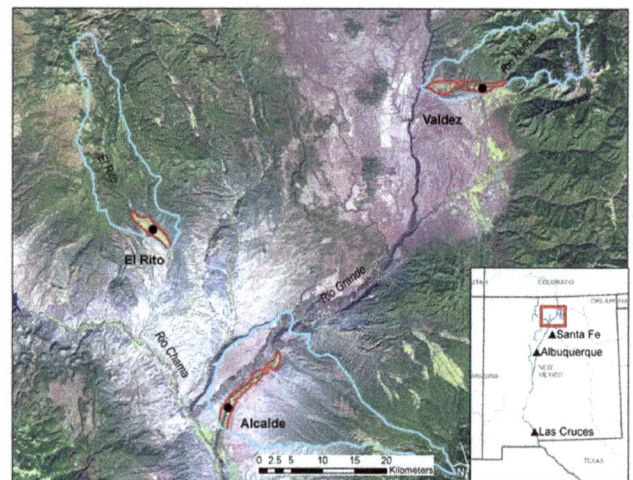

Figure 2. Northern New Mexico acequia study communities (black dots), their associated irrigated valleys (red lines), and contributing watersheds (blue lines) (Fernald et al., 2012).

Farming on the alluvial floodplains is dependent on the connectivity between surface water and shallow groundwater, particularly in the forms of precipitation, runoff, and infiltration processes in the upper watershed and their linkages with streamflow, irrigation, and aquifer recharge in the lower valleys. Results from our combined intensive field monitoring and modeling in communities of northern New Mexico indicate that there is a strong hydrologic connectivity between snowmelt-driven runoff in the headwaters and recharge of the shallow aquifer in the valleys, mainly driven by the use of traditional irrigation systems (Ochoa et al., 2013a). Figure 3 presents results from one of our study sites, illustrating that during the irrigation season (April–October), shallow groundwater levels rise in response to irri-

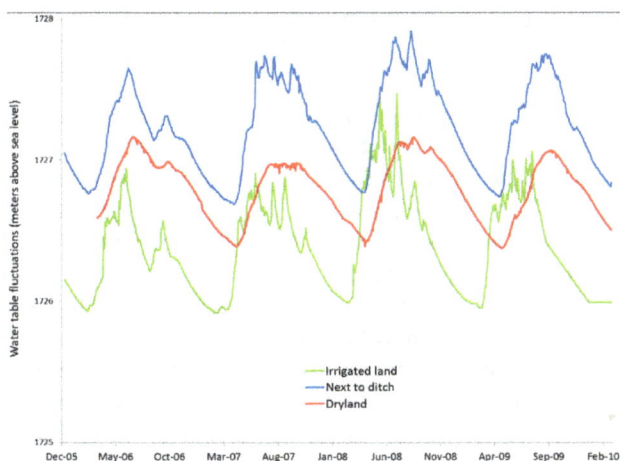

Figure 3. Seasonal water table fluctuations in response to irrigation inputs in one transect of wells in an acequia-irrigated agricultural valley in northern New Mexico.

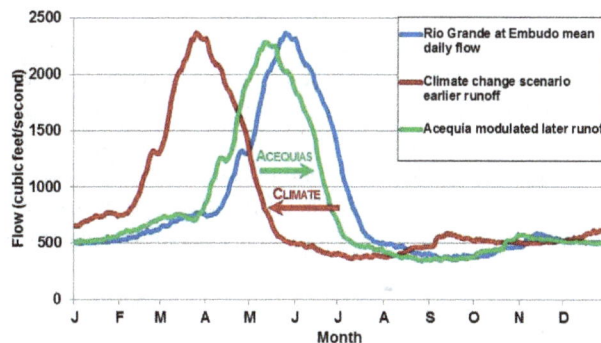

Figure 4. At the regional scale, acequia surface water–groundwater interactions may ameliorate effects of climate change by delaying the spring runoff that is projected to be earlier in the year.

gation percolation and canal seepage. Then in the late season, without irrigation, the river acts as a drain and starts receiving delayed return flow from groundwater that was temporarily stored in the shallow aquifer during the irrigation season. Also, similar patterns in seasonal shallow aquifer recharge have been observed in wells located in dry land at distances of about 1 km from the main irrigation canal and from any irrigated fields (Ochoa et al., 2013b). Conservation of this seasonal aquifer recharge provides several ecosystem functions including water quality enhancement, riparian habitat support, and river connection to the groundwater. It also supports important economic and ecological functions downstream through temporary storage and release processes. Return flow to rivers from groundwater ameliorates the impacts of climate change by retransmitting to later in the year those snowmelt hydrographs that are earlier and shorter due to increasing hydrographs (Fernald et al., 2010).

The hydrologic connectivity between upland water sources and irrigated valleys through the shallow groundwater system can be important for understanding the hydrologic resilience of agroecosystems in the face of climate variability. Human-induced changes (e.g., changes in land use or in technology) and natural processes (e.g., severe drought) can modify the spatial and temporal patterns of hydrologic connectivity in a given landscape. For instance, a significant change in land use from agricultural to residential and/or a big shift in irrigation technology that favors drip irrigation over the traditional use of flood may severely affect the recharge of the local aquifer. Irrigation efficiency at the farm scale can lead to increased crop yields and less seepage past the plant rooting zone. Flood irrigation is used because it is inexpensive and traditional. Seepage is an unintentional byproduct of the flood irrigation. The seepage, groundwater recharge, and delayed return flow back to the river support important economic and ecological functions.

Climate warming and changes in the quantity and temporal frequency of precipitation threaten the accumulation of snow, timing of melt, and thus the timely delivery of water for irrigation. In a case study of the effects of climate warming on streamflow in the El Rito watershed, it was determined that warmer temperatures predicted by the end of the 21st Century are likely to cause peak runoff to occur approximately 15 days earlier than it now does.

Although peak streamflow is predicted to occur earlier in the year and the volume of water delivered during peak streamflow is potentially greater than historical years, the connectivity between snowmelt-driven runoff and aquifer recharge provided by acequia flood irrigation provides a means for mitigating the potential for flooding and/or loss of water. The temporary storage and release of water provided by flood irrigation is important locally for ensuring water supply longevity through the growing season within the acequia valley itself. The effects of runoff modulation by acequias are also anticipated to extend beyond the local scale to the regional scale as illustrated by a conceptual diagram of the Embudo Station stream gauge hydrograph on the Rio Grande (Fig. 4).

2.2 Climate change effects on biodiversity and related ecosystem services

Biological diversity is threatened by land use change, yet acequia system contributions of water maintain riparian areas and so help mitigate the loss of these habitats in adjacent landscapes. Acequia irrigation provides much of the water for habitat in our semiarid study area and the climate changes that impact water and agriculture may negatively impact ecosystem health. Biodiversity and ecosystem services can be used to provide perspective on the ecological integrity or health of an ecosystem. A measure often used for biodiversity, and now for ecosystem services, is species richness. This is not the only measure, but it is one that can be calculated across time and space at a reasonable cost. Recent literature has identified that acequias benefit biodiversity and provide ecosystem services. To measure this impact we used biodiversity metrics created from a species habitat model (Boykin et al., 2007). These metrics reflect ecosystem services directly or components of biodiversity (Boykin et al., 2013). For example, the ecosystem services concept can be connected to metrics such as bird richness. There is an economy tied to avid bird-watchers who travel to view species. These people often go to species-rich areas to see a wide variety of avifauna.

We looked at the northern Rio Grande watershed. The regional focus provided the context for further analysis of biodiversity metrics and ecosystem services within the smaller acequia study areas. These perspectives allow us to understand where the most species-rich areas are regionally. We can also start to understand how broadscale land use changes may affect the finer-scaled acequia areas and vice versa using land use scenarios consistent with the Intergovernmental Panel on Climate Change (IPCC) global greenhouse gas emission storylines (Bierwagen et al., 2010). All future land use scenarios predicted losses in bird richness in each richness class. Species richness losses ranged from greater than 24 % for high development scenarios to less than 4.4 % for low development scenarios. A large percentage of habitat, including acequia-irrigated fields and riparian areas that support species diversity, are lost to urban development. The scenarios driven by economic forces lose a large percentage of habitat, while those driven by environmental forces lose the least amount of habitat. These data for our acequia community study region show that water and climate alone are not likely future determinants of ecosystem health. These results indicate that minimizing habitat loss from expanding urban footprints along with maximizing riparian habitat, is likely to reduce biodiversity loss along river corridors.

3   Place-based adaptation to changing land use and economics

If communities can stay connected to the acequia landscape, grazing, farming, economics, and land use changes allow them to adapt and survive. Region-wide livestock numbers have fallen somewhat since the mid-1980s, yet strong links between livestock raising and irrigated farming contribute to the cohesion of acequia communities. Streamflow is highly variable and the recent past includes very dry years, but added income independent of farming provides a coping strategy to weather periods of drought. Acequia farming has lost its role as a main provider of food for the community, but regional urban area demand for local food along with alternative crop availability to acequia communities provides options for adaptive responses. Drought and climate change threaten continued crop production, but a history of adaptations as well as maintenance of acequia infrastructure and farm plot arrangements indicates flexibility to transform and meet future challenges.

3.1   Historical changes in settlement morphology and agriculture at county and local community levels

In the future, acequia communities may need to respond to water demands in urban centers and drought through adaptation of land use and agricultural practices at the community scale. In order to understand the resilience and adaptive capacity of acequia-based community systems, we focused research on how acequia settlements have adapted over time in response to economic and environmental factors. The construction of Los Alamos National Laboratory in 1943 transformed the economy from agropastoral subsistence farming to wage-based off-farm employment. For Rio Arriba County in 1935, there were 3500 farms documented, 1400 of which were between 3 and 9 acres, and another 600 farms ranging from 10 to 19 acres accounting for 57 % of the farms in that year (USDC/USDA, 2014). In subsequent decades, agricultural parcels would increase in size; 1200 farms were reported in 1964, 380 of which were between 1 and 9 acres in size, and another 400 farms were classified as having from 10 to 49 acres, making up 65 % of the farms in that year (USDC/USDA, 2014). In 2007, of the nearly 1400 farms reported to the US Census, the range in parcel size remained consistent with that of the 1960s (Fig. 5). The overall decline in small-size farms from 1935 to 1964 suggests a shift in the agricultural system from a subsistence form of agriculture to one of supplemental income.

At the county scale, there has also been a major transformation from a sheep economy to a cattle-based economy, which may have also resulted in a change in cropping patterns in acequia communities over the last 75 years. In the 1930s, sheep and chickens significantly outranked cattle numbers. However, by 2007 cattle numbers in Rio Arriba County far exceeded any other livestock type (Fig. 6).

At the community scale, findings on historical land use change in Alcalde resulted in a decrease of total acreage of field crops from 84.5 to 30 acres, urban development into the irrigated farm land increased 42.3 acres, and average agricultural parcel size decreased from 3.6 to 1.9 acres (Fig. 7). Also, the channelization of the Rio Grande increased the bosque (riparian) vegetation in the study area by approximately 14.9 acres during the same period. Initial findings from mapping community scale morphology changes over the past 75 years revealed an increase in farm plot size, a dispersed settlement pattern, and less crop diversity. In addition our findings reveal a restricted channelized riparian condition along a portion of Alcalde (Fig. 8).

3.2   Upland grazing and recent trends in livestock inventories

Despite longer term changes in the agricultural landscape of the valleys, livestock raising and irrigation farming continue to be tightly intertwined in traditional Hispanic rural communities of northern New Mexico. Owning livestock continues

Change in Farm Size, 1935-2007

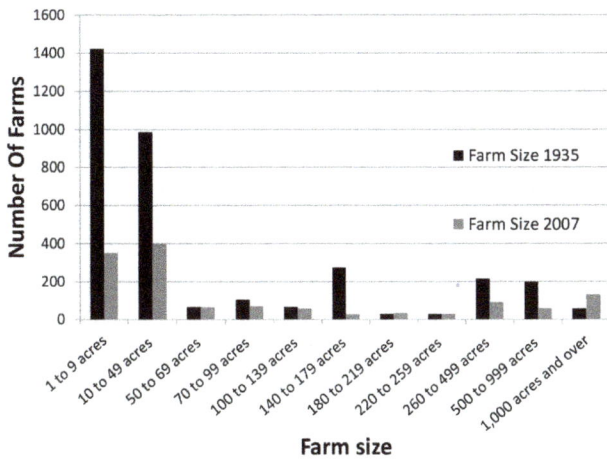

Figure 5. Agricultural parcel size for Rio Arriba County in 1935 and 2007 (NASS, 2012; USDC/USDA, 2014).

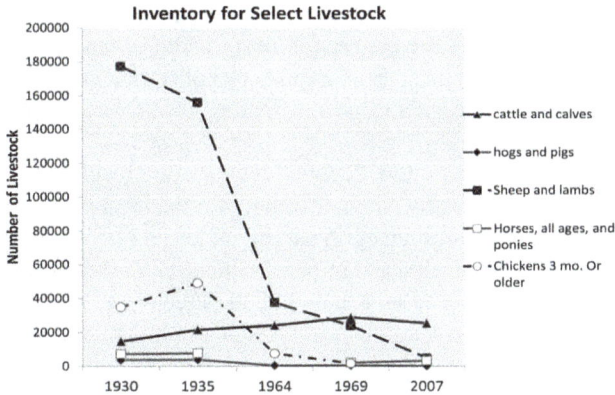

Figure 6. Inventory of livestock 1930–2007, Rio Arriba County (NASS, 2012; USDC/USDA, 2014).

Alcalde, NM

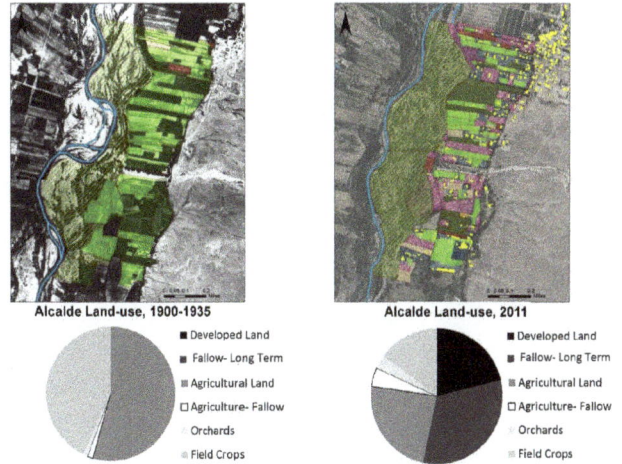

Figure 7. Community level changes in land use for Alcalde (Earth Data Analysis Center, 2010; Rio Arriba County Assessor's Office, 2013).

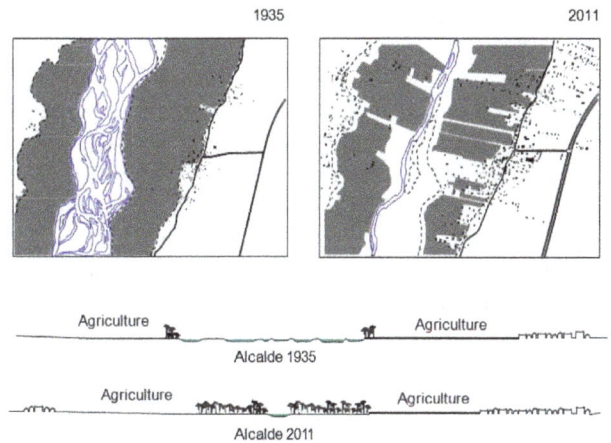

Figure 8. Changes in landscape morphology of the community of Alcalde (Earth Data Analysis Center, 2010; Rio Arriba County Assessor's Office, 2013).

to be vital to acequia community families; it is a way of re-connecting to their heritage (Eastman and Gray, 1987) and "an essential component of [their] historic persistence and self-reliance" (Cox, 2010; p. 65). Livestock raising, however, is only possible if access to uplands surrounding irrigated valleys is available for grazing during the growing season. Spanish deposition of farming lands included grazing rights on adjoining uplands (now regulated by the federal government), which constitutes a testament to the historical importance ascribed to livestock raising as a means of community financial stability (E. Gomez, personal communication, 2012).

Region-wide sheep and goat grazing permits on upland forests administered by the United States Forest Service were reduced dramatically beginning in the early 1900s (Fig. 9). The reduction in cattle grazing permits on forested uplands has become more pronounced since the mid-1980s (Fig. 9).

These changes have been mirrored by trends in livestock inventories in the counties surrounding our study sites (Fig. 6). Although this phenomenon varies locally (Raish and Mc-Sweeney, 2003; Cox, 2010; Lopez, 2014), and even though acequia communities have the ability to adapt to change, these trends could point to the weakening of an activity that has historically provided a natural link between valleys and uplands and has contributed to the cohesion of acequia communities (Cox, 2010; E. Gomez, personal communication, 2012).

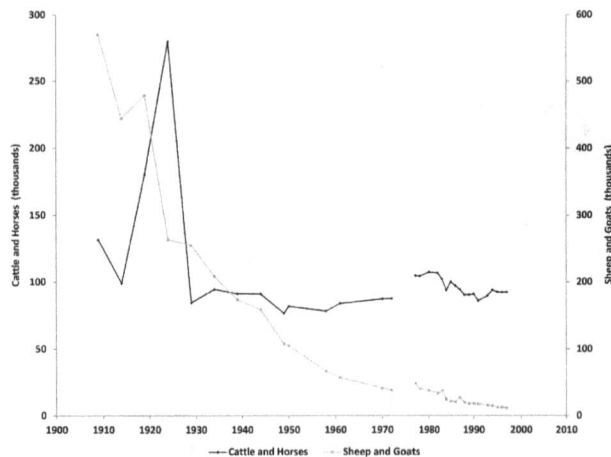

Figure 9. Grazing permits on national forests of New Mexico granted by the United States Forest Service from 1909 to 1997 (Source: USDA – Forest Service, 1991 and 1998).

3.3   Acequia resilience in the face of economic and land use changes

The land use change in Alcalde has been similarly documented for other communities in Rio Arriba County. Since WWII, acequia communities moved away from small tract subsistence agriculture to an agricultural system based on cattle production in order to maintain the irrigation system despite economic shifts in the region. The move from field crop diversity to pasture forages for livestock feed provided economic gain for farmers to participate in the post-WWII wage economy while maintaining irrigable acequia farmland. The challenge today of prolonged drought and climate change coupled with the reduction of grazing animal units on US forest land will increase pressure on the grazing system. However, acequia communities have the opportunity to develop strategies for adapting the existing land use configuration of irrigable farmland into a system capable of dealing with the potential impacts of prolonged drought and climate variability. Possible strategies include value added crop production, use of drip irrigation, and the use of alternative agricultural technology. The pre-WWII acequia community was resilient in that the farmers promoted crop diversity, utilized low energy inputs from farm machinery, and maintained sustainable practices on small parcel farms. Overall, land use conditions in Alcalde have been altered, but the acequia infrastructure and farm plot arrangements have endured and have adapted over time, indicators of resilience with flexibility to transform and meet climatic and other challenges in the future.

3.4   Economic trends and adaptation opportunities analyzed in terms of community balance sheets

The community capacity for resilience depends on preparedness and adaptation through accumulation of various types of human, financial, natural resource, and social–cultural assets. These assets act broadly as indicators of community wealth and strength, representing a type of community balance sheet. Each asset category is strengthened or diminished in a community balance sheet by changes in the status or performance of a community's economic, demographic, and environmental systems. For example, changes in regional net income from employment and productivity will change the community's aggregate financial wealth. As another example, natural and environmental resources, including fertile agricultural crop, grazing and forest lands, and water resources, are seen as contributing to the essential character of an acequia community. Many of these resource assets are renewable and their quantity and quality are subject to fluctuation and change, depending on variable factors such as climate, natural events and cycles, rates of extraction and transformation (including conversion of agricultural land to development), and changes in regulatory policy that can erode or strengthen the asset value by affecting resource access and use.

Mayagoitia et al. (2012) found that acequia residents exhibit a strong identity with and affinity for the local land, water, and cultural resources. A survey of acequia irrigators in northern New Mexico and in-person interviews asked irrigators about both economic and cultural perceptions and their views of acequia strengths and challenges. A particular survey focused on the current state of preparedness to endure stress from adverse economic conditions, continued drought (under climate change), and regional population growth. 77 percent of respondents cited connectivity and mutual relationships as factors contributing most to the acequia community's preparedness (i.e., adaptive capacity) (Fig. 10).

The concept of a community balance sheet also conforms well to notions of sustainable development as defined by the Brundtland Commission as "development that meets the needs of the present without compromising the ability of future generations to meet their own needs" (United Nations, 1987). Long-term sustainability and resilience would be consistent with a community's overall "balance sheet" remaining healthy and resistive to excessive degradation and long-term net loss. Although reductions in non-renewable resource capital are expected, it would also be consistent to expect an associated accretion in economic capital as a result of the transformation of natural capital into economic capital. For example, for a mining community to remain economically viable in the long term as mineral resources are depleted, its economy must invest in transitional economic development. Otherwise it would risk long-term economic decline and consequential losses and erosion in the other categories

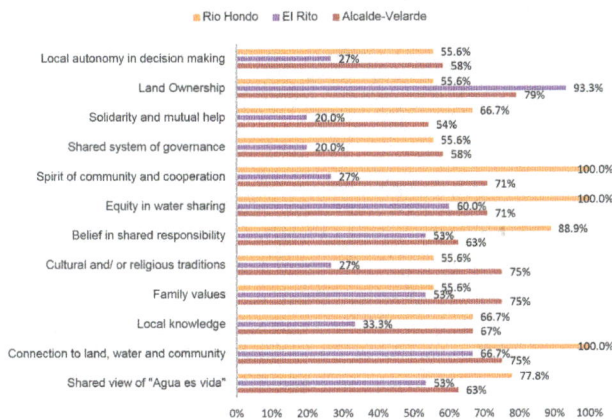

Figure 10. Acequia characteristics perceived to "best contribute" to acequia adaptive capacity, past adaption, and resilience.

of community capital, including population and sociocultural resources.

In the 42 years preceding 2011 there was a steady transformation of the employment base for the communities in Taos and Rio Arriba Counties, the two counties in Northern New Mexico with the largest concentrations of acequia communities. As Fig. 11 shows, these communities are both transforming from primary extractive and resource-based economies to service, professional, and government-service centered economies. Between the years 1970 and 2000 the services sector shares have risen from about 50 % to nearly 70 % in Taos County and similarly from about 40 % to over 50 % in Rio Arriba County. Agricultural employment has remained relatively flat at approximately 4 % for Taos and 8 % for Rio Arriba. These trends suggest a greater dependence on regional employment centers, greater off-farm income support, and higher commuting rates. Such regionalization can affect community "balance-sheets" in several, sometimes offsetting, ways. For example, diversification from traditional extractive economic activities can strengthen incomes and raise financial wealth and capacity; however, this may come at some cost to sociocultural strength and well-being, with less time and commitment to community-centered activities and relationships.

Trends and changes in farm income and the agricultural economy of the acequia region, coupled with changes and variations in the water supply situation, are shown in Fig. 12. Roughly consistent with the long-term steady employment picture in farming and ranching, there does not appear to be a strong positive or negative trend in farm and ranch incomes over this period. However, there is a relatively high degree of income variability that is largely independent of regional water supply conditions.

4 Sociocultural perspectives to understand resilience

In the acequia culture of the historic Rio Arriba bioregion (northern Rio Grande), attachment to place is strongly held by multiple generations of irrigators who have long historical connection to their landholdings. This sense of place is exhibited in many ways that promote resilience related to acequias and water management.

4.1 Sociocultural knowledge and the hydrosocial cycle

As with other forms of traditional agriculture around the world, acequia irrigation in the Rio Arriba is knowledge intensive in terms of understanding and responding to the local hydrological and environmental conditions upon which the system depends. The complete system is carried collectively in the local knowledge of the irrigators, particularly with regard to the distinctive micro-region of their community: soils, climate conditions, crops, and water requirements for every niche suitable for agriculture (Glick, 2006). The mutuality of the irrigators derives from the values encoded in the operational rules of water sharing, namely equity, justice, and local control (Maass, 1998). This knowledge is derived from and expressed in practical, experiential terms and, after repeated cycles, is embedded in the culture and passed on to new generations as part of the social and institutional memory of the community (Folke et al., 2003).

4.2 Acequia resilience: the role of the New Mexico Acequia Association

The New Mexico Acequia Association (NMAA) conducted an assessment survey in 2008 (before the current research project reported here) to document the concerns of its membership in order to generate effective governance program planning and strategies to benefit acequias. Topic areas included needs, crop and water management trends, infrastructure conditions, and overall health of acequias. The survey was distributed to acequia officers as the targeted group due to their knowledge of their local acequias. In some cases, individual member irrigators completed the surveys when the governing body was not available. A synopsis report by the NMAA highlighted the survey responses in the two example charts below (Fig. 13a and b). The results show that the cited values are of high importance while low community participation, few members irrigating their crops or raising livestock, and problems with the irrigation works were noted as important challenges. The NMAA took action to address these needs and concerns by way of programs, workshops, and leadership development projects intended to assist acequias in recruiting, train the next generation of acequia managers, strengthen acequia administration, and recruit community leaders to take active roles in policy development. With respect to infrastructure repairs, the NMAA holds workshops

(a) Taos County

(b) Rio Arriba County

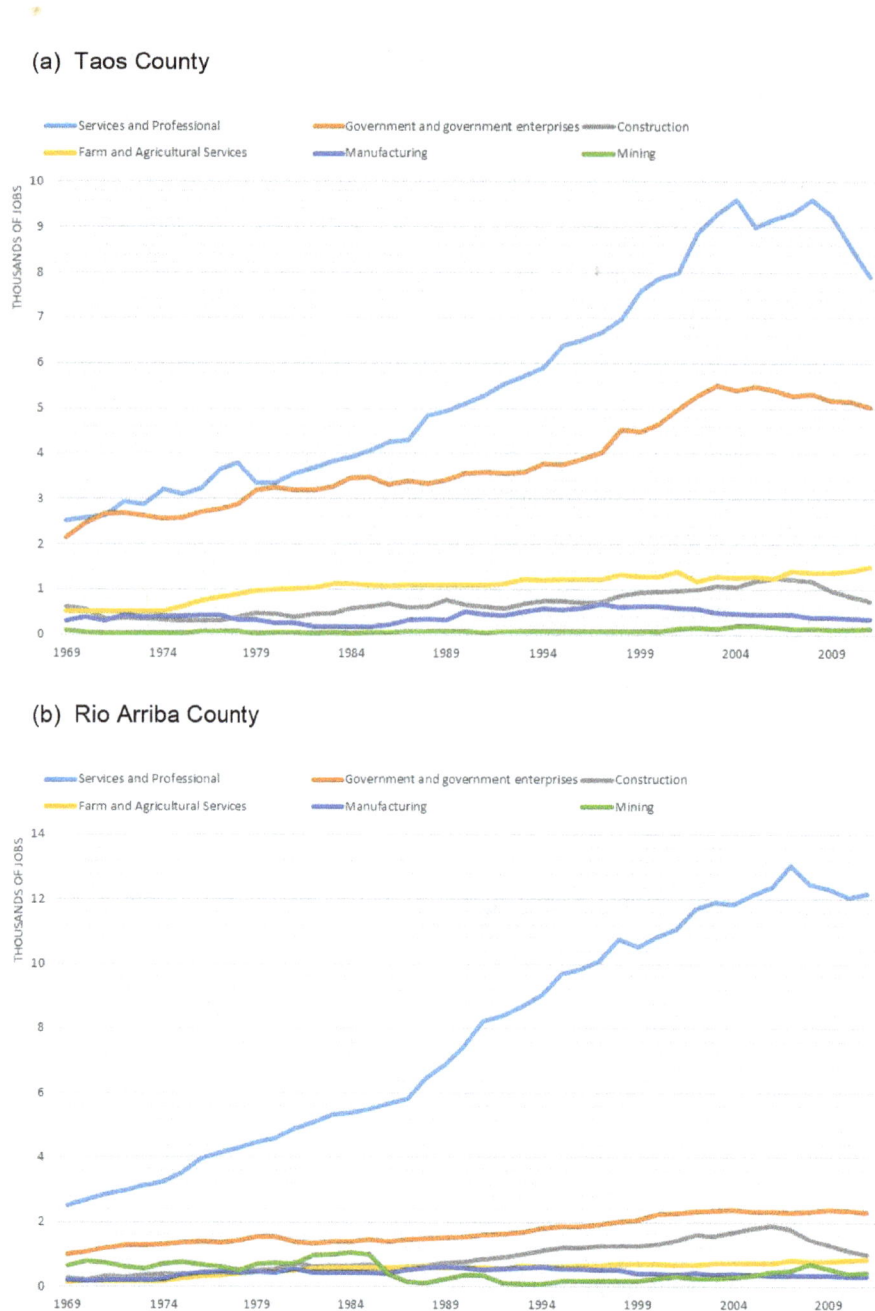

Figure 11. Employment shares and trends by sector in Taos and Rio Arriba Counties, New Mexico (US Bureau of Economic Analysis, 2013a).

on how member acequias can qualify for the ditch rehabilitation financing available from state and federal agencies.

4.3   Acequia resilience: views of officials and irrigators

Traditional local knowledge about the hydrosocial cycle of acequia operations is a key factor in acequia resilience when the irrigators are confronted with disturbances, unexpected events, or changing climate that affects water supply. Adaptability in times of stressors is self-evident by the fact that ace-

quias as human and social institutions still operate and have not disappeared even after political administration under four sovereigns and their shifting water law regimes: (1) Spanish colonial, (2) Mexican period, (3) US Territorial, and (4) New Mexico statehood in 1912. Focus group sessions with acequia officials and irrigators, along with supporting evidence from similar case studies in other arid regions of the world, suggest a number of intriguing propositions with regard to resilience factors of acequia irrigation systems:

(a) Taos County, New Mexico

(b) Rio Arriba County, New Mexico

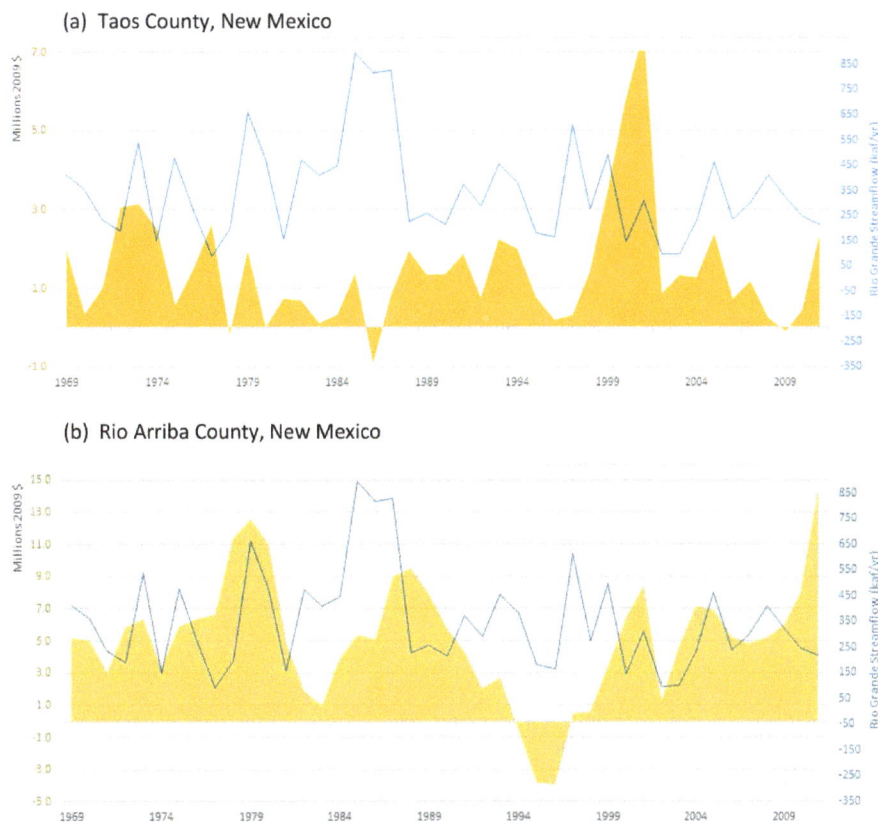

Figure 12. Trends and changes in agricultural incomes (orange areas) and streamflow (blue lines) for Taos and Rio Arriba Counties in New Mexico (US Bureau of Economic Analysis, 2013b; US Geological Survey, 2013).

1. Attachment to land and place develop a collective identity with a set of shared values and cultural norms, producing what can be called an "acequia imaginary".

2. Mutual networks and social density result in cohesion and solidarity of community when confronted with change or stressors from outside the community.

3. Leadership by key individuals such as the acequia officers maintains and retains the customary rules and local management practices of the system.

4. Social memory embedded in the culture instructs acequia leaders and irrigators on how to respond to and withstand disturbances or year-to-year changes.

5. Ecological knowledge of local conditions and environment is carried collectively and transmitted to new generations.

6. Local control of resources increases the capacity to adapt in times of scarcity such as cycles of prolonged drought.

7. Autonomy of decision-making structure and discretionary authority permit rapid adjustments in opera-

tional rules and practices when warranted by changing or unexpected conditions.

5 Discussion

5.1 Identifying essential variables to model future scenarios

Information gathered from the local communities was placed into the framework of a causal loop diagram. The causal loop diagram is a tool to represent interacting variables of a system, and it is a precursor to simulation modeling. Causal loops were also created for hydrology, economic, and environmental variables as discussed in Fernald et al. (2012). Community members were invited to a workshop with researchers to refine the causal loop diagrams based on their own understanding. The data that researchers collected were used to identify and assess the key variables among all of those identified in a collaborative community water research process (Guldan et al., 2013).

Ongoing system dynamics modeling based on the causal loop diagrams will be used to turn narratives into future scenarios that identify thresholds and tipping points of sustainability (Fernald et al., 2012). Our ongoing approach is to

a)

- Part of our culture
- Needed to grow crops and raise livestock
- Creates green valleys that make the landscape beautiful
- Creates more habitat for wildlife
- Contributes to replenishment of groundwater or aquifer
- Responsible as a local government for taking care of the water
- Other

b)

- Low parciante participation in the acequia
- Few people irrigating their pasture/crops or raising livestock
- Irrigation works (presa, ditch, headgates) are difficult to operate or are in disrepair.
- Mayordomo or Commissioners lack time or knowledge to properly manage acequia
- Disputes between parciantes or landowners regarding easements
- Disputes between parciantes regarding irrigation schedule

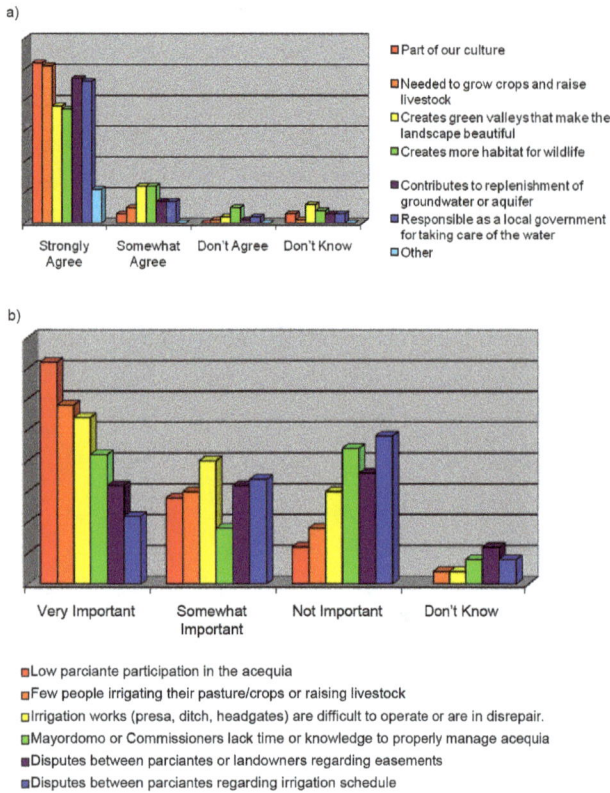

Figure 13. (a) Responses to the question "what is your opinion about the following statements about the importance of your acequia to your community?" and **(b)** Responses to the question "what do you believe are the important challenges within your acequia?"

distill out the essential variables in the system using our field data from the multiple disciplines and incorporating them into a simulation model. When we put together the causal loop diagrams for all disciplines and work with community members, key variables emerge as shown in the essential causal loop diagram (Fig. 14).

In order to identify resilience, sustainability, thresholds, tipping points, and future directions for hydrologic and community health, our ongoing work is developing a model that brings together all scenarios to help identify higher levels of interaction than are obtainable with disciplinary approaches.

We will use cross-cutting scenarios within system dynamics modeling to test tipping point hypotheses. System dynamics modeling uses stocks of key variables and parameterized flows between them to recreate the systems under consideration. In our case, we will bring together the multiple threads and model scenarios based on our field data from the multiple disciplines. We will establish a dynamic hypothesis specific to the system dynamics modeling exercise. We will iteratively put the hypothesis in front of the different researchers as the model is developed based on the causal loop diagrams that were drawn with community help. Then the model itself

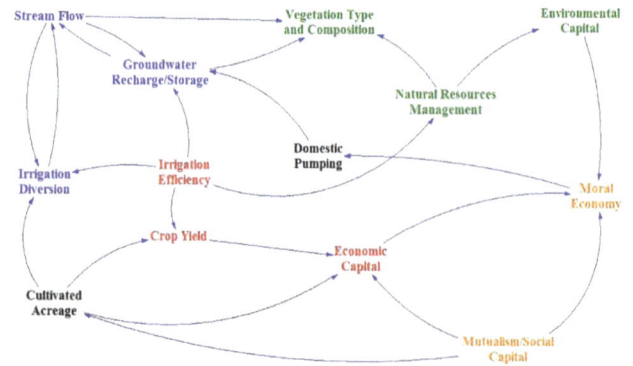

Figure 14. Essential causal loop diagram with key variables for modeling. Black variables are critical elements integrating across multiple subsystems, while colors are primary to individual disciplines.

will be tested and validated with community members and researchers.

5.2   Example model development

The causal loop diagrams that we have created in collaboration with the communities are allowing us to model scenarios of interacting water, agriculture, society, and economics. We present here a theoretical case of model development showing interacting prices, crops, and water use. Fig. 14 shows a number of variables. We select two here, cultivated acreage and crop yield, that relate closely to crop production (irrigated agriculture) for the Acequia de Alcalde, an acequia along the main stem of the Rio Grande.

Each of these two variables is affected by more variables than the essential causal loop diagram indicates (see Fernald et al., 2012), for example labor availability, crop price, and market. Perennial forages and tree fruit orchards are the two primary, yet very different, crop types that make up the majority of the irrigated land that the Acequia de Alcalde provides water for. Currently, there are on the order of 50 acres of tree fruit orchards and 620 acres of hay forage (Table 1).

The proportion of irrigated acres in orchards vs. forage crops can potentially be influenced by changes to various factors/drivers. Decrease in availability of farm labor could lead to a decline in orchard acres, whereas an increased demand for local fresh food might convert some forage fields into orchards. Climate change may exacerbate the common problem of tree fruit crop loss due to late spring frosts. As has been the trend, any factor that decreases acreage of orchard crops may increase forage acreage or result in less overall cultivated acres.

A scenario of decreased irrigation water supply, such as with a long-term drought, might shift acres from forages to apples or other specialty or higher value crops that can more easily be adapted to micro-sprinkler or drip irrigation (a difficult resilience strategy for producers only having equipment

Table 1. Approximate cultivated acreage (D. Archuleta, personal communication, 2012), yield, and gross revenue ranges (Forage, Currier et al., 1995; Lauriault et al., 2004; Apple/Orchard crops, S. Yao, personal communication, 2012) of two crops and potential effects from various factors (Acequia de Alcalde).

Cultivated area	×	Yield/area	=	Total yield	Gross revenue
Forage 620 acres		4–6 tons acre^{-1}		2500–3700 tons	0.5–1 million USD
Apple/orchard crops					
50 acres		6–13 tons acre^{-1}		300–650 tons	0.3–0.65 million USD*

* Through direct marketing at roadside stands or farmers markets, apples and other orchard crops can potentially produce significantly higher revenues per acre than indicated.

and experience with forage crops). Forage producers might also shift to forages that use less water, but that would also yield less, or a loss in overall cultivated acres may occur as hay fields are left idle.

Cropping patterns could also change due to socioeconomic, policy, and cultural factors, in extreme cases having an effect on the hydrology of the system. For example, grazing restrictions or increased costs of grazing permits in uplands due to policy changes could impact local demand for hay: an increase in demand could occur to meet herd feed needs, or a decrease in demand could occur if herds are reduced, possibly leading to a shift to other crops or idling of acres. The latter could reduce aquifer recharge and groundwater return flow to the river due to lack of seepage and deep percolation (Fernald et al., 2010).

Along many acequias, demand for housing and the associated increase in land prices have often caused pressure to subdivide fields into residential lots. This decrease in irrigated acres has likely reduced seepage and percolation in the system and thus reduced aquifer recharge and groundwater return flow. In the case of one county (Rio Arriba), a resilient response to this has been an ordinance passed several years ago that stipulates agricultural fields be developed in a manner that maintains 70 % as open field, with 30 % allowed for development as cluster housing (Rio Arriba Agricultural Protection and Enhancement Ordinance, adopted 31 January 2002; prepared by the Rio Arriba County Planning and Zoning Department, Rio Arriba County).

Our approach enables a quantification of system connections as illustrated by the grazing example. A preliminary analysis of farmer/rancher surveys and historical records of public land grazing in areas adjoining irrigated valley study sites suggests a tight connection between irrigated hay production and year-to-year variation in upland livestock numbers (Lopez, 2014). Although it is difficult to isolate the influence of hay production from factors such as public land use policies that determine the length of the summer grazing season (and therefore the number of winter feeding days), availability of valley-grown forages appears to be an important driver of livestock herd dynamics in the local farming/ranching communities (Lopez, 2014). Thus, projected changes in snowmelt regimes and irrigation water availability

could indirectly affect traditional livestock-raising activities and weaken local economies and ancestral valley upland sociocultural connections.

We have identified resilience as well as susceptibility to change. We contend that multiple lines of evidence enable us to construct meaningful future scenarios to test the limits of these systems. We envision an interacting causal model based on our causal loop diagrams and field evidence that sheds light on the tipping point hypothesis. What will happen, for example, if acequia farmers sell their land, and water is transferred off the land to regional urban centers? Based on the integrated model, we might find that under this scenario farming is reduced, impacting the timing and distribution of flow, reducing seepage and groundwater return flow, reducing riparian function, and reducing river flow in late summer and fall. Reduced farming and grazing result in changing vegetation structure and increased density and cover, which leads to increased wildfire in a warmer future, further exacerbating pressures on grazing and farming. We have shown that these systems adapt and change, but there are also signs that key components of the acequia systems have limits to resilience.

5.3 Integrating elements

Although a more comprehensive integration of elements affecting resilience in acequia systems will be provided upon completion of the system dynamics model (Fernald et al., 2012), we highlight in this section some key hydrosocial interactions that have a bearing on resilience. Seepage from the acequia canals and flooded fields studied recharges the aquifer. This provides resilience for the local community – for farmer and non-farm rural residents alike who rely on either shallow domestic wells or community water systems. Recharged aquifers appear to provide delayed groundwater return flow to the river, providing resilience for the basin as a whole by benefitting downstream irrigators and other users.

A drastic reduction in seepage through wide-spread adoption of practices meant to increase irrigation efficiency at the field scale (e.g., drip irrigation) or a reduction in area under irrigated agriculture due to urbanization or greater movement away from on-farm employment can decrease these resiliencies. On the other hand, practices such as drip irrigation can allow irrigated agriculture to continue if water becomes significantly less available because of long-term drought or climate change. With increasing pressures and policy development to facilitate the transfer of agricultural water to growing cities, drip irrigation in combination with partial lease of irrigation water rights to the cities could provide a trade-off that helps a farmer stay in agriculture. This conversion, on the other hand, would possibly entail a shift from alfalfa pastures to row crops, which would require irrigators to modify their livestock-raising enterprises to adopt off-site winter feeding options.

Whereas urbanization puts pressure on agricultural land and water resources, a certain critical mass of nearby urban consumers can provide demand for local, high-value agricultural products. The high-value products can increase incomes for small-scale acequia farmers, again increasing chances that agriculture continues and seepage benefits continue.

Acequia group leaders discuss the lack of community participation in operating, governing, and maintaining acequias. Our conjecture is that this does impact resilience, but more in terms of community cohesiveness and not so much in terms of hydrologic regimes, because acequias continue to flow. It is interesting to note that although individual acequias may not have the active participation of all irrigators, participation of regional acequia associations at the annual meeting of acequias (*Congreso de las Acequias*) has increased greatly since the meetings began 15 years ago.

Socioeconomic data indicate that in some acequia communities, income is not strongly influenced by regional water supply conditions. This can be viewed in different ways. For example, are acequias not important for the community? Or, because acequias continue to operate regardless of water supply, does this indicate they play a role in or are a measure of community resilience? Survey results (Mayagoitia et al., 2012) indicate acequia residents continue to exhibit strong identity with and affinity for the local land, water, and cultural resources. This cultural aspect, as well as the policy and socioeconomic influences mentioned above, supports a key principle – understanding people and society is critical to understanding the hydrology of a system.

6 Conclusions

Keys to resilience are found in hydrologic and human system connections. Seepage from acequia systems supports a host of hydrologic and riparian resilience functions. Hydrologically, seepage recharges groundwater and provides attenuated return flow to rivers and streams. Riparian areas support most of the biodiversity in these regions. Reduction in water may reduce riparian areas and acequias can provide additional refugia in times of low water. Community cohesion has resilience in the value of attachment to place derived from acequia and local farming culture. Livestock raising contributes to strengthening the economic and social resilience of traditional acequia irrigation communities of northern New Mexico. The ability to grow irrigated forages appears to be critical to the persistence of this well-established agricultural activity. Changes in snowmelt regime and water availability for irrigation could cause further reductions in herd numbers and severely weaken the cohesion of acequia farming communities.

Acequias are resilient because they are in step with the scope and scale of variability in the natural systems. The roots of sustainability are the intricate linkages that have developed over generations, connecting human and hydrologic systems. For example, acequia water is distributed in keeping with the highly variable precipitation of the region. Unlike priority water law that gives the oldest water rights the water in times of scarcity, acequias share the water. In wet times everyone gets more, and in dry times everyone gets less. Irrigated lands were established to match the wet and dry years, with vital lands near the river irrigated in dry years and lands farther from the river added to the irrigated footprint in wet years. Adaptation to semiarid system variability and connection to place are at the heart of an acequia's ability to adapt.

Our data have demonstrated that acequia systems have very high resilience and adaptive capacity but also show susceptibility to major upheaval. Thus shocks to the system such as climate change and land use change that impact water and ties to the land are particularly disruptive. Tipping points may be reached when external drivers push these systems beyond their historic limits. It is widely acknowledged that a regional megadrought in what is now the US southwest pushed Pueblo peoples beyond their capacity to adapt and caused widespread migration and cultural upheaval. Signs of tipping points are showing now in the Taos valley where developers have paved over acequias, blocking downstream users access to water. In 2013, after 10 years of drought, river water was significantly low, and acequia communities were on the verge of filing lawsuits against upstream users until seasonal monsoon rains allowed irrigation to resume. Resilience and tipping points can be propagated both upstream and downstream due to hydrologic connections and trans-basin water movement. Although timely rains occurred in 2013, if droughts of historic depth and duration occur, acequia systems appear vulnerable to upheaval.

Acknowledgements. This study was funded in part by the New Mexico Agricultural Experiment Station and National Science Foundation grants no. 814449 New Mexico EPSCoR and no. 1010516 Dynamics of Coupled Natural and Human Systems. We thank the special issue editor Murugesu Sivapalan, and Christopher Scott and the anonymous reviewers for incredibly helpful comments.

Edited by: M. Sivapalan

References

Barnett, T. P., Pierce, D. W., Hidalgo, H. G., Bonfils, C., Santer, B. D., Das, T., Bala, G., Wood, A. W., Nozawa, T., Mirin, A. A., Cayan, R. R., and Dettinger, M. D.: Human-induced changes in the hydrology of the western United States, Science, 319, 1080–1083, 2008.

Boykin, K. G., Thompson, B. C., Deitner, R. A., Schrupp, D., Bradford, D., O'Brien, L., Drost, C., Propeck-Gray, S., Rieth, W., Thomas, K., Kepner, W., Lowry, J., Cross, C., Jones, B., Hamer, T., Mettenbrink, C., Oakes, K. J., Prior-Magee, J., Schulz, K., Wynne, J. J., King, C., Puttere, J., Schrader, S., and Schwenke, Z.: Predicted animal habitat distributions and species richness, in: Southwest Regional Gap Analysis Final Report, edited by: Prior-Magee, J. S., U.S. Geological Survey, Gap Analysis Program, Moscow, Idaho, USA, 2007.

Boykin, K. G., Kepner, W. G., Bradford, D. F., Guy, R. K., Kopp, D. A., Leimer, A. K., Samson, E. A., East, N. F., Neale, A. C., and Gergely, K. J.: A national approach for mapping and quantifying habitat-based biodiversity metrics across multiple spatial scales, Ecol. Indic., 33, 139147, doi:10.1016/j.ecolind.2012.11.005, 2013.

Bierwagen, B. G., Theobald, D. M., Pyke, C. R., Choate, A., Groth, P., Thomas, J. V., and Morefield, P.: National housing and impervious surface scenarios for integrated climate impact assessments, P. Natl. Acad. Sci., 107, 20887–20892, 2010.

Cox, M.: Exploring the dynamics of social-ecological systems: The Case of the Taos Valley acequias, PhD dissertation, Indiana University, United States of America, 129 pp., 2010.

Cox, M. and Ross, J. M.: Robustness and vulnerability of community irrigation systems: The case of the Taos valley acequias, J. Environ. Econ. Manag., 61, 254–266, 2011.

Currier, C., Henning, J., Townsend, S., Barnes, C., Mayernak, J., Nelson, S., Nuñez, A., Gregory, E. J., McGarrah, C., and Ward, C.: Alfalfa variety trials in New Mexico, New Mexico Agric. Exp. Sta. Research Report 695, Las Cruces, New Mexico, United States of America, 1995.

Earth Data Analysis Center: 1935 aerial photography of Rio Arriba County, Image Archive and Data Service, University of New Mexico, Albuquerque, New Mexico, USA, 2010.

Eastman, C. and Gray, J. R.: Community Grazing: Practice and Potential in New Mexico, University of New Mexico Press, Albuquerque, New Mexico, United States of America, 1987.

Fernald, A. G., Cevik, S. Y., Ochoa, C. G., Tidwell, V. C., King, J. P., and Guldan, S. J.: River hydrograph retransmission functions of irrigated valley surface water–groundwater interactions, J. Irrig. Drain. E.-ASCE, 136, 823–835, 2010.

Fernald, A., Tidwell, V., Rivera, J., Rodríguez, S., Guldan, S., Steele, C., Ochoa, C., Hurd, B., Ortiz, M., Boykin, K., and Cibils, A.: Modeling sustainability of water, environment, livelihood, and culture in traditional irrigation communities and their linked watersheds, Sustainability, 4, 2998–3022, 2012.

Folke, C., Colding, J., and Berkes, F.: Synthesis: Building resilience and adaptive capacity in socialecological systems, in: Navigating Social-Ecological Systems, edited by: Berkes, F., Colding, J., and Folke, C., Cambridge University Press, West Nyack, New York, United States of America, 352–387, 2003.

Glick, T. F.: Historical status and cultural meaning of historic hydraulic landscapes, unpublished paper: *Primer Congreso Internacional Oasis y Turismo Sostenible*, Elche, Spain, 14–16 December 2006.

Guldan, S. J., Fernald, A. G., Ochoa, C. G., and Tidwell, V. C.: Collaborative community hydrology research in northern New Mexico, J. Contemporary Water Research & Education, 152, 49–54, 2013.

Lauriault, L. M., Ray, I. M., Pierce, C. A., McWilliams, D. A., English, L. M., Flynn, R. P., Guldan, S. J., and O'Neill, M. K.: The 2004 New Mexico alfalfa variety test report, available at: http://aces.nmsu.edu/pubs/variety_trials/var04.pdf (last access: 19 December 2014), 2004.

Lopez, S. C.: The Role of Livestock in Suppressing Rangeland Weeds and Sustaining Traditional Agropastoral Communities in Northern New Mexico, M.S. Thesis, New Mexico State University, United States of America, 108 pp., 2014.

Maass, A., and Anderson, R. L.: ... and the Desert Shall Rejoice: Conflict, Growth and Justice in Arid Environments, Robert E. Krieger Publishing Co., Malabar, Florida, United States of America, 1–10, 366–376, 1978 (Reprint Edition 1986).

Mayagoitia, L., Hurd, B., Rivera, J., and Guldan, S.: Rural community perspectives on preparedness and adaptation to climate-change and demographic pressure, J. Contemporary Water Res. Education, 147, 49–62, 2012.

National Agricultural Statistics Service: www.nass.usda.gov, last access: December 2012.

Ochoa, C. G., Fernald, A. G., Guldan, S. J., Tidwell, V. C., and Shukla, M. K.: Shallow aquifer recharge from irrigation in a semiarid agricultural valley in New Mexico, J. Hydrologic Engineering, 18, 1219–1230, 2013a.

Ochoa, C. G., Guldan, S. J., Cibils, A., Lopez, S., Boykin, K., Tidwell, V. C., and Fernald, A. G.: Hydrologic connectivity of head waters and floodplains in a semiarid watershed, J. Contemporary Water Res. Education, 152, 69–78, 2013b.

Ortiz, M., Brown, C., Fernald, A., Baker, T. T., Creel, B., and Guldan, S.: Land use change impacts on *Acequia* water resources in northern New Mexico, J. Contemporary Water Res. Education, 137, 47–54, 2007.

Raish, C. and McSweeney, A. M.: Economic, social, and cultural aspects of livestock ranching on the Espanola and Canjilon Ranger Districts of the Santa Fe and Carson National Forests: A pilot study, Gen. Tech. Rep. RMRS-GTR-113, United States Department of Agriculture, Rocky Mountain Research Station, Fort Collins, Colorado, United States of America, 2003.

Rio Arriba County Assessor's Office: Rio Arriba County GIS parcel data, retrieved from County Assessor's Office in Española, New Mexico, July 2013..

Rivera, J. A.: Acequia Culture: Water, Land, and Community in the Southwest, University of New Mexico Press, Albuquerque, New Mexico, USA, 1998.

United Nations: Report of the World Commission on Environment and Development: Our Common Future, General Assembly Resolution 42/187, 11 December 1987.

US Bureau of Economic Analysis: Regional Economic Accounts, Tables CA25 and CA25N, Total Employment by Major SIC (NAICS) Industry, available at: http://bea.gov/regional/downloadzip.cfm (accessed 18 October 2013), 2013a.

US Bureau of Economic Analysis: Regional Economic Accounts, Table CA45, Farm Income and Expenses, available at: http://bea.gov/regional/downloadzip.cfm (accessed 18 October 2013), 2013b.

US Department of Agriculture (USDA) Forest Service: Timeless Heritage: A History of the Forest Service in the Southwest, FS-409, 1991

US Department of Agriculture (USDA) Forest Service: Region III, Permitted Livestock and Paid Livestock, Years 1909–1997, Data Compilation from Annual Statistical Grazing Reports, 1998.

US Department of Commerce/US Department of Agriculture: USDA Census of Agriculture Historical Archive, available at: http://agcensus.mannlib.cornell.edu/AgCensus/homepage.do, last access: 19 December 2014.

US Geological Survey (USGS): National Streamflow Information Program USGS Current Conditions for the Nation, URL: http://waterdata.usgs.gov/nwis/uv, last access: 18 October 2013.

12

From days to decades: numerical modelling of freshwater lens response to climate change stressors on small low-lying islands

S. Holding and D. M. Allen

Department of Earth Sciences, Simon Fraser University, 8888 University Drive, Burnaby, British Columbia, V5A 1S6, Canada

Correspondence to: S. Holding (sholding@sfu.ca)

Abstract. Freshwater lenses on small islands are vulnerable to many climate change-related stressors, which can act over relatively long time periods, on the order of decades (e.g., sea level rise, changes in recharge), or short time periods, such as days (storm surge overwash). This study evaluates the response of the freshwater lens on a small low-lying island to various stressors. To account for the varying temporal and spatial scales of the stressors, two different density-dependent flow and solute transport codes are used: SEAWAT (saturated) and HydroGeoSphere (unsaturated/saturated). The study site is Andros Island in the Bahamas, which is characteristic of other low-lying carbonate islands in the Caribbean and Pacific regions. In addition to projected sea level rise and reduced recharge under future climate change, Andros Island experienced a storm surge overwash event during Hurricane Francis in 2004, which contaminated the main wellfield. Simulations of reduced recharge result in a greater loss of freshwater lens volume (up to 19 %), while sea level rise contributes a lower volume loss (up to 5 %) due to the flux-controlled conceptualization of Andros Island, which limits the impact of sea level rise. Reduced recharge and sea level rise were simulated as incremental instantaneous shifts. The lens responds relatively quickly to these stressors, within 0.5 to 3 years, with response time increasing as the magnitude of the stressor increases. Simulations of the storm surge overwash indicate that the freshwater lens recovers over time; however, prompt remedial action can restore the lens to potable concentrations up to 1 month sooner.

1 Introduction

Small islands are particularly vulnerable to stressors associated with climate change. The freshwater lens is generally sensitive to hydrological disturbances, as a consequence of the low hydraulic gradient and limited thickness of the lens (Vacher, 1988; Falkland, 1991; Robins and Lawrence, 2000; White and Falkland, 2010). As groundwater recharge is the primary source of freshwater to a freshwater lens, an adequate amount of recharge is critical for maintaining the lens morphology (Falkland, 1991). Changes in groundwater recharge due to climate change are likely to result from increases in temperature and changes in the spatial distribution, frequency and magnitude of precipitation (Green et al., 2011). Conditions of reduced recharge disturb the balance of freshwater outflow necessary to maintain the extent of the freshwater lens, and may lead to loss of freshwater volume due to saltwater intrusion (Oude Essink, 2001; Ranjan et al., 2009).

Sea level rise may result in inundation and a landward shift of the saltwater interface, particularly on low-lying islands (Bear et al., 1999). This would result in a loss of freshwater lens volume, either by a reduction in areal extent and/or a thinning of the lens (Oude Essink, 2001). Projected changes in the frequency of hurricanes and tropical storms are uncertain (IPCC – Intergovernmental Panel on Climate Change, 2014); however, there is evidence to suggest that storms may become more intense, increasing the likelihood of storm surge occurrence (Biasutti et al., 2012). Storm surge overwash can lead to salt contamination of the freshwater lens and a temporary loss of freshwater (Anderson, 2002; Illangasekare et al., 2006; Terry and Falkland, 2010). Due to topography, low-lying islands are more susceptible to saltwater inundation from sea level rise and storm surge overwash.

Previous modelling studies have investigated aspects of climate change impacts on the freshwater lenses of islands or coastal aquifers. Simulations of decreased recharge resulted in more saltwater intrusion and impact to water supply infrastructure than simulations of sea level rise alone (Rasmussen et al., 2013). However, for regions with future projected increases in recharge, the impact of sea level rise and other stresses (i.e., increases in pumping) may be counteracted by increased recharge (Sulzbacher et al., 2012). Analytical and numerical models of sea level rise indicate that the degree of saltwater intrusion (or loss of freshwater lens volume) resulting from sea level rise depends on many factors. Whether the hydrogeological system is recharge-limited or topography-limited (Michael et al., 2013) influences whether or not the water table rise that accompanies sea level rise can be accommodated by the system. Werner and Simmons (2009) showed that less saltwater intrusion is expected when the system is recharge-limited (flux-controlled). Unsurprisingly, the degree of land surface inundation was found to control the amount of saltwater intrusion (Ataie-Ashtiani et al., 2013), and the impact of sea level rise on saltwater intrusion is enhanced by groundwater extraction from coastal wellfields (Bobba, 2002; Langevin and Zygnerski, 2013).

Models of storm surge overwash events have been developed to evaluate their impact on the freshwater lens. Most of these models used codes that neglect the surface domain. However, Yang et al. (2013) used a fully coupled subsurface and surface approach that simulated tidal activity, coastal flow dynamics, and a hypothetical storm surge on a coastal aquifer. All models indicate initial salt contamination of the freshwater lens, which recovers to fresh concentrations over time due to freshwater recharging at surface and density-driven downward migration of salt water (Terry and Falkland, 2010). The occurrence of multiple storm surges (Anderson, 2002) and accumulations of salt water at the surface in low depressions (Chui and Terry, 2012) may increase the time for recovery of the lens. Where the vadose zone becomes thinner under conditions of sea level rise (because the freshwater lens has risen in the subsurface), the impact of storm surge alongside sea level rise may result in less salt contamination of the freshwater lens (Chui and Terry, 2013). However, the salt contamination that does occur under sea level rise conditions remains close to the surface of the lens (Terry and Chui, 2012). Wider islands generally result in less freshwater lens contamination than narrow islands, as a result of their thicker lens morphology (Chui and Terry, 2013).

Although many aspects of climate change impacts on freshwater lenses have been modelled previously, few studies have investigated both the spatial and temporal response of the freshwater lens to the stressors. Climate change related stressors operate on various spatial and temporal scales: island-wide impacts due to sea level rise and changes in recharge occur over long time periods, on the order of decades, whereas local-scale impacts due to storm surge overwash occur over short time periods, on the order of days.

This study evaluates the spatial and temporal response of an island freshwater lens to various climate change stressors using a numerical modelling approach. To account for the varying temporal and spatial scales of the stressors, two different density-dependent flow and transport modelling codes are used. SEAWAT (Langevin et al., 2007) models were developed on an island scale to simulate long-acting stressors, including sea level rise and change in recharge. HydroGeoSphere (Therrien et al., 2010) models were developed on a local scale to simulate storm surge, which is a short-acting stressor. The study aims to identify critical factors and stressors that may affect freshwater resources of small, low-lying islands, using Andros Island in the Bahamas as a representative island. The results of the study are intended to be applicable to other islands with similar hydrogeological settings.

2 Site description

The study site is Andros Island in the Bahamas. Andros Island has undergone limited development and groundwater exploitation; therefore, the hydrogeological data collected in the 1970s (Little et al., 1973) are considered generally representative of current conditions and can be used for baseline model calibration. Andros Island is representative of other low-lying carbonate islands with thin freshwater lenses commonly found throughout the Caribbean and Pacific regions (Falkland, 1991; Vacher and Quinn, 1997).

Andros Island is the largest island in the Bahamas, and is located 200 km southeast of Florida (Fig. 1). It is $14\,000\,\text{km}^2$ in area and is comprised of several smaller islands and cays. The highest elevation on the island is 20 m a.s.l. (metres above sea level) along a ridge that parallels the eastern coast, whereas lower elevations (< 1 m a.s.l.) are common towards the west. The western coastline is largely composed of wetlands and saltwater marshes, and, therefore, most settlements are along the eastern coast of the island (Fig. 1). The remainder of the island is largely covered in pine forest. There are no permanent surface water bodies on the island.

Andros Island is located on the Great Bahama carbonate bank (Fig. 1). The geology of the island is a predominantly Pleistocene Lucayan limestone formation, which is around 40 m thick (Beach and Ginsburg, 1980). Discontinuity surfaces (unconformities) within the limestone are present as layers of paleosols recurring in the upper stratigraphy (Beach and Ginsburg, 1980). These layers represent episodes of subaerial exposure and are largely concentrated within the top 20 m (Beach and Ginsburg, 1980; Boardman and Carney, 1997). Underlying the Lucayan is a cavernous, highly karstic, and relatively more permeable unit termed the pre-Lucayan, which is present from 43 m b.g.s. (metres below ground surface) to at least 75 m b.g.s. (Boardman and Carney, 1997). The geology below this depth has not been observed as most studies focus on the shallow, freshwater-bearing units; how-

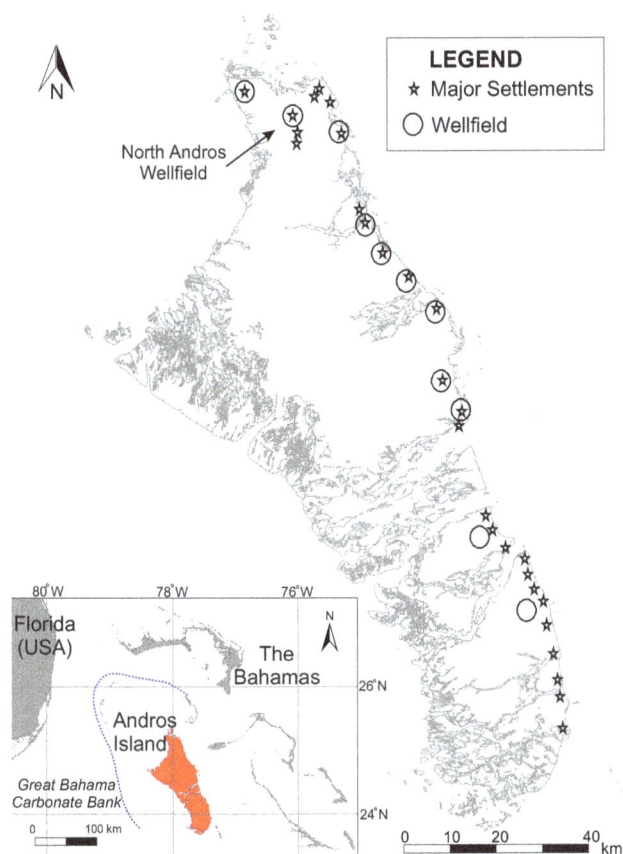

Figure 1. Andros Island, indicating the location of settlements and wellfields.

ever, deposits of carbonates on the Great Bahama bank are estimated to be up to 7 km thick (Cant and Weech, 1986).

Due to its large size, the freshwater lens on Andros Island represents the principal source of natural freshwater for the Bahamas. Most local residents rely on the municipal potable water supply ($< 0.4\,g\,L^{-1}$ salt concentration), which extracts groundwater from the lens via 11 wellfields distributed across the island (Fig. 1). The local drinking water guidelines define potable water as having salt concentrations of less than $0.4\,g\,L^{-1}$. The largest of these wellfields is the North Andros Wellfield. As is common with many freshwater lenses, there is potential for upconing of the underlying saltwater and degradation of the lens if wells are deep and the lens thin (Werner et al., 2009; White and Falkland, 2010). Therefore, the wellfields on Andros employ horizontal trench-based groundwater extraction or a series of interconnected shallow boreholes pumped at low rates. Typical depth of the wellfields is between 1 and 5 m b.g.s. Water flows within the trench-based wellfields under a very low gradient, towards a central low sump where water is pumped to storage reservoirs.

The hydrogeology of Andros Island is based on previous studies, most of which were conducted around the wellfields and other developed areas. The principal aquifer is the un-

confined Lucayan limestone, as the older (deeper) geological units are too permeable and thus are not able to prevent freshwater from mixing with the surrounding saltwater (Cant and Weech, 1986; Schneider and Kruse, 2003). Soil zones are sparse, and minimal runoff occurs during precipitation events (Little et al., 1973; Tarbox, 1987). The freshwater lens is recharged solely through infiltrating precipitation, which generally occurs during the wet season from May to October (Bukowski et al., 1999). Average annual precipitation in the south is 39 % less than average annual precipitation in the north of Andros Island (Cant and Weech, 1986; Bahamas Department of Meteorology, Climate Averages 1979–2000). Based on resistivity surveys conducted in the north of the island, the thickness of the freshwater lens ranges from 3 to 20 m (Wolfe et al., 2001); however, previous studies cite the maximum thickness as 34 m (Cant and Weech, 1986) and borehole salinity profiles indicate that the maximum thickness of the lens is up to 39 m (Little et al., 1973). The lens is generally shallower in the southern regions of Andros Island compared to the northern regions, with a measured thickness of at least 15 m b.g.s. (municipal water supply managers, Bahamas Water and Sewerage Corporation, personal communication, 2013). The elevation of the lens inland is approximately 2 m a.s.l. (Ritzi et al., 2001) with typical depth to water of 1–2 m b.g.s., although it is deeper (up to 5 m b.g.s.) under the high topography ridge along the eastern coast (Little et al., 1973; Boardman and Carney, 1997). The hydraulic conductivity of the principal aquifer (Lucayan limestone) is estimated to range from 86 to $8640\,m\,day^{-1}$ based on short-duration, single-well specific capacity pumping tests conducted in the 1970s (Whitaker and Smart, 1997). The hydraulic gradient (ranging from 0.0005 to 0.001) was determined from historic field observations and estimates of the freshwater lens morphology (Little et al., 1973). Porosity ranges from 10 to 20 % (Bukowski et al., 1999). Sparse hydrogeological field data are available for the majority of the island; therefore, in the past, the morphology of the freshwater lens was largely inferred based on vegetation patterns, geological setting and anecdotal observations. Because Andros Island is composed of several small islands and cays, the freshwater lens is also composed of multiple lenses present on the different land masses. Lenses are anticipated to be present across most of the island, except in areas that are heavily intersected by saltwater marshes and wetlands.

In September 2004, Hurricane Frances caused a storm surge on the western coast of Andros Island, which resulted in extensive salinization of the North Andros Wellfield (Fig. 2). The hurricane ranged from a Category 4 to Category 2 on the Saffir Simpson Hurricane Scale while it travelled across the Bahamas from the southeast to northwest (Franklin et al., 2006). The surge occurred 3–4 September 2004, while Hurricane Frances passed near Andros Island. The exact time of occurrence of the storm surge and the actual extent of the overwash are unknown because the western coast of Andros Island is largely unpopulated. However,

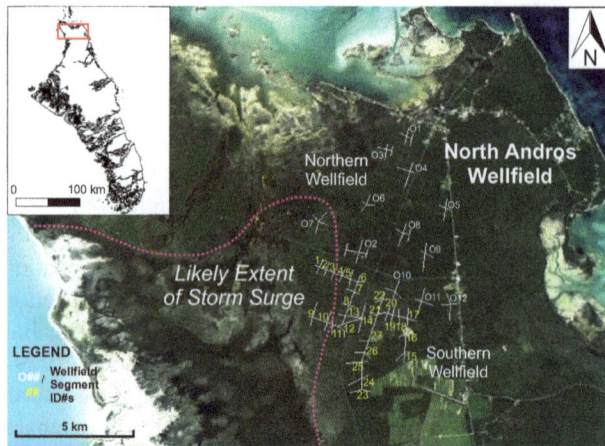

Figure 2. Layout of the North Andros Wellfield indicating the likely extent of the 2004 Hurricane Frances storm surge overwash.

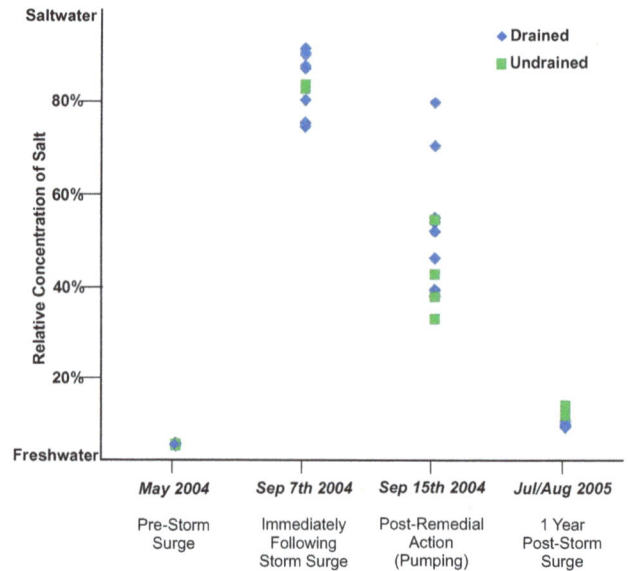

Figure 3. Salinity monitoring data before and after the 2004 Hurricane Frances storm surge. Data are shown for the southern trench segments of the North Andros Wellfield only (shown in yellow on Fig. 2).

after the hurricane had passed, evidence of the overwash was observed, such as flooded ground and the presence of marine fish at inland locations (Bowleg and Allen, 2011). The likely extent of the overwash is thus based on observations of damage following the surge (e.g., water marks on trees, presence of seaweed and marine organisms, etc.), and is shown in Fig. 2.

Salinity concentration data from the southern wellfield (Fig. 2) were provided from the water managers for the dates May 2004 (pre-storm), 7 September (immediately post-storm surge), 15 September (following remedial action) and July/August 2005 (approximately 1 year post-storm surge). These data are presented in Fig. 3, which illustrates the abrupt increase in salinity within the trenches following the storm surge and the eventual recovery to pre-storm concentrations. As a form of remedial action, the contaminated trenches were pumped to remove the ponded seawater beginning on 8 September (approximately 4 days following the storm surge). Salinity in the affected trenches improved, reducing by up to 88 % on 15 September, relative to the maximum recorded concentrations in each trench. However, remedial pumping of the trenches was not completed because freshwater was required to support post-hurricane relief efforts on other islands. Therefore, some of the contaminated trenches were closed off from the wellfield system to allow for extraction of freshwater from the unaffected parts of the freshwater lens and were not drained. Several of these contaminated trenches remained closed for 2 years due to poor water quality. Trenches that were drained are distinguished from those that were not in Fig. 3. The wellfield eventually recovered to normal salinity concentrations between 1 and 2 years post-storm, with all trenches recovered by 2009.

3   Methodology

3.1   SEAWAT model: long-acting stressors

3.1.1   Baseline model setup

A three-dimensional numerical density-dependent groundwater flow and solute transport model was developed using SEAWAT. The island was simulated using two separate models, a northern and southern model, to allow for refined grid resolution and a reasonable run time for each simulation. Each model was run for 100 years, during which time the freshwater lenses developed; both models reached steady state (i.e., no further change in lens morphology) within 20–25 years. Specified head boundaries were defined along the perimeter of the domain to simulate sea level, with density specified at $1.025 \, \mathrm{kg \, L^{-1}}$, representative of typical seawater composition. Specified concentration boundaries were assigned to the same grid cells as the specified head boundaries with concentrations of $35 \, \mathrm{g \, L^{-1}}$ salt. The initial concentration of the entire model domain was specified at $35 \, \mathrm{g \, L^{-1}}$ salt. The ground surface for the model was based on a digital elevation model for Andros Island (90 m resolution). The model grid was uniform in plan view, with each grid cell 500 m by 500 m. In the vertical dimension, the model included 44 layers, with individual layer thicknesses of 1 m in the upper 20 m of the model domains, which transitioned to 2.5, 5 and then 10 m thickness from a depth of 60 m to the base of the domain (200 m b.g.s.).

Hydraulic conductivity of the principal aquifer was based on field data (Little et al., 1973) and a sensitivity analysis was

Table 1. SEAWAT model parameters.

Parameter	Value
Model domain	200 m deep; lateral extent based on island area
Lucayan/pre-Lucayan interface	40 m b.g.s.
Paleosol depths	9–10 and 14–15 m b.g.s.
Hydraulic conductivity	Lucayan: $864 \, \text{m day}^{-1}$ – paleosols: $8640 \, \text{m day}^{-1}$ – pre-Lucayan: $86\,400 \, \text{m day}^{-1}$
Specific storage/specific yield	$1 \times 10^{-5} \, \text{m}^{-1}/0.15$
Effective porosity	0.15
Dispersivity	Longitudinal 1.0 m; transverse (vertical & horizontal) 0.1 m
Specified head boundary	0 m a.s.l. along model domain periphery; specified density $1.025 \, \text{kg L}^{-1}$
Concentration at specified head boundary	$35 \, \text{g L}^{-1}$ along model domain periphery
Initial concentration	$35 \, \text{g L}^{-1}$ throughout model domain
Recharge	$877 \, \text{mm yr}^{-1}$ (north) and $426 \, \text{mm yr}^{-1}$ (south); concentration $0 \, \text{g L}^{-1}$
Time steps	Initial: 14 min; maximum: 1 day

conducted to identify the optimal configuration and hydraulic properties of the layers to simulate the observed freshwater lens thickness on Andros Island. Previous studies had characterized the paleosols as low hydraulic conductivity layers (Ritzi et al., 2001); however, anecdotal evidence indicates that the layers are very weathered and may be highly conductive. In this study, the paleosols are represented by relatively high hydraulic conductivity layers (interbeds) within lower permeability limestone. This layer configuration with the assigned layer hydraulic properties is supported by model calibration. Assigning a low conductivity to the paleosols resulted in the lens being perched, which is not observed in the field. Whereas, representing the paleosols as high conductivity layers within lower conductivity limestone resulted in thin lenses being developed, similar to field observations. This approach is consistent with other studies based in the Bahamas, which have suggested that layers of high hydraulic conductivity in the subsurface are responsible for thin freshwater lenses (Wallis et al., 1991). The optimal configuration of aquifer layers and hydraulic conductivities are provided in Table 1.

Recharge was applied to the top layer of the model with concentration of $0 \, \text{g L}^{-1}$ salt to simulate the average annual recharge to the aquifer. Recharge is the only input of freshwater to the hydrogeological system and, therefore, is the main mechanism by which the simulated freshwater lens develops in the model. The annual recharge amount for Andros Island was estimated using the United States Environmental Protection Agency's software HELP (Hydrologic Evaluation of Landfill Performance) (Schroeder et al., 1994). HELP utilizes a storage routing technique based on hydrological water balance principles. It accounts for soil moisture storage, runoff, interception, and evapotranspiration. HELP has been used to estimate recharge for a variety of climatic and physiographic settings (Scibek and Allen, 2006; Jyrkama and Sykes, 2007; Toews and Allen, 2009; Allen et al., 2010).

Within HELP, a representative vertical percolation profile was defined for the unsaturated zone. The depth of the profile was 2 m, based on a sensitivity analysis using the minimum and maximum observed depths to the water table on Andros Island. No soil zone was specified due to the generally thin/absent soils on Andros Island (Little et al., 1973). The lithology was homogeneous (representing limestone), with a saturated hydraulic conductivity ($864 \, \text{m day}^{-1}$) based on the mean value from field studies (Little et al., 1973) and the calibrated value from the baseline SEAWAT model. Vegetation cover was assigned to the highest class in the software (a leaf area index of 5) based on the large proportion of pine forests. The surface was assigned zero slope given that minimal runoff is observed. The wilting point was assigned 0.05 and field capacity 0.1 in the absence of measured values.

Two 100 year climate data series were generated using the embedded stochastic weather generator; one for North Andros and one for South Andros because the historical climate differs between the two regions. The average annual precipitation on North Andros is $1442 \, \text{mm yr}^{-1}$, while on South An-

Table 2. Projected climate shifts for the 2090s, and the resulting projected values for monthly mean temperature and monthly mean precipitation for North and South Andros.

Parameter	D	J	F	M	A	M	J	J	A	S	O	N
Temperature shift (°C)		+2.8			+3.0			+3.2			+3.2	
Projected monthly mean temperature (°C) North/south	25.2	24.3	24.6	25.3	26.8	28.5	30.4	31.3	31.3	30.9	32.7	27.7
Precipitation shift (mm)		−2			−18			−24			+12	
Projected monthly mean precipitation (mm) North	45	48	50	47	66	90	189	138	210	190	176	98
Projected monthly mean precipitation (mm) South	51	34	37	24	27	89	81	40	57	112	138	103

dros, it is $889 \, mm \, yr^{-1}$. Temperature averages were not available for South Andros; therefore, the monthly averages for North Andros were applied to both models. Other climate parameters (e.g., windspeed and relative humidity) were identical for both models. The historical statistical parameters for climate were based on values for the nearest climate station (Miami, Florida, USA) in the weather generator database.

The average annual recharge for the north was estimated at $877 \, mm \, yr^{-1}$, with a minimum monthly average of $24 \, mm \, month^{-1}$ in December and a maximum monthly average of $163 \, mm \, month^{-1}$ in August. The average annual recharge for the south was estimated at $426 \, mm \, yr^{-1}$, with a minimum monthly average of $17 \, mm \, month^{-1}$ in February and a maximum monthly average of $70 \, mm \, month^{-1}$ in October. These values were used as input for the northern and southern SEAWAT models, respectively.

The hydrogeological parameters assigned to the SEAWAT model, based on field data and sensitivity analyses, are summarized in Table 1. Storage parameters were based on common values for the aquifer lithology (Younger, 1993). The wellfields were not simulated in the baseline model in order to represent natural historical conditions. Given their small size, the wellfields are not anticipated to affect the freshwater lens response. If the system were head-controlled, however, on a local scale a rise in water table could result in more loss of freshwater from the top of the lens.

3.1.2 Climate change simulations

Future climate for this study was based on published climate change projections for the Bahamas (UNDP – United Nations Development Programme, 2010). The projections were derived from 15 global climate models (GCMs) simulating three emissions scenarios (SRES A2, A1B, and B1). Summaries of projected changes were compiled as seasonal shifts for 3 month groupings (McSweeney et al., 2010). For each grouping, a range in values (minimum, median, and max-

imum) for each emissions scenario were provided for the 2030s, 2060s and 2090s. The median seasonal shift in temperature and precipitation projected for the 2090s for the A2 scenario (expected to result in the greatest change) was selected for Andros Island, as summarized in Table 2. Average daily temperature for the 2090s is projected to increase during all seasons (between 2.8 and 3.2 °C). Changes to precipitation are projected to occur primarily during the summer (up to 42 % reduction relative to current conditions). Overall, the projected climate shifts represent conditions with less precipitation and higher temperatures – a drier and hotter climate state.

Changes to groundwater recharge were determined by re-modelling recharge in HELP using the projected 2090s climate. The seasonal climate shifts (applied evenly to each month according to season) were applied to the monthly normals for temperature and precipitation in the weather generator, and a new stochastic weather series was generated to represent the projected future climate. This approach is consistent with that used in other studies (e.g., Scibek and Allen, 2006). The adjusted climate data series was then used as input to the vertical percolation profile to determine the annual average groundwater recharge expected under projected climate change. As in the baseline recharge modelling, recharge estimates were produced for North and South Andros, and these values were applied to the SEAWAT models for each region, respectively. The predicted average annual recharge for the north was $777 \, mm \, yr^{-1}$, with a minimum monthly average of $18 \, mm \, month^{-1}$ in March and a maximum monthly average of $130 \, mm \, month^{-1}$ in August. The predicted average annual recharge for the south was $360 \, mm \, yr^{-1}$, with a minimum monthly average of $4 \, mm \, month^{-1}$ in July and a maximum monthly average of $82 \, mm \, month^{-1}$ in November.

Sea level rise was simulated by increasing the elevation of the specified head boundaries in the model domain. Loss of land surface due to inundation associated with sea level rise was not simulated, as the grid resolution of the model is larger than the inundation anticipated based on ground surface elevation. Therefore, the boundaries at the edge of the model domain are anticipated to remain at the same model grid cell, only representing a higher specified head value. Although sea level rise has been already observed over the last several decades (White et al., 2005), there is uncertainty as to the rate that it will occur in the future (Rahmstorf, 2007). Geographic variability in the rates of sea level rise is also expected (White et al., 2005). Therefore, a predicted mean sea level increase of 0.6 m by the 2090s (relative to 1980) was selected as an average estimate based on global and regional projections of sea level rise (IPCC, 2007; Rahmstorf, 2007; Obeysekera, 2013). The hydrogeological system of Andros Island is considered recharge-limited rather than topography-limited, because there is some capacity for the freshwater lens to rise in the unsaturated zone without leading to surface flooding (Werner and Simmons, 2009).

Both the reduction in recharge and sea level rise were simulated in the models as incremental instantaneous shifts. Three models were run: one for recharge reduction alone, one for sea level rise alone, and one including both stressors. The baseline model was run for 50 years to allow the freshwater lens to develop. The recharge and specified head boundary values were then adjusted every 10 years until reaching the projected values for the 2090s. This assumes uniform rates of change throughout the 100 year simulation.

Observation wells were defined in the models to capture a discrete record of simulated concentration for every time step. The observation wells were located within the center and at the edge of the freshwater lens to represent areas that are anticipated to be, respectively, most resilient and most vulnerable to stressors. The northern model consists of one landmass and, therefore, one principal lens, whereas the southern model consists of multiple landmasses. As discussed below, two principal lenses form in the southern model. Therefore, two observation wells were assigned in the northern model and four observation wells were assigned in the southern model, representing central and peripheral wells for each anticipated freshwater lens. The wells are identified as A and B to distinguish between the two principal lenses in the southern model. Each well was screened from the ground surface to 5 m b.g.s., corresponding to the maximum depth of most wells/wellfields on Andros Island.

In order to evaluate changes to the freshwater lens morphology in response to climate change, the SEAWAT model island-scale results were quantitatively evaluated using a geographic information system (GIS). The volume and area of the lens were calculated based on a threshold salt concentration 0.4 g L^{-1} or less (representing local potable water guidelines) and porosity. Although there are inaccuracies inherent in this approach, it provides an estimate of the lens morphology that allows for quantitative comparison of the changes in freshwater lens morphology between different stressors applied in the island-scale model. This threshold concentration is based on the water quality guidelines for salinity in the

municipal supply on Andros Island. It also falls within common definitions of freshwater containing less than 1.0 g L^{-1} of total dissolved solids (Freeze and Cherry, 1977; Barlow, 2003). The World Health Organisation (WHO) drinking-water guidelines do not stipulate a maximum threshold for salt in water, except as it relates to unacceptable taste. The WHO recognizes that water that tastes fresh often has a salt concentration of less than 0.25 g L^{-1}; however, in regions where there is naturally more salt in the water, there may be a higher taste threshold (WHO, 2011).

3.2 HydroGeoSphere model: short-acting stressor

Modelling the impact of storm surge overwash on a hydrogeological system involves simulating density-dependent flow and solute transport across the surface, the vadose zone and the saturated domain. HydroGeoSphere (HGS) was identified as the most suitable tool to simulate these coupled processes because it is a fully integrated surface and variably saturated subsurface model that is capable of simulating these processes across all domains. By solving the surface and subsurface flow equations simultaneously, HGS provides more realistic representations of the major processes than simpler or independently coupled models (Goderniaux et al., 2009).

One of the mechanisms of aquifer contamination following storm surge is from open wells or trenches that provide direct access to the water table and collect the salt water during inundation (Terry and Falkland, 2010). In addition, salt water trapped within a borehole, or other direct pathway into the aquifer, may lead to prolonged release of salt water into the surrounding aquifer over time (Illangasekare et al., 2006). These features may delay recovery of the aquifer and, therefore, are an important component to include in modelling studies of storm surge impacts (Chui and Terry, 2013). Major consequences to water supply are likely to result when storm surge waves strike trench-based wellfields or open boreholes, as occurred on Andros Island in 2004. Notwithstanding this risk, trench-based wellfields are commonly used on low-lying islands to limit upconing. The models developed for this study aim to characterise aquifer damage and recovery from a storm surge overwash, specifically in the context of a trench-based wellfield and the impact on water supply.

The model domain represents a highly discretized, two-dimensional cross-section of one of the trenches in the North Andros Wellfield (Fig. 4). The size of the model domain had to be made as small as possible for computational reasons. Therefore, several different model configurations were tested by varying the model domain width and the hydraulic conductivity distribution (limestone and paleosols) to identify the optimal combination of parameters that best approximates observed conditions. The physically based seawater boundaries are important components in simulating flow within a freshwater lens. In reality, these boundaries are located along the coastline; however, the coastline is far from the North Andros Wellfield. Therefore, local-scale models

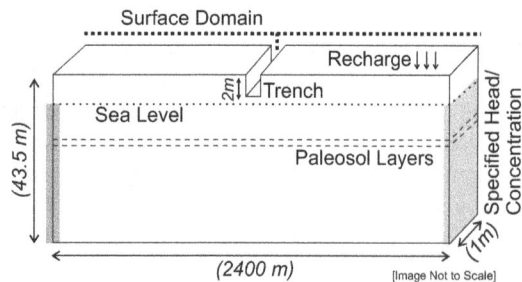

Figure 4. HydroGeoSphere model domain and boundary conditions.

were developed using boundary conditions assigned in such a way as to simulate a realistic flow field surrounding the trench. The local-scale models were calibrated based on critical factors that are expected to affect freshwater lens contamination and recovery. These critical factors include recharge, thickness of the vadose zone, aquifer hydraulic conductivity, geological heterogeneity (e.g., paleosols), water table gradient, and thickness of the freshwater lens. Field data for each of these factors (as presented earlier) comprise the calibration criteria as summarized in Table 3.

With increasing model domain width, the elevation of the water table and gradient both increase, whereas the thickness of the lens decreases. The opposite response was observed when hydraulic conductivity was increased. The model setup that satisfied the calibration criteria with the smallest domain width was selected as the baseline model for this study (Fig. 4).

The model uses block elements that range from 0.35 to 1.0 m. Grid refinement was done in order to optimise simulation of flow and transport across the three hydrologic domains and to allow for the evaluation of small-scale changes in response to overwash. The model domain covers a horizontal extent of 2400 m and a vertical extent of 43.5 m, with sea level assumed to be 3.5 m b.g.s. The vertical extent of the domain was determined to represent the Lucayan limestone. The model domain was 1 unit thickness, with a uniform horizontal grid spacing of 1 m. Vertical grid refinement varied from 1 m thick in the lower 20 m, to 0.5 m thick in the overlying 20 m, and 0.35 m thick in the uppermost 3.5 m. Paleosols were simulated in the subsurface as 1 m thick zones at 9 and 14 m b.g.s. (corresponding to field observations). The hydraulic conductivity was defined as isotropic at $86.5 \, \text{m day}^{-1}$ for the portion of the domain representing the Lucayan limestone and $865 \, \text{m day}^{-1}$ for the paleosols. The hydraulic conductivities lie within the observed range, although they are lower than that used in the SEAWAT model in order to calibrate the freshwater lens morphology on the local scale surrounding the trench. The underlying high conductivity pre-Lucayan limestone was not included in the model, as it was observed not to have a significant impact on the freshwater lens morphology on the scale of the model.

Table 3. Observed conditions used for calibrating the HydroGeoSphere model.

Parameter	Value
Vadose zone thickness	1.5–2 m
Water table elevation	2 m a.s.l.
Gradient	0.0005–0.001
Average velocity	$0.3 \, \text{m day}^{-1}$
Thickness of lens	15–20 m

The trench itself extends 2 m b.g.s., intersecting the top of the water table. Most trench-based wellfields rely on gravity flow; therefore, water tends to move very slowly within the trenches and is observed to be almost stagnant unless the trench is actively being drained. Therefore, lateral flow within the trench was assumed to have a negligible impact on the storm surge impact and recovery of the aquifer. The model provides a snapshot of the impact of trench-based wellfields in terms of salt water capture and transport into the aquifer, which may be scaled up to represent the whole wellfield.

Specified head with associated concentration boundaries were assigned to both sides of the model to represent the surrounding seawater (Fig. 4). Recharge was applied to the surface domain as an annual average quantity based on the HELP recharge modelling, presented earlier. Recharge provides the only input of freshwater that enables the freshwater lens to develop. The boundary conditions and hydrogeological parameters assigned to the HGS model are summarized in Table 4.

The simulation of storm surge overwash required three separate modelling phases: (1) development of the freshwater lens to steady state conditions; (2) short temporal-scale modelling of the rise in salt water height accompanying the overwash; and (3) recovery of the freshwater lens. The heads and concentrations at the end of each phase are used as initial conditions for the subsequent phases; however, the boundary conditions are changed to reflect the different scenarios. The three phases are required to accommodate the different temporal scales (i.e., decades for lens development and minutes for storm surge occurrence) as well as to assign the time-varying boundary conditions. All model simulations used the same initial steady state freshwater lens (Phase 1) and simulation of the storm surge overwash (Phase 2). Different scenarios of remedial action were simulated for Phase 3 and compared to the baseline recovery scenario.

3.2.1 Phase 1: freshwater lens development

Phase 1 is a model spin-up period during which the freshwater lens develops. The initial concentration in the baseline model domain was salty ($35 \, \text{g L}^{-1}$), with the only source of freshwater being recharge. The model was run for 50 years to reach steady state.

Table 4. HydroGeoSphere model parameters.

Parameter	Value
Model domain	2400 m model domain width; 43.5 m domain depth (representing Lucayan limestone)
Paleosol depths	9–10 and 14–15 m b.g.s.
Trench dimensions	1 m wide, 2 m deep
Hydraulic conductivity	Lucayan limestone: 86.4 m day^{-1}; paleosols: 864 m day^{-1}
Effective porosity	0.15
Specific storage	1×10^{-5} m^{-1}
Dispersivity	Longitudinal 1.0 m; transverse horizontal 0.1 m; transverse vertical 0.01 m
Specified head boundary	0 m a.s.l. along model domain periphery; specified density 1.025 kg L^{-1}
Concentration at specified head boundary	35 g L^{-1} along model domain periphery
Initial concentration	35 g L^{-1} throughout model domain
Recharge	877 mm yr^{-1}; concentration 0 g L^{-1}
Time steps	Initial time step: 0.8 s Maximum time step: 1 day

3.2.2 Phase 2: storm surge inundation

Phase 2 simulates the occurrence of a storm surge overwash event. The surface domain was inundated with up to 1 m of water, based on observations following the 2004 storm surge on Andros Island. Flooding was simulated at a gradual rate of 0.1 m per 10 min stress period to satisfy model convergence criteria. Once full inundation was reached (1.5 h after start of flooding), the maximum flood level was held constant for 2 h. The actual period of inundation is not known, so this period was estimated to allow for sufficient salt water to enter the system. The salt concentration of the flood water was assigned as 35 g L^{-1} to represent seawater.

3.2.3 Phase 3: recovery of the freshwater lens

Phase 3 involved simulating the recovery of the freshwater lens. Several different scenarios were tested to enable comparison of recovery times when different factors are varied. All scenarios are based on the output from Phase 2, with the head and concentration boundaries of the surface domain unconstrained to allow release of the salty flood water. All other boundaries remained the same as the initial Phase 1 model.

A baseline recovery scenario was simulated for 10 years following the storm surge to allow the salt water to be flushed out of the system under the influence of recharge. In the baseline recovery model, the freshwater lens returns naturally to its original morphology.

Several other scenarios were simulated to represent different remedial actions. Following a storm surge event when the trenches are filled with salt water, a common remedial action is to drain out the trenches to remove the captured salt water (Illangasekare et al., 2006; Terry and Falkand, 2010; Chui and Terry, 2012). Draining, or pumping out the trenches, is meant to improve the recovery time and assist with removal of the salt water from the system. However, draining may often be delayed due to access constraints or due to lack of coordination and emergency response following the storm surge. Therefore, models were developed where the trenches are drained at different times and for different durations to evaluate the impact of draining protocol on recovery times of the freshwater lens and impact to water supply. Scenarios were modelled whereby draining was delayed by 1, 2, 3, or 4 days after the storm surge (to reflect a delay in action). Other scenarios modelled draining initiated 1 day after the storm surge, whereby the duration of draining was 1, 2, or 3 days (to investigate the effect of sustained periods of draining).

For all recovery simulations, observation points were assigned within and immediately below the trench to monitor salt concentrations during recovery. This allowed for the comparison of recovery times between different scenarios, specifically the number of days for potable water to return to the trench and aquifer.

4 Results

4.1 Long-acting stressors

4.1.1 Baseline model

The simulated freshwater lens in the baseline model provides a snapshot of the average annual freshwater lens morphology. The model results indicate that a lens is present throughout most of the model domain (not shown); however, this study focuses on areas considered viable to provide a sustainable water supply, which are defined as having a lens thickness of greater than 2 m and a concentration of less than 0.4 g L^{-1} (Fig. 5). The shape of the lens is relatively symmetrical in cross-section with an average hydraulic head of 1.8 m a.s.l., which corresponds to typical elevations observed of 2 m a.s.l. The estimated total area of the viable freshwater lens on Andros Island is around 2000 km^2 with a freshwater volume of 5.9×10^9 m^3.

The baseline model was calibrated to observations, where available, although these were sparse and based on varying time periods (from the 1970s to early 2000s). The extent of the lens generally corresponds to observations of freshwater occurrence (i.e., the presence of wells and wellfields) and the results of previous studies (Little et al., 1973; Cant and Weech, 1986; Wolfe et al., 2001). The freshwater lens in the northern model is composed of a single lens that is much larger than the smaller, separate lenses in the southern model. Along the coastlines, particularly in the southern regions of the island, the simulated freshwater lens tends to be situated further inland than is observed; however, the depth

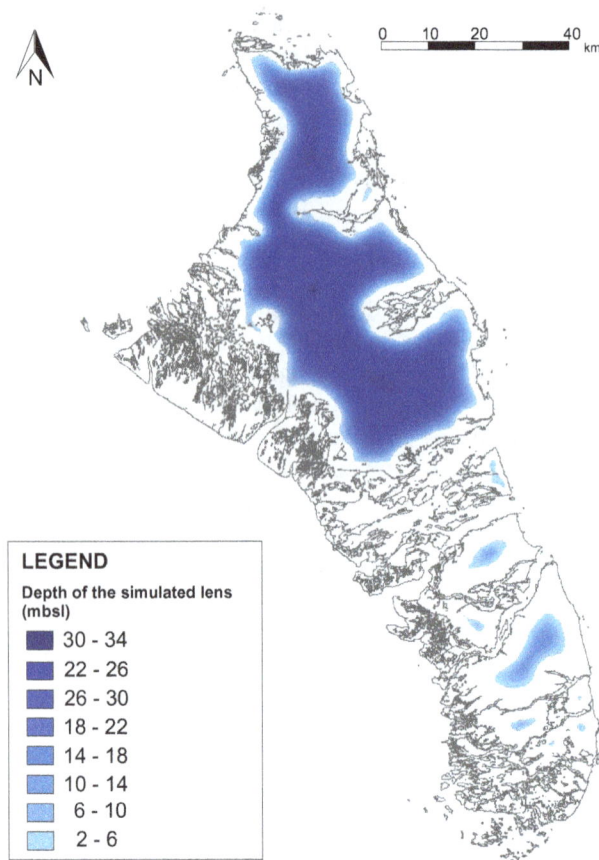

Figure 5. Baseline freshwater lens representing current conditions.

Table 5. Percent change in freshwater lens morphology relative to the baseline model for the combined effect of reduced recharge and sea level rise.

Modelled region	% change area	% change volume
Northern	−4.1	−5.9
Southern	−16.8	−24.2

tion. Although the worst case scenario (e.g., lowest recharge during the dry season) is not accounted for in this study, other studies have shown that there is little seasonal variation in groundwater levels for islands of similar hydrogeological settings (Momi et al., 2005). Overall, the simulated lens is within the range of observed depths, although it represents a slight over-estimation of the freshwater resources in the northern region of Andros Island. The model provides a generalized estimate of the freshwater lens morphology and serves as a reasonable baseline for investigating the impacts due to climate change stressors.

4.1.2 Climate change models

As noted above, the HELP model utilizes site-specific climate averages so that predictions can be made regarding the impact of future climate conditions on recharge. Recharge is projected to decrease by 11 % in the northern model and decrease by 15 % in the southern model by the 2090s relative to baseline (current) recharge. This is due largely to decreases in average annual precipitation, and slight increases in evapotranspiration rates. Minimal changes in soil storage were simulated in the HELP model.

The results of the climate change modelling, including a reduction in recharge and a rise in sea level, indicate that the freshwater lens will reduce in areal extent and volume under future climate change conditions. The percent change in freshwater lens area and volume relative to the baseline values are presented in Table 5. The change in area and volume of the lens indicate that the lens shrinks and thins in response to the stressors. For both the northern and southern models, simulations of reduced recharge alone result in the majority of freshwater lens reduction, with sea level rise contributing a smaller proportion of lens reduction. The freshwater lens in the southern model is predicted to incur a greater percentage of loss of lens compared to the northern model under climate change conditions. In the southern model, the results indicate a 19 % volume loss due to reduced recharge compared to 5 % volume loss due to sea level rise relative to baseline morphology. Whereas, in the northern model, 5 % of volume loss is due to reduced recharge with 0.9 % volume loss due to sea level rise. The simulated lens at the end of the 100 year simulation is presented, illustrating areal loss of lens relative to the baseline model (Fig. 6).

of the simulated lens in the south is consistent with field observations. The depth of the simulated lens in the northern regions of Andros Island falls within the range of maximum observed lens depth (up to 39 m b.g.s.), although it is slightly deeper than typical observations of around 15 to 20 m b.g.s. Because most of the model parameters are based on field data and sensitivity analyses, the deeper simulated lens is likely the result of slight over-estimation of recharge in the HELP model. HELP applies daily precipitation to the lithology profile evenly over a 24 h period, when in reality, precipitation events occur within shorter time intervals (hourly) and leads to some pooled water on the ground surface. Given that the intensity of the precipitation events is not accounted for in HELP, the resulting recharge estimates may be slightly overestimated. However, there is no clear basis upon which the recharge estimates can be adjusted to achieve better model calibration, due to the lack of field data for actual evapotranspiration and recharge.

Some local-scale variations are neglected in the model due to the limitations of the large grid cell size required to cover the area of the island, which resulted in a low resolution of the ground surface elevation. In addition, the model was developed to represent the average annual freshwater lens morphology and, therefore, does not include seasonal varia-

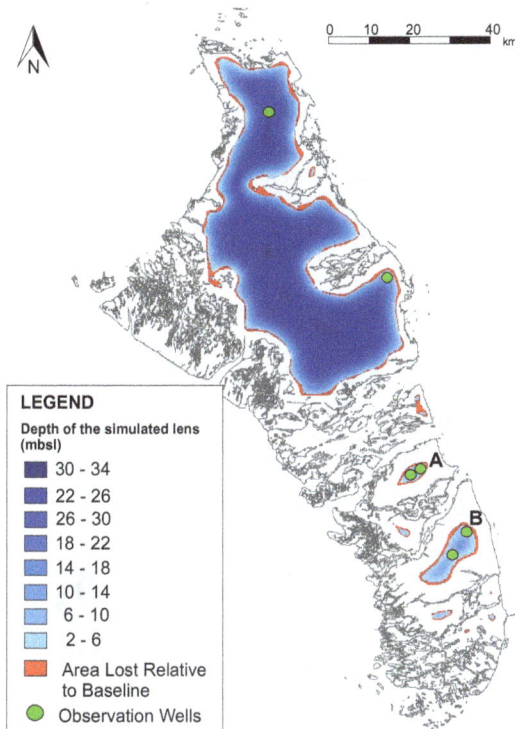

Figure 6. Model result for climate change simulations for the combined effect of reduced recharge and sea level rise, indicating area lost relative to baseline conditions.

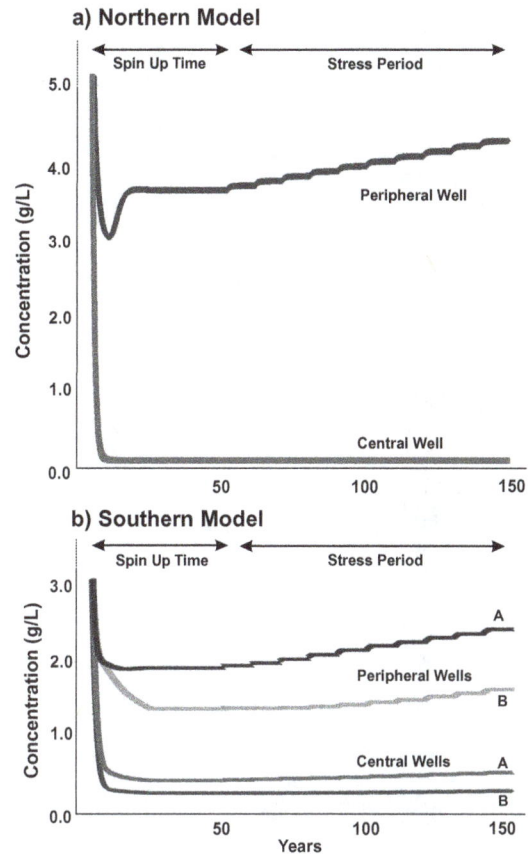

Figure 7. Simulated dissolved salt concentrations over time at the observation wells for climate change models. (**a**) Northern model, (**b**) southern model with two observation wells for each landmass shown.

The simulated time-varying dissolved salt concentrations in the observation wells are shown in Fig. 7. The simulated concentrations at most observation wells indicate that salinity in the lens progressively increases in response to the climate change shifts applied every 10 years starting at 50 years. Prior to 50 years, the model is spinning up from a fully salty state. Dissolved salt concentrations in all of the observation wells reach near steady state between stress periods (only very small changes continue to occur on the order of $10^{-10}\,\mathrm{g\,L^{-1}\,day^{-1}}$). The time to reach steady concentrations is relatively similar in all wells, ranging from 0.5 to 3 years and increasing as the simulation progresses. This indicates that even though the climate change shifts in each stress period are the same magnitude, the freshwater lens takes longer to adjust to the shifts as the cumulative magnitude of climate change increases.

The central wells were placed in areas that were anticipated to be more resilient to stressors, and the peripheral wells in areas that were anticipated to be more vulnerable to stressors (thereby showing a more immediate lens thinning). The simulation results are consistent with the anticipated behaviour. The peripheral observation wells have higher dissolved salt concentrations than the central wells because they are situated in the thinner part of the freshwater lens, and therefore, are more likely to intersect the base of the lens. The highest dissolved salt concentrations are in the periph-

eral well in the northern model, which is closer to the coast than the peripheral wells in the southern model. This is because the edge of the northern freshwater lens extends further coastward than the southern freshwater lens. Greater changes in dissolved salt concentration are also observed in the peripheral wells compared to the central wells, as would be expected.

4.2   Short-acting stressor

4.2.1   Freshwater lens development and storm surge inundation

The morphology of the freshwater lens reaches steady state within 25 years at a maximum depth of 23 m b.s.l. (metres below sea level) (Phase 1; Fig. 8a). The model is calibrated to observed conditions outlined in Table 3. The maximum elevation of the freshwater lens is observed in the trench at 1.8 m a.s.l. The vadose zone surrounding the trench is approximately 1.7 m thick. The gradient across the model domain is 0.0015, with an average horizontal groundwater velocity of 0.87 m day^{-1}. The inflections on the sides of the

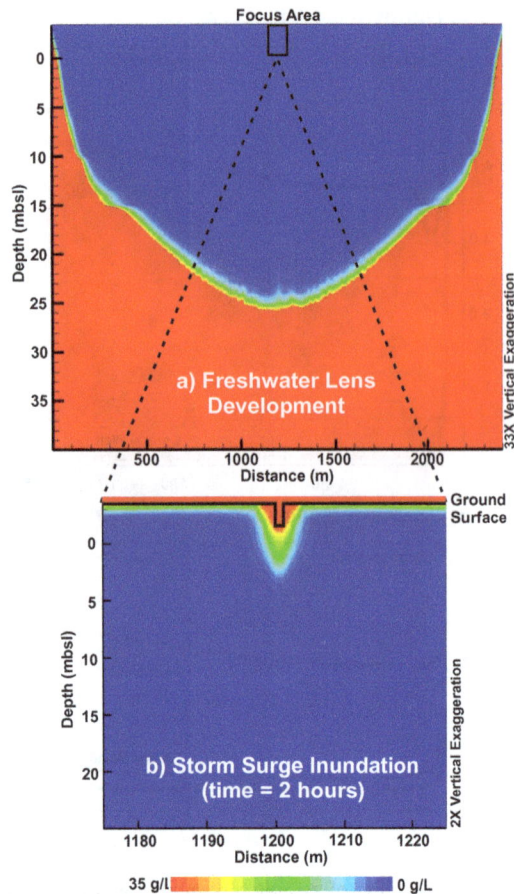

Figure 8. (**a**) Freshwater lens development after 50 years (Phase 1); (**b**) storm surge inundation in the focus area at 2 h (Phase 2).

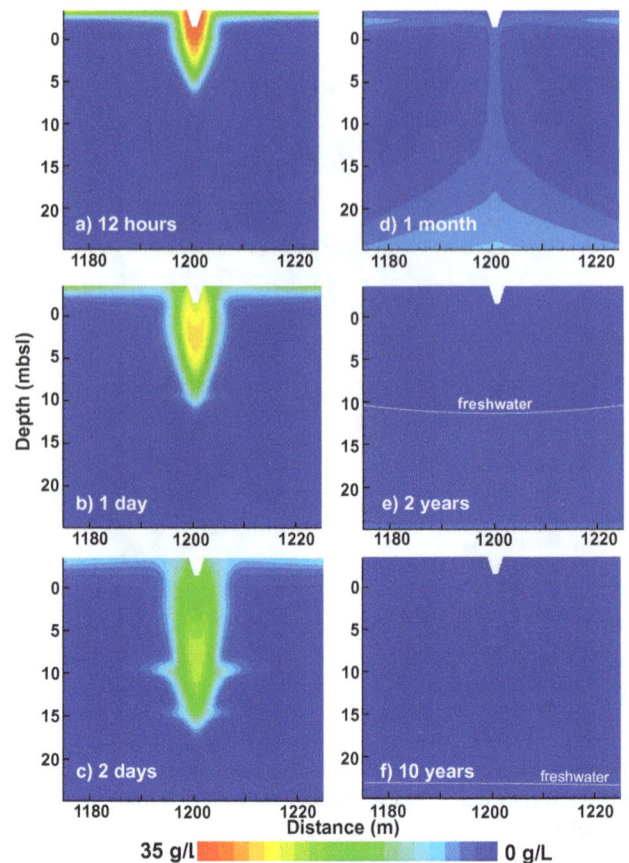

Figure 9. Baseline recovery of the freshwater lens post-storm surge at (**a**) 12 h, (**b**) 1 day, (**c**) 2 days, (**d**) 1 month, (**e**) 2 years, and (**f**) 10 years.

lens at 9 and 14 m b.g.s. reflect the high hydraulic conductivity paleosol layers.

Simulation of storm surge inundation (Phase 2) resulted in high salt concentrations at the surface of the model up to 1 m above ground surface (Fig. 8b). The results of the inundation model are shown for a focus area within 25 m of the trench (focus area indicated in Fig. 8a). Within the 2 h inundation period, the salt water had already been transported into the vadose zone due to the hydraulic gradient associated with the surface flood, and had also filled the trench with salt water (Fig. 8b).

4.2.2 Aquifer recovery

The baseline recovery of the freshwater lens (natural recovery) is shown for six times post-storm surge (Fig. 9): 12 h, 1 day, 2 days, 1 month, 2 years, and 10 years. The baseline recovery scenario indicates that the freshwater lens returns to its original morphology approximately 10 years post-storm surge. The salt water is transported from the surface domain into the aquifer system, where it forms a salt plume within the subsurface. This plume is flushed out over time due to

the infiltrating freshwater recharge. Salt concentration within the trench returns to levels below the potable water threshold within 149 days following the storm surge for the baseline recovery scenario.

The results of the different draining scenarios are shown in Fig. 10, alongside the baseline recovery scenario, as relative concentration data over time, where 1.0 represents salt water and 0.0 represents freshwater. The number of days to reach potable concentration in the trenches is indicated for each scenario. Observed concentration data for the North Andros Wellfield trenches are also presented in Fig. 10. Trenches that were drained following the storm surge, and those that were isolated from the system and not drained, are distinguished by different symbols.

There is little difference in observed concentrations when comparing the trenches that were drained and those that were not. The observed concentrations are similar to the simulated concentrations immediately following the overwash event; however, at 1 year post-storm surge, the observed concentrations are slightly above the potable water threshold. By 2 years post-storm surge (not shown), the observed concentra-

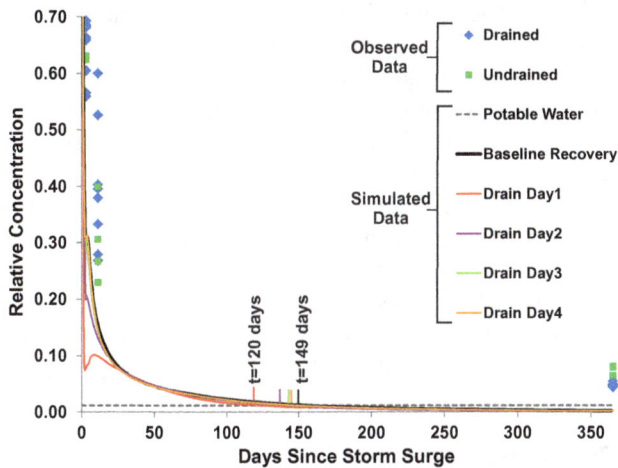

Figure 10. Observed and simulated concentrations within the trench. The times for concentrations to reach potable water threshold are indicated by the small vertical bars for the baseline recovery scenario (149 days) and the various scenarios of draining on different days following the surge (120 days for draining on day 1). The increase in concentration observed for Scenario Drain Day 1 represents the end of the draining period, when high concentration water re-enters the trench from the surrounding aquifer and vadose zone.

tions are similar to the simulated concentrations, and below the potable threshold.

Draining of the trenches generally results in a faster recovery. If draining occurs within 1 day of the storm surge, potable water returns to the trench by about 120 days (Fig. 10), approximately 1 month sooner compared to the baseline recovery simulation (149 days). With every day that draining is delayed, it takes longer for potable water to return to the trench (corresponding to the small vertical lines in Fig. 10 for each scenario). After a delay of 3 days, the recovery time for potable water to return is the same as the case when no remedial action is undertaken. Therefore, the improvements in recovery time are dependent on the timing of draining.

In contrast, the duration of draining (not shown) does not significantly improve recovery times. Draining that occurs for multiple days results in slightly longer times for potable water to return compared to short-duration draining (i.e., over a single day).

As mentioned earlier, the observed data at 1 year post-storm surge are higher than the model results, indicating that the trenches on Andros Island recovered slower than the model results. This is likely the result of several factors:

1. The amount of salt water entering the aquifer system largely depends on the time of inundation. As this was unknown, it was assumed to be a 2 h inundation. However, the inundation may have lasted much longer, as no observations of the area were made until 3 days after the storm surge. To account for this uncertainty, a Phase 2 model was run with a longer inundation period

of 2 days. The recovery from this storm surge scenario took at least 2 months longer, with higher concentrations at 1 year post-storm surge. However, the freshwater lens morphology recovered at the same time as the baseline scenario.

2. The amount of recharge that specifically occurred on Andros Island may have been different during 2004–2005. Alternate recovery simulations were run where recharge was applied as monthly averages based on the 2004 and 2005 rainfall data for Andros Island. These simulations resulted in longer recovery times, up to 6 weeks more than baseline recovery.

3. As previously discussed, the HELP recharge results may over-estimate actual recharge to the freshwater lens. Therefore, additional models were run with recharge applied at half the baseline amount. These simulations indicated that recovery was delayed by 2 months.

4. Additional factors may impact the calibration to observed data. The models were developed based on field studies that were not all specific to the North Andros Wellfield area; therefore, hydrogeological conditions (such as porosity or hydraulic conductivity) at the wellfield may differ from those on the island scale.

5. The exact timing, duration and method of draining utilised on Andros Island are also unclear. While the best possible information was obtained from the Bahamas Water and Sewerage Corporation, it is likely that the details of operations were inexact.

6. Lastly, other hurricanes passed near to Andros Island in the weeks and months following Hurricane Frances; however, it is unknown whether any of these caused an additional storm surge event (NOAA – National Hurricane Center, National Oceanic and Atmospheric Administration, 2014). Regardless, the close passage of other storms would have attributed to atypical rainfall events. In addition, the concentration of recharging freshwater may be higher than $0\,g\,L^{-1}$ during storms due to salt spray, thereby introducing higher salt concentrations at the surface and delaying recovery.

Although many factors contribute to the uncertainty in the calibration, the recovery models are likely reasonable representations that allow for comparison of the impact of remedial actions on recovery.

5　Discussion

5.1　Long-acting stressors

The volume and area of the freshwater lens are reduced under stressed conditions, indicating that the lens both shrinks and thins. A significant impact is observed in areas where the lens shrinks (i.e., along the periphery), as most settlements and the related infrastructure are typically near the coast on small islands (Ranjan et al., 2009; Cashman et al., 2010). As a result, any changes in the freshwater lens morphology within the coastal zone may affect access and availability of freshwater near the population centres.

The loss of freshwater lens area and extent under climate change conditions is attributed more to the impact of changes to groundwater recharge than the impact of sea level rise. Although loss of land surface due to sea level rise was not simulated in the models, estimates based on ground surface elevation suggest loss of land surface (and resulting loss of freshwater lens volume) is limited. On islands with lower topography and/or smaller land area, inundation would have a greater effect on loss of freshwater lens volume. The model results for Andros Island are supported by other studies, which show that conditions of reduced groundwater recharge (or prolonged drought, which results in reduced recharge) disturb the balance of freshwater outflow necessary to maintain the extent and thickness of the freshwater lens, thereby leading to loss of freshwater resources due to saltwater intrusion (Ranjan et al., 2009; White and Falkland, 2010; Mollema and Antonellini, 2013). In addition, the hydrogeological system on Andros Island is recharge-limited, meaning that the freshwater lens is able to rise in the subsurface in response to sea level rise. Therefore, it is less vulnerable to sea level rise because the freshwater lens is able to maintain a balance between the hydraulic gradient of the fresh and salt water (Michael et al., 2013). This assumption is only valid to a point; for higher magnitudes of sea level rise, the freshwater lens would likely become topographically limited and, therefore, have a larger response (i.e., loss of lens) due to sea level rise. Although sea level rise appears not to be a significant factor for saltwater intrusion on Andros Island, it may increase the island's vulnerability to other events, such as extreme high tides and storm surge overwash. These events have the potential to result in significant impacts to the freshwater lens, as is discussed below.

The northern regions of Andros Island appear to be more resilient to climate change stressors than the southern regions. Several factors contribute to the difference in response between the northern and southern regions: (1) the south is composed of smaller landmasses, resulting in smaller areas for the freshwater lenses to develop; (2) significantly less rainfall occurs in the south, meaning that there is less recharge to sustain the freshwater lenses; and (3) lower recharge results in a thinner lens developing, leading to lower hydraulic gradient of the freshwater lens. The combined impact of these factors is that the southern region of Andros Island has smaller freshwater lenses that are more vulnerable to damage from stressors.

The simulated freshwater lens on Andros takes longer to respond to climate change stressors as the magnitude of the cumulative stress increases (i.e., lower recharge and higher

sea level). The implication is that as climate change progresses over time, the ability of the freshwater lens to respond to these changes decreases. Because recharge is the main driver of lens formation and maintenance, when the rate of recharge is reduced, the response time of the hydrogeological system is also reduced. This has been observed in laboratory experiments (Stoeckl and Houben, 2012) whereby the lens takes longer to reach steady state when there is reduced input (i.e., specified flux or concentration boundaries) to the system. Therefore, areas where there is less recharge, such as the southern regions of Andros Island, are expected to take longer to react and adapt to stresses to the hydrogeological system.

5.2 Short-acting stressor

Trench-based wellfields result in large salt plumes that develop in the aquifer following a storm surge overwash. This is because the trench provides direct access for inundating salt water to travel into the aquifer. The salt plume remains larger surrounding the trench than in the rest of the aquifer throughout recovery, and takes 3 months longer to recover than the surrounding aquifer. This is supported by other studies, where it was observed and modelled that areas where salt water pools or is collected during inundation (such as open boreholes or depressions) result in longer recovery times (Terry and Falkland, 2010; Chui and Terry, 2012).

The timing of remedial action (specifically draining of the trenches) is more critical than the duration of draining. It is critical to drain the trenches as soon as possible following a storm surge overwash in order to remove the initial salt load to the aquifer before it is transported deeper into the aquifer. After a certain period of delay, there is no improvement in recovery achieved by draining. This is illustrated in the simulation results as well as the observation data, where there is little improvement in recovery for trenches on Andros Island that were drained after a 4 day delay. The time of this delay threshold, where there is still benefit to be gained in draining the trenches, will depend on many factors, such as the hydraulic conductivity, the groundwater velocity, and recharge rates. For most typical low-lying islands, the delay threshold is likely quite soon after storm surge due to the high hydraulic conductivity of geological materials normally found on low-lying islands (Ayers and Vacher, 1986). Coarser aquifer material may allow for faster salt transport into the aquifer (Chui and Terry, 2012). Although this effect may also speed up recovery, it means that there is a limited time in which to perform remedial action to remove the salt water. On Andros Island, the delay threshold is 3 days. The duration of draining should also be short, because longer durations of draining may result in slower recovery times. This is likely due to the fact that draining of the trenches removes the recharging freshwater, along with the salt water.

6 Conclusions

Stressors act over varying spatial and temporal scales to impact the freshwater lenses of low-lying islands. Both short and long-acting stressors may result in significant loss of freshwater resources. The model results are inherently uncertain due to uncertainty associated with the input data, model conceptualization, and stressor scenarios. The greatest uncertainty lies in the simplification of the hydrogeology and the associated parameters. This is largely due to limited studies having been conducted on Andros. However, small islands often have limited capacity for hydrogeological investigations. Therefore, this study was not predictive, but rather aimed to identify the likely response based on the hydrogeological setting and the mean projected climate state derived from multiple climate change model scenarios. To address uncertainty rigorously, a series of models with a range of input parameters and climate scenarios would be required; however, this was beyond the scope of the current study. Within these limitations, the results of the study provide the following conclusions:

1. The impacts of stressors on the freshwater lens are predicted to occur primarily in areas where the freshwater lens is smaller or thinner, such as the periphery of the lens. As most settlements are concentrated within the coastal zone, even small-scale changes to the freshwater lens morphology in these areas may have significant implications for freshwater sustainability.

2. Change to groundwater recharge is identified as a key stressor to Andros Island, where greater impacts on the freshwater lens are observed in areas with lower recharge.

3. The response time of the freshwater lens (time to reach steady state) increases as the magnitude of the stressors increase. With increasing magnitude of change to the hydrogeological system, the freshwater lens takes longer to adjust to the new state.

4. The freshwater lens is generally able to recover from storm surge inundation over time as fresh recharge flushes the salt plume out of the aquifer. Eventually, the freshwater lens returns to the original morphology.

5. Trench-based wellfields may increase the potential storm surge impacts on the freshwater lens, depending on the hydraulic conductivity, the vadose zone thickness, and land cover. However, they also allow for remedial action (such as draining of the trenches) to be undertaken, which can improve recovery times. The sooner draining occurs, the more improvement in recovery, because, if draining is delayed by too long (in this case, 3 days or more), there is no improvement in recovery. The duration of draining has less effect on recovery and only needs to occur for a short period of time.

Acknowledgements. Funding for this research was provided by the Natural Sciences and Engineering Research Council (NSERC) through a Discovery Grant to Diana Allen, and a grant to Simon Fraser University by The Nature Conservancy through the Royal Bank of Canada's Blue Water Project[TM]. The authors also acknowledge the contribution of the Bahamas Water and Sewerage Corporation in providing data for model calibration.

Edited by: M. Bakker

References

Allen, D. M., Cannon, A. J., Toews, M. W., and Scibek, J.: Variability in simulated recharge using different GCMs, Water Resour. Res., 46, W00F03, doi:10.1029/2009WR008932, 2010.

Anderson Jr., W. P.: Aquifer salinization from storm overwash, J. Coast. Res., 18, 413–420, 2002.

Ataie-Ashtiani, B., Werner, A. D., Simmons, C. T., Morgan, L. K., and Lu, C.: How important is the impact of land-surface inundation on seawater intrusion caused by sea-level rise?, Hydrogeol. J., 21, 1673–1677, doi:10.1016/j.advwatres.2012.03.004, 2013.

Ayers, J. F. and Vacher, H. L.: Hydrogeology of an atoll island: A conceptual model from detailed study of a Micronesian example, Ground Water, 24, 185–198, doi:10.1111/j.1745-6584.1986.tb00994.x, 1986.

Barlow, P. M.: Ground Water in Freshwater-Saltwater Environments of the Atlantic Coast, US Department of the Interior, US Geol. Surv. Circular 1262, Reston, Virginia, USA, 2003.

Beach, D. K. and Ginsburg, R. N.: Facies succession of Pliocene-Pleistocene carbonates, northwestern Great Bahama Bank, AAPG Bull.-Am. Assoc. Petr. Geol., 64, 1634–1642, 1980.

Bear, J., Cheng, A. H. D., Sorek, S., Herrera, I., and Ouazar, D. (Eds.): Seawater Intrusion in Coastal Aquifers, Kluwer Academic Publishers, Dordrecht, the Netherlands, 1999.

Biasutti, M., Sobel, A. H., Camargo, S. J., and Creyts, T. T.: Projected changes in the physical climate of the gulf coast and caribbean, Climatic Change, 112, 819–845, doi:10.1007/s10584-011-0254-y, 2012.

Boardman, M. R. and Carney, C.: Influence of sea level on the origin and diagenesis of the shallow aquifer of Andros Island, Bahamas, in: Proceedings of the Eighth Symposium on the Geology of the Bahamas, Bahamian Field Station, edited by: Carew, J. L., San Salvador, Bahamas, 13–32, 1997.

Bobba, A. G.: Numerical modeling of salt-water intrusion due to human activities and sea-level change in the Godavari Delta, India, Hydrolog. Sci. J., 47, 67–80, 2002.

Bowleg, J. and Allen, D. M.: Effects of storm surges on groundwater resources, North Andros Island, Bahamas, in: Climate Change Effects on Groundwater Resources: A Global Synthesis of Findings and Recommendations, IAH International Contributions to Hydrogeology, editd by: Treidgel, H., Martin-Bordes, J. L., and Gurdak, J. J., CRC Press, London, UK, 2011.

Bukowski, J. M., Carney, C., Ritzi Jr., R. W., and Boardman, M. R.: Modeling the fresh-salt water interface in the Pleistocene aquifer on Andros Island, Bahamas, in: Proceedings of the Ninth Symposium on the Geology of the Bahamas, Bahamian Field Station, edited by: Curran, H. A. and Mylroie, J. E., San Salvador, Bahamas, 1–13, 1999.

Cant, R. V. and Weech, P. S.: A review of the factors affecting the development of Ghyben-Herzberg lenses in the Bahamas, J. Hydrol., 84, 333-343, doi:10.1016/0022-1694(86)90131-9, 1986.

Cashman, A., Nurse, L., and Charlery, J.: Climate change in the Caribbean: the water management implications, J. Environ. Develop., 19, 42–67, doi:10.1177/1070496509347088, 2010.

Chui, T. F. M. and Terry, J. P.: Modeling fresh water lens damage and recovery on atolls after storm-wave washover, Ground Water, 50, 412–420, 2012.

Chui, T. F. M. and Terry, J. P.: Influence of sea-level rise on freshwater lenses of different atoll island sizes and lens resilience to storm-induced salinization, J. Hydrol., 502, 18–26, 2013.

Falkland, A. (Ed.): Hydrology and water resources of small island: a practical guide, United Nations Educational, Scientific, and Cultural Organization – UNESCO, Paris, 1991.

Franklin, J., Pasch, R., Avila, L., Beven, J., Lawrence, M., Stewart, S., and Blake, E.: Atlantic hurricane season of 2004, Mon. Weather Rev., 134, 981–1025, 2006.

Freeze, R. A. and Cherry, J. A.: Groundwater, Prentice-Hall, Upper Saddle River, NJ, USA, 1977.

Goderniaux, P., Brouyere, S., Fowler, H. J., Blenkinsop, S., Therrien, R., Orban, P., and Dassargues, A.: Large scale surface-subsurface hydrological model to assess climate change impacts on groundwater reserves, J. Hydrol., 373, 122–138, 2009.

Green, T. R., Taniguchi, M., Kooi, H., Gurdak, J. J., Allen, D. M., Hiscock, K. M., and Aureli, A.: Beneath the surface of global change: Impacts of climate change on groundwater, J. Hydrol., 405, 532–560, doi:10.1016/j.jhydrol.2011.05.002, 2011.

Illangasekare, T., Tyler, S. W., Clement, T. P., Villholth, K. G., Perera, A. P. G. R. L., Obeysekera, J., Gunatilaka, A., Panabokke, C. R., Hyndman, D. W., Cunningham, K. J., Kaluarachchi, J. J., Yeh, W. W. G., van Genuchten, M. T., and Jensen, K.: Impacts of the 2004 tsunami on groundwater resources in Sri Lanka, Water Resour. Res., 42, W05201, doi:10.1029/2006WR004876, 2006.

IPCC: Climate Change 2007: The Physical Science Basis, in: Contribution of Working Group I to the Fourth Assessment Report of the Intergovernmental Panel on Climate Change, edited by: Solomon, S., Qun, D., Manning, M., Chen, Z., Marquis, M., Averyt, K. B., Tignor, M., and Miller, H. L., Cambridge University Press, Cambridge, UK, 2007.

IPCC: Climate Change 2014: Impacts, Adaptation, and Vulnerability, Part B: Regional Aspects, in: Contribution of Working Group II to the Fifth Assessment Report of the Intergovernmental Panel on Climate Change, edited by: Barros, V. R., Field, C. B., Dokken, D. J., Mastrandrea, M. D., Mach, K. J., Bilir, T. E., Chatterjee, M., Ebi, K. L., Estrada, Y. O., Genova, R. C., Girma, B., Kissel, E. S., Levy, A. N., MacCracken, S., Mastrandrea, P. R., and White, L. L., Cambridge University Press, Cambridge, UK and New York, NY, USA, 2014.

Jyrkama, M. I. and Sykes, J. F.: The impact of climate change on spatially varying groundwater recharge in the grand river watershed (Ontario), J. Hydrol., 338, 237–250, doi:10.1016/j.jhydrol.2007.02.036, 2007.

Langevin, C. D. and Zygnerski, M.: Effect of sea-level rise on salt water intrusion near coastal well field in southeastern Florida, Ground Water, 51, 781–803, 2013.

Langevin, C. D., Thorne, D. T., Dausman, A. M., Sukop, M. C., and Guo, W.: SEAWAT Version 4: A Computer Program for Simulation of Multi-Species Solute and Heat Transport, US Geol. Surv. Techniques and Methods Book 6, Chapter A22, US Geological Survey, Florida, USA, 2007.

Little, B. G., Buckley, D. K., Jefferiss, A., Stark, J., and Young, R. N.: Land resources of the commonwealth of the Bahamas, Volume 4 Andros Island, Land Resources Division, Tolworth Tower, Surrey, England, 1973.

McSweeney, C., New, M., Lizcano, G., and Lu, X.: The UNDP Climate Change Country Profiles, B. Am. Meteorol. Soc., 91, 157–166, doi:10.1175/2009BAMS2826.1, 2010.

Michael, H. A., Russoniello, C. J., and Byron, L. A.: Global assessment of vulnerability to sea-level rise in topography-limited and recharge-limited coastal groundwater systems, Water Resour. Res., 49, 2228–2240, doi:10.1002/wrcr.20213, 2013.

Mollema, P. N. and Antonellini, M.: Seasonal variation in natural recharge of coastal aquifers, Hydrogeol. J., 21, 787-797, doi:10.1007/s10040-013-0960-9, 2013.

Momi, K., Shoji, J., and Nakagawa, K.: Observations and modeling of seawater intrusion for a small limestone island aquifer, Hydrol. Process., 19, 3897–3909, doi:10.1002/hyp.5988, 2005.

NOAA: http://www.nhc.noaa.gov/, last access: 10 June 2014.

Obeysekera, J., Park, J., Irizarry-Ortiz, M., Barnes, J., and Trimble, P.: Probabilistic projection of mean sea level and coastal extremes, J. Waterw. Ports Coast. Ocean Eng., 139, 135–141, doi:10.1061/(ASCE)WW.1943-5460.0000154, 2013.

Oude Essink, G. H. P.: Improving fresh groundwater supply – problems and solutions, Ocean Coast. Manage., 44, 429–449, doi:10.1016/S0964-5691(01)00057-6, 2001.

Rahmstorf, S.: A semi-empirical approach to projecting future sea-level rise, Science, 315, 368–370, doi:10.1126/science.1135456, 2007.

Ranjan, P., Kazama, S., Sawamoto, M., and Sana, A.: Global scale evaluation of coastal fresh groundwater resources, Ocean Coast. Manage., 52, 197–206, doi:10.1016/j.ocecoaman.2008.09.006, 2009.

Rasmussen, P., Sonnenborg, T. O., Goncear, G., and Hinsby, K.: Assessing impacts of climate change, sea level rise, and drainage canals on saltwater intrusion to coastal aquifer, Hydrol. Earth Syst. Sci., 17, 421–443, doi:10.5194/hess-17-421-2013, 2013.

Ritzi, R., Bukowski, J., Carney, C., and Boardman, M.: Explaining the thinness of the fresh water lens in the Pleistocene carbonate aquifer on Andros Island, Bahamas, Ground Water, 39, 713–720, doi:10.1111/j.1745-6584.2001.tb02361.x, 2001.

Robins, N. and Lawrence, A.: Some hydrogeological problems peculiar to various types of small islands, Water Environ. J., 14, 341–346, 2000.

Schneider, J. and Kruse, S.: A comparison of controls on freshwater lens morphology of small carbonate and siliciclastic islands: Examples from barrier islands in Florida, USA, J. Hydrol., 284, 253–269, doi:10.1016/j.jhydrol.2003.08.002, 2003.

Schroeder, P. R., Dozier, T. S., Zappi, P. A., McEnroe, B. M., Sjostrom, J. W., and Peyton, R. L.: The Hydrologic Evaluation of Landfill Performance (HELP) model: Engineering documentation for Version 3, Rep. EPA/600/R-94/168b, US Environmental Protection Agency, Washington, D.C., USA, 1994.

Scibek, J. and Allen, D. M.: Modeled impacts of predicted climate change on recharge and groundwater levels, Water Resour. Res., 42, W11405, doi:10.1029/2005WR004742, 2006.

Stoeckl, L. and Houben, G.: Flow dynamics and age stratification of freshwater lenses: Experiments and modeling, J. Hydrol., 458, 9–15, doi:10.1016/j.jyhrol.2012.05.070, 2012.

Sulzbacher, H., Wiederhold, H., Siemon, B., Grinat, M., Igel, J., Burschil, T., Günther, T., and Hinsby, K.: Numerical modelling of climate change impacts on freshwater lenses on the North Sea Island of Borkum using hydrological and geophysical methods, Hydrol. Earth Syst. Sci., 16, 3621–3643, doi:10.5194/hess-16-3621-2012, 2012.

Tarbox, K. L.: Occurrence and development of water resources in The Bahamas, in: Proceedings of the Third Symposium on the Geology of the Bahamas, Bahamian Field Station, edited by: Curran, H. A., San Salvador, Bahamas, 139–144, 1987.

Terry, J. P. and Chui, T. F. M.: Evaluating the fate of freshwater lenses on atoll islands after eustatic sea-level rise and cyclone-driven inundation: A modeling approach, Global Planet. Change, 88–89, 76–84, doi:10.1016/j.gloplacha.2012.03.008, 2012.

Terry, J. P. and Falkland, A. C.: Responses of atoll freshwater lenses to storm-surge overwash in the Northern Cook Islands, Hydrogeol. J., 18, 749–759, 2010.

Therrien, R., McLaren, R., Sudicky, E. and Panday, S.: Hydro-GeoSphere – A three-dimensional numerical model describing fully-integrated subsurface and surface flow and solute transport, University of Waterloo and Université Laval, Waterloo, Canada, 2010.

Toews, M. W. and Allen, D. M.: Evaluating different GCMs for predicting spatial recharge in an irrigated arid region, J. Hydrol., 374, 265–281, doi:10.1016/j.jhydrol.2009.06.022, 2009

UNDP – United Nations Development Programme: Climate Change Country Profiles, The Bahamas, 2010.

Vacher, H. L.: Dupuit-Ghyben-Herzberg analysis of strip-island lenses, Bull. Geol. Soc. Am., 100, 580–591, doi:10.1130/0016-7606(1988)100<0580:DGHAOS>2.3.CO;2, 1988.

Vacher, H. L. and Quinn, T. M. (Eds.): Geology and hydrogeology of carbonate island, in: Developments in Sedimentology Vol. 54, Elsevier, Tampa Bay, Florida, USA, 1997.

Wallis, T. N., Vacher, H. L., and Stewart, M. T.: Hydrogeology of freshwater lens beneath a Holocene strandplain, Great Exuma, Bahamas, J. Hydrol., 125, 93–109, doi:10.1016/0022-1694(91)90085-V, 1991.

Werner, A. D. and Simmons, C. T.: Impact of sea-level rise on sea water intrusion in coastal aquifers, Ground Water, 47, 197–204, doi:10.1111/j.1745-6584.2008.00535.x, 2009.

Werner, A. D., Jakovovic, D., and Simmons, C. T.: Experimental observations of saltwater up-coning, J. Hydrol., 373, 230–241, doi:10.1016/j.jhydrol.2009.05.004, 2009.

Whitaker, F. and Smart, P.: Climatic control of hydraulic conductivity of Bahamian limestones, Ground Water, 35, 859–868, doi:10.1111/j.1745-6584.1997.tb00154.x, 1997.

White, I. and Falkland, A.: Management of freshwater lenses on small Pacific islands, Hydrogeol. J., 18, 227–246, doi:10.1007/s10040-009-0525-0, 2010.

White, N. J., Church, J. A., and Gregory, J. M.: Coastal and global averaged sea level rise for 1950 to 2000, Geophy. Res. Lett., 32, L01601, doi:10.1029/2004GL021391, 2005.

WHO – World Health Organisation: Guidelines for drinking-water quality, 4th Edn., Geneva, Switzerland, 2011.

Wolfe, P. J., Adams, A. L., and Carney, C. K.: A resistivity study of the freshwater lens profile across North Andros Island, The Bahamas, in: Proceedings of the Tenth Symposium on the Geology of the Bahamas and Other Carbonate Regions, edited by: Greenstein, B. J. and Carney, C. K., Gerace Research Center, San Salvador, Bahamas, 31–40, 2001.

Yang, J., Graf, T., Herold, M., and Ptak, T.: Modeling the effects of tides and storm surges on coastal aquifers using a coupled surface-subsurface approach, J. Contam. Hydrol., 149, 61–75, 2013.

Younger, P. L.: Simple generalized methods for estimating aquifer storage parameters, Q. J. Eng. Geol. Hydrogeol., 26, 127–135, doi:10.1144/GSL.QJEG.1993.026.02.04, 1993.

Permissions

List of Contributors

C.-H. Su
Department of Infrastructure Engineering, University of Melbourne, 3010 Victoria, Australia

D. Ryu
Department of Infrastructure Engineering, University of Melbourne, 3010 Victoria, Australia

R. G. Knox
Massachusetts Institute of Technology, Cambridge, Massachusetts, USA

M. Longo
Harvard University, Cambridge, Massachusetts, USA

A. L. S. Swann
University of Washington, Seattle, Washington, USA

K. Zhang
Harvard University, Cambridge, Massachusetts, USA

N. M. Levine
Harvard University, Cambridge, Massachusetts, USA

P. R. Moorcroft
Harvard University, Cambridge, Massachusetts, USA

R. L. Bras
Georgia Institute of Technology, Atlanta, Georgia, USA

G. Balsamo
European Centre for Medium-Range Weather Forecasts (ECMWF), Reading, UK

C. Albergel
European Centre for Medium-Range Weather Forecasts (ECMWF), Reading, UK

A. Beljaars
European Centre for Medium-Range Weather Forecasts (ECMWF), Reading, UK

S. Boussetta
European Centre for Medium-Range Weather Forecasts (ECMWF), Reading, UK

E. Brun
Météo-France, Toulouse, France

H. Cloke
University of Reading, Reading, UK

D. Dee
European Centre for Medium-Range Weather Forecasts (ECMWF), Reading, UK

E. Dutra
European Centre for Medium-Range Weather Forecasts (ECMWF), Reading, UK

J. Muñoz-Sabater
European Centre for Medium-Range Weather Forecasts (ECMWF), Reading, UK

F. Pappenberger
European Centre for Medium-Range Weather Forecasts (ECMWF), Reading, UK

P. de Rosnay
European Centre for Medium-Range Weather Forecasts (ECMWF), Reading, UK

T. Stockdale
European Centre for Medium-Range Weather Forecasts (ECMWF), Reading, UK

F. Vitart
European Centre for Medium-Range Weather Forecasts (ECMWF), Reading, UK

K. Džubáková
Institute of Environmental Engineering, ETH Zurich, Switzerland
Department of Physical Geography and Geoecology, Comenius University in Bratislava, Slovakia

P. Molnar
Institute of Environmental Engineering, ETH Zurich, Switzerland

K. Schindler
Institute of Geodesy and Photogrammetry, ETH Zurich, Switzerland

M. Trizna
Department of Physical Geography and Geoecology, Comenius University in Bratislava, Slovakia

T. Erfani
School of Mechanical, Aerospace and Civil Engineering, University of Manchester, Manchester, M13 9PL, UK
Department of Civil, Environmental and Geomatic Engineering, University College London, Chadwick Building, Gower Street, London, WC1E 6BT, UK

O. Binions
Department of Civil, Environmental and Geomatic Engineering, University College London, Chadwick Building, Gower Street, London, WC1E 6BT, UK

J. J. Harou
School of Mechanical, Aerospace and Civil Engineering, University of Manchester, Manchester, M13 9PL, UK
Department of Civil, Environmental and Geomatic Engineering, University College London, Chadwick Building, Gower Street, London, WC1E 6BT, UK

X. Han
Key Laboratory of Remote Sensing of Gansu Province, Cold and Arid Regions Environmental and Engineering Research
Forschungszentrum Jülich, Agrosphere (IBG 3), Leo-Brandt-Strasse, 52425 Jülich, Germany
Centre for High-Performance Scientific Computing in Terrestrial Systems: HPSC TerrSys, Geoverbund ABC/J, Leo-Brandt-Strasse, 52425 Jülich, Germany

H.-J. H. Franssen
Forschungszentrum Jülich, Agrosphere (IBG 3), Leo-Brandt-Strasse, 52425 Jülich, Germany
Centre for High-Performance Scientific Computing in Terrestrial Systems: HPSC TerrSys, Geoverbund ABC/J, Leo-Brandt-Strasse, 52425 Jülich, Germany

R. Rosolem
Department of Civil Engineering, University of Bristol, Bristol BS8 1TR, UK

R. Jin
Key Laboratory of Remote Sensing of Gansu Province, Cold and Arid Regions Environmental and Engineering Research
CAS Center for Excellence in Tibetan Plateau Earth Sciences, Chinese Academy of Sciences, Beijing 100101, PR China

X. Li
Key Laboratory of Remote Sensing of Gansu Province, Cold and Arid Regions Environmental and Engineering Research
CAS Center for Excellence in Tibetan Plateau Earth Sciences, Chinese Academy of Sciences, Beijing 100101, PR China

H. Vereecken
Forschungszentrum Jülich, Agrosphere (IBG 3), Leo-Brandt-Strasse, 52425 Jülich, Germany
Centre for High-Performance Scientific Computing in Terrestrial Systems: HPSC TerrSys, Geoverbund ABC/J, Leo-Brandt-Strasse, 52425 Jülich, Germany

R. G. Anderson
USDA, Agricultural Research Service, San Joaquin Valley Agricultural Sciences Center, Water Management Research Unit, Parlier, California, USA
USDA, Agricultural Research Service, U.S. Salinity Laboratory, Contaminant Fate and Transport Unit, Riverside, California, USA

D. Wang
USDA, Agricultural Research Service, San Joaquin Valley Agricultural Sciences Center, Water Management Research Unit, Parlier, California, USA

R. Tirado-Corbalá
USDA, Agricultural Research Service, San Joaquin Valley Agricultural Sciences Center, Water Management Research Unit, Parlier, California, USA
Crops and Agro-Environmental Science Department, University of Puerto Rico, Mayagüez, Puerto Rico, USA

H. Zhang
USDA, Agricultural Research Service, San Joaquin Valley Agricultural Sciences Center, Water Management Research Unit, Parlier, California, USA
USDA, Agricultural Research Service, Water Management Research Unit, Fort Collins, Colorado, USA

J. E. Ayars
USDA, Agricultural Research Service, San Joaquin Valley Agricultural Sciences Center, Water Management Research Unit, Parlier, California, USA

M. Guderle
Friedrich Schiller University, Institute for Geosciences, Burgweg 11, 07749 Jena, Germany
Max Planck Institute for Biogeochemistry, Biogeochemical Processes, Hans-Knöll-Str. 10, 07745 Jena, Germany
International Max Planck Research School for Global Biogeochemical Cycles, Hans-Knöll-Str. 10, 07745 Jena, Germany

A. Hildebrandt
Friedrich Schiller University, Institute for Geosciences, Burgweg 11, 07749 Jena, Germany
Max Planck Institute for Biogeochemistry, Biogeochemical Processes, Hans-Knöll-Str. 10, 07745 Jena, Germany

V. Srinivasan
Ashoka Trust for Research in Ecology and the Environment, Royal Enclave Sriramapura, Jakkur Post, Bangalore, Karnataka, India

J. H. Gao
Ministry of Education Key Laboratory for Coast and Island Development, Nanjing University, Nanjing 210093, China

J. Jia
State Research Centre for Island Exploitation and Management, Second Institute of Oceanography, State Oceanic Administration, Hangzhou 310012, China

Y. P. Wang
Ministry of Education Key Laboratory for Coast and Island Development, Nanjing University, Nanjing 210093, China

Y. Yang
Ministry of Education Key Laboratory for Coast and Island Development, Nanjing University, Nanjing 210093, China

J. Li
Qingdao Institute of Marine Geology, Qingdao 266071, China

F. Bai
Qingdao Institute of Marine Geology, Qingdao 266071, China

X. Zou
Ministry of Education Key Laboratory for Coast and Island Development, Nanjing University, Nanjing 210093, China

S. Gao
Ministry of Education Key Laboratory for Coast and Island Development, Nanjing University, Nanjing 210093, China

A. Fernald
College of Agricultural, Consumer and Environmental Sciences, New Mexico State University, Las Cruces, New Mexico, USA

S. Guldan
Sustainable Agriculture Science Center at Alcalde, New Mexico State University, Alcalde, New Mexico, USA

K. Boykin
College of Agricultural, Consumer and Environmental Sciences, New Mexico State University, Las Cruces, New Mexico, USA

A. Cibils
College of Agricultural, Consumer and Environmental Sciences, New Mexico State University, Las Cruces, New Mexico, USA

M. Gonzales
Community & Regional Planning Program, University of New Mexico, Albuquerque, New Mexico, USA

B. Hurd
College of Agricultural, Consumer and Environmental Sciences, New Mexico State University, Las Cruces, New Mexico, USA

S. Lopez
College of Agricultural, Consumer and Environmental Sciences, New Mexico State University, Las Cruces, New Mexico, USA

C. Ochoa
Department of Animal and Rangeland Sciences, Oregon State University, Corvallis, Oregon, USA

M. Ortiz
New Mexico Acequia Association, Santa Fe, New Mexico, USA

J. Rivera
Center for Regional Studies, University of New Mexico, Albuquerque, New Mexico, USA

S. Rodriguez
Department of Anthropology, University of New Mexico, Albuquerque, New Mexico, USA

C. Steele
College of Agricultural, Consumer and Environmental Sciences, New Mexico State University, Las Cruces, New Mexico, USA

S. Holding
Department of Earth Sciences, Simon Fraser University, 8888 University Drive, Burnaby, British Columbia, V5A 1S6, Canada

D. M. Allen
Department of Earth Sciences, Simon Fraser University, 8888 University Drive, Burnaby, British Columbia, V5A 1S6, Canada